ESTABLISHING THE NEW SCIENCE

Frontispiece. The Royal Society's coat of arms granted in its second charter, as shown on the leaf facing the title-page of Thomas Sprat's History of the Royal Society *(1667). Reproduced by permission of the British Library.*

Establishing the New Science

The Experience of the Early Royal Society

Michael Hunter

THE BOYDELL PRESS

First published 1989
The Boydell Press, Woodbridge

Transferred to digital printing

ISBN 978-0-85115-506-7

The Boydell Press is an imprint of Boydell & Brewer Ltd
PO Box 9, Woodbridge, Suffolk IP12 3DF, UK
and of Boydell & Brewer Inc.
668 Mt Hope Avenue, Rochester, NY 14620, USA
website: www.boydellandbrewer.com

A CiP catalogue record for this book is available from the British Library

This publication is printed on acid-free paper

CONTENTS

DOCUMENTS AND NOTES

LIST OF ILLUSTRATIONS

LIST OF ABBREVIATIONS

Bacon, *Works*	James Spedding, R. L. Ellis and D. D. Heath, eds., *The Works of Francis Bacon* (London, 1857 – 74)
Birch, *History*	Thomas Birch, *The History of the Royal Society of London* (London, 1756 – 7)
Boyle, *Works*	Thomas Birch, ed., *The Works of the Hon. Robert Boyle* (2nd ed., London, 1772)
Cl.P.	Royal Society Classified Papers
D.M.	Royal Society Domestic Manuscripts
DNB	*Dictionary of National Biography*
E.L.	Royal Society Early Letters
Evelyn, *Diary*	E.S. de Beer, ed., *The Diary of John Evelyn* (Oxford, 1955)
Evelyn, *Diary and Correspondence*	William Bray [and John Forster], eds., *The Diary and Correspondence of John Evelyn* (London, 1852)
Grew, *Musaeum*	Nehemiah Grew, *Musaeum Regalis Societatis* (London, 1681)
Hooke, *Diary*	H. W. Robinson and W. Adams, eds., *The Diary of Robert Hooke, 1672 – 1680* (London, 1935)
Hooke, *Posthumous Works*	Richard Waller, ed., *The Posthumous Works of Robert Hooke* (London, 1705)
Hunter, *Royal Society*	Michael Hunter, *The Royal Society and its Fellows, 1660 – 1700: the Morphology of an Early Scientific Institution* (Chalfont St Giles, 1982)
Hunter, *Science and Society*	Michael Hunter, *Science and Society in Restoration England* (Cambridge, 1981)
Oldenburg	A. R. and M. B. Hall, eds., *The Correspondence of Henry Oldenburg* (Madison and Milwaukee and London, 1965 – 86)
Notes and Records	*Notes and Records of the Royal Society of London*
OED	*Oxford English Dictionary*
PRO	Public Record Office
Pepys, *Diary*	Robert Latham and William Matthews, eds., *The Diary of Samuel Pepys* (London, 1972 – 83)
Phil. Trans.	*Philosophical Transactions*
The Record	*The Record of the Royal Society of London* (4th ed., London, 1940)
Sprat, *History*	Thomas Sprat, *The History of the Royal Society of London, 1667*, ed. J. I. Cope and H. W. Jones (London, 1959)

PREFACE

The study of seventeenth-century English science, its context and its organization, exerts a continuing fascination on historians. It might be thought that, after decades of scholarly activity, this topic had been 'done', and that attention would shift to other, hitherto neglected, periods. In fact, however, interest in this field remains as lively as ever, as is illustrated by the abundance of recent writings on the subject surveyed in the Bibliographical Essay that I have written for this volume (below, pp. 356 – 68). Moreover, though it might be considered that the Royal Society itself had been particularly fully scrutinized, as the first example of a type of organization that has since emerged as crucial in the Western World, the amount that still remains to be said about this seminal institution is indicated by the fact that this volume is entirely devoted to new findings concerning various facets of the Society's early development.

The justification of such research lies first and foremost in the profuse untapped documentation that survives, in the Society's own archives, among the papers of prominent early Fellows or in such repositories as the Public Record Office. There have been times when I have come to feel that every stone I turned hid a teeming mass of often unexpected detail. The result is that in the essays that follow, a prominent role in the argument is frequently played by previously unknown material. Indeed, I have tried to make a virtue of this by incorporating into the studies, or appending to them, hitherto unpublished materials which throw important light on the Society's rationale, activities and problems in its early years. In my view, one of the advantages of archival research is its ability to revivify debate by adding to the corpus of available texts: it is to be hoped that the documents made available here will enrich future scholarship in this way.

Quite apart from the material on which they are based, however, the studies presented here should prove of value in their own right. Each of them offers a definitive revaluation of some facet of the Society's early history. In addition, it may be argued that the whole represents more than the sum of its parts. For one thing, almost all of the Society's early institutional initiatives are dealt with in the pages that follow, while consideration is also given to such crucial themes as its relations with its principal employee, Robert Hooke, and the surprising range of proposals for reforming the Society put forward in these years by Hooke and others. More important is the cumulative implication of such detailed analysis, which forces us to look afresh at the formative years of this seminal institution, and especially at the earliest stages of its development, the distinctive and embryonic nature of which can easily be overlooked.

Up to a point, the picture offered is one of naivety, disappointment and opportunities missed, but it would be wrong to take an unduly negative view of what comes to light. For such problems were merely symptoms of the fact that the task of organizing the new science in its formative years proved much more difficult than those who set about the task expected: the result is that a process which is often taken for granted turns out to be worthy of study in its own right. The chapters that follow illustrate the disagreements that existed among the Society's leading members as to the proper aims of science and the most effective strategy for a scientific institution; the constraints imposed by practical difficulties and by the disappointing response that the new science received in contemporary society as a whole; and, above all, the extent to which the experience of trying to run a scientific institution for the first time proved educative in itself, illustrating some initiatives as more fruitful than others in a manner that could not have been predicted in advance. The book thus argues that the institutionalization of science in the late seventeenth century can only be properly understood if the interplay of ideals and realities is appreciated, and that the process had a momentum of its own which explains much about the way in which organized science was to develop.

Such broad reflections are pursued further in the third section of the Introduction that follows, 'The Fruits of Experience', which attempts to draw out some of the implications of the content of the book as a whole. This section is, however, preceded by two others which are intended to contribute to our understanding of the process by which the new science was 'established', considering, first, the issue of *why* institutionalization seemed so desirable as a way of forwarding the new science in the years around 1660 — which is not as self-evident as is sometimes presumed — and secondly just how the Society's organizers went about constituting the Society in its earliest years: this involves a detailed study of the charters, statutes and other measures implemented at that stage.

This Introduction is wholly new, and so are chapters 2, 3, 9, and the Bibliographical Essay. Chapter 2, which is based on a paper given at a conference on 'Latitudinarianism, Science and Society' held in Los Angeles in April 1987, presents a fresh reading of Thomas Sprat's famous *History of the Royal Society* (1667). It stresses the element of naivety that underlay the enterprise of producing an apologia on the Society's behalf, and the embarrassment that followed the book's hostile reception; it also draws broader conclusions about the political and religious affiliations of the Society's membership.

Chapter 3 deals with the committees which the Society set up in 1664 to deal with various aspects of its activities. It is accompanied by the complete text of a fascinating series of minutes and reports on the work of these groups, here published for the first time. Apart from making a substantial addition to our knowledge of the Society's early activities, this hitherto neglected material has interesting implications concerning the evolution of the Society's research programme, its attitude to technology and the nature of its members' involvement in its work: all these are drawn out in the essay which precedes the fully annotated text of the minutes themselves.

In chapter 9, hitherto unknown manuscript material is deployed to throw a quite new light on Robert Hooke's relations with his patron, Sir John Cutler. This lengthy study not only illustrates the extent to which the Society tried to manipulate Cutler's intentions for his benefaction for purposes of its own which were slightly at odds with what Cutler himself apparently wanted. It also shows the effect that the endowment had on Hooke, particularly in forcing him to a candid statement of his priorities concerning science and technology during the protracted negotiations with Cutler that were to ensue. It is also argued that the Lectureship inspired Hooke to a series of programmatic statements about the aims of natural philosophy and how they might be realized which he might not have made without this stimulus.

Chapter 4 gives a full account of a little-known episode and its implications, the setting up of a 'repository' or museum under the Society's auspices. Though this is virtually a new essay, it overlaps with and largely supersedes an earlier study of this subject that I published in 1985, 'The Cabinet Institutionalized: the Royal Society's Repository and its Background', in *The Origins of Museums: The Cabinet of Curiosities in Sixteenth and Seventeenth-century Europe*, eds. Oliver Impey and Arthur MacGregor (Oxford, 1985), pp. 156 – 68. In that paper I was under fairly tight constraints of space; my remit entailed dealing with other institutional collections in seventeenth-century England as well as the Royal Society's; and I also wanted to include a brief disquisition on early English usage of the word 'museum'. The result was that the coverage of the Royal Society's collection suffered, even by comparison with the longer version of the paper that I had given at the conference on which the volume was based, held at Oxford in July 1983 to celebrate the tercentenary of the Ashmolean Museum.

I have therefore decided to publish here a full-length essay specifically devoted to the Royal Society's repository, including much more detail than there was room for in my earlier paper, and drawing out some important implications concerning the lessons that the Royal Society learnt from putting together and administering such a collection about the function that such a resource might serve in advancing understanding of the natural world. On the other hand, I have omitted the material dealing with other institutional collections, which seemed out of place here, and for which the reader should consult my 1985 essay.

The remaining essays in the book, chapters 5 – 8 and the Appendix, have already appeared in print. Chapter 5 comprises a study of the Society's plan to erect a 'college' in 1667 – 8 and the difficulties which ultimately killed this project. Chapter 7 reprints my evaluation of the achievement of Henry Oldenburg, written to mark the completion of the edition of his correspondence by A. R. and M. B. Hall. Chapter 8 gives an account of the strategies by which the Society funded a research post for the plant anatomist, Nehemiah Grew, in the 1670s. Chapter 6 is an analysis of the contrasting priorities for scientific research and organization revealed by discussions as to how the Society might best be restructured which occurred at various points in the late seventeenth century: I originally wrote this in collaboration with Paul B. Wood, but he has generously allowed me to reprint it here because it naturally complements the remainder of the book's content. Lastly, there is my critique of the approach to the early Royal Society of Lotte and Glen Mulligan, which I have consigned to an Appendix because it is more of an occasional piece than the other essays: it seemed worth including, however, because it complements them in giving a sense of how the Society actually worked in its early years and of which figures within it were responsible for the initiatives that were taken.

I have myself always found it annoying to discover that a paper exists in print in two different versions, and I have therefore refrained from inserting second thoughts into these articles. Instead, they are reprinted almost exactly as they originally appeared, except that I have corrected a handful of misprints and minor errors of fact and have altered cross-references to other material reprinted here so that the reader is directed to the version of it to be found within these covers raıher than to the original.

A minor complication arises in connection with citations of my detailed analysis of the Society's early membership, to which reference is repeatedly made in the pages that follow, since this exists in two recensions: the first, which covered the period 1660 – 1685 only, was published in the Royal Society's *Notes and Records* in 1976, while the second, which revises this and extends its coverage to 1700, appeared as a monograph published by the British Society for the History of Science in 1982 (it is herinafter cited as Hunter, *Royal Society*: see List of Abbreviations). Two studies reprinted here — chapter 8 and the Appendix — appeared before the revised version did, and, whereas in the case of my study of Grew, which cited this work only for factual information, I have silently altered references from the 1976 to the 1982 version, in the case of the Appendix, 'Reconstructing Restoration Science', it seemed more appropriate to leave references to the passages originally cited, which were sometimes slightly revised and reorganized in the 1982 version: I have therefore there cited the revised version which was 'forthcoming' when that article originally appeared only for sections which did not appear in the original.

I might note parenthetically that, though chapter 5 appears exactly as it was originally published, I have taken the opportunity to improve the quality of one of the illustrations to it, while I have annexed a further pair of illustrations in the form of John Evelyn's hitherto unknown drawings of the 'college' that he projected in 1659, an interesting precursor of the Royal Society to which reference is also made in the first section of the Introduction; these are accompanied by a brief explanatory note, as is some of the other illustrative material that is included.

Individual acknowledgments in the footnotes to papers already published have been retained, and, for the sake of consistency, it seemed best to deal in a similar manner with acknowledgments relating to each of the hitherto unpublished chapters. In addition, however, I would like to thank the following. John Henry, Steven Shapin and Paul B. Wood read a draft of the Introduction, and they have also helped me to develop my conception of the book as a whole. Richard Barber at Boydell & Brewer has been patient and accommodating as this volume has grown and altered in shape: I hope that he will feel that his temerity in signing a contract on trust in the first place has been justified. References and advice have been provided by Jeremy Black, Giles Mandelbrote, Simon Schaffer, Christopher Whittick and David Wootton. The British Library, the Trustees of the British Museum, the President and Council of the Royal Society, the Department of Plant Sciences, University of Oxford, and the Trustees of the Will of Major Peter George Evelyn have given me permission to reproduce material in their custody. In the latter case, I am also indebted to the Trustees for allowing me to consult and cite the Evelyn MSS at Christ Church, Oxford, while I am grateful to the Earl of Shelburne for comparable facilities in connection with the microfilms of the Petty MSS at Bowood House now on deposit at the Bodleian Library, Oxford.

Lastly, I must thank the publishers and editors of the following journals for permission to reproduce material which originally appeared in their pages as follows: 'Early Problems in Professionalizing Scientific Research: Nehemiah Grew and the Royal Society, with an Unpublished Letter to Henry Oldenburg' *Notes and Records of the Royal Society of London*, 36 (1982), 189 – 209; 'A "College" for the Royal Society: the Abortive Plan of 1667 – 8', ibid., 38 (1984), 159 – 86; 'Towards Solomon's House: Rival Strategies for Reforming the Early Royal Society', *History of Science*, 24 (1986), 49 – 108; 'Promoting the New Science: Henry Oldenburg and the Early Royal Society', ibid., 26 (1988), 165 – 81; and 'Reconstructing Restoration Science: Problems and Pitfalls in Institutional History', *Social Studies of Science*, 12 (1982), 451 – 66.

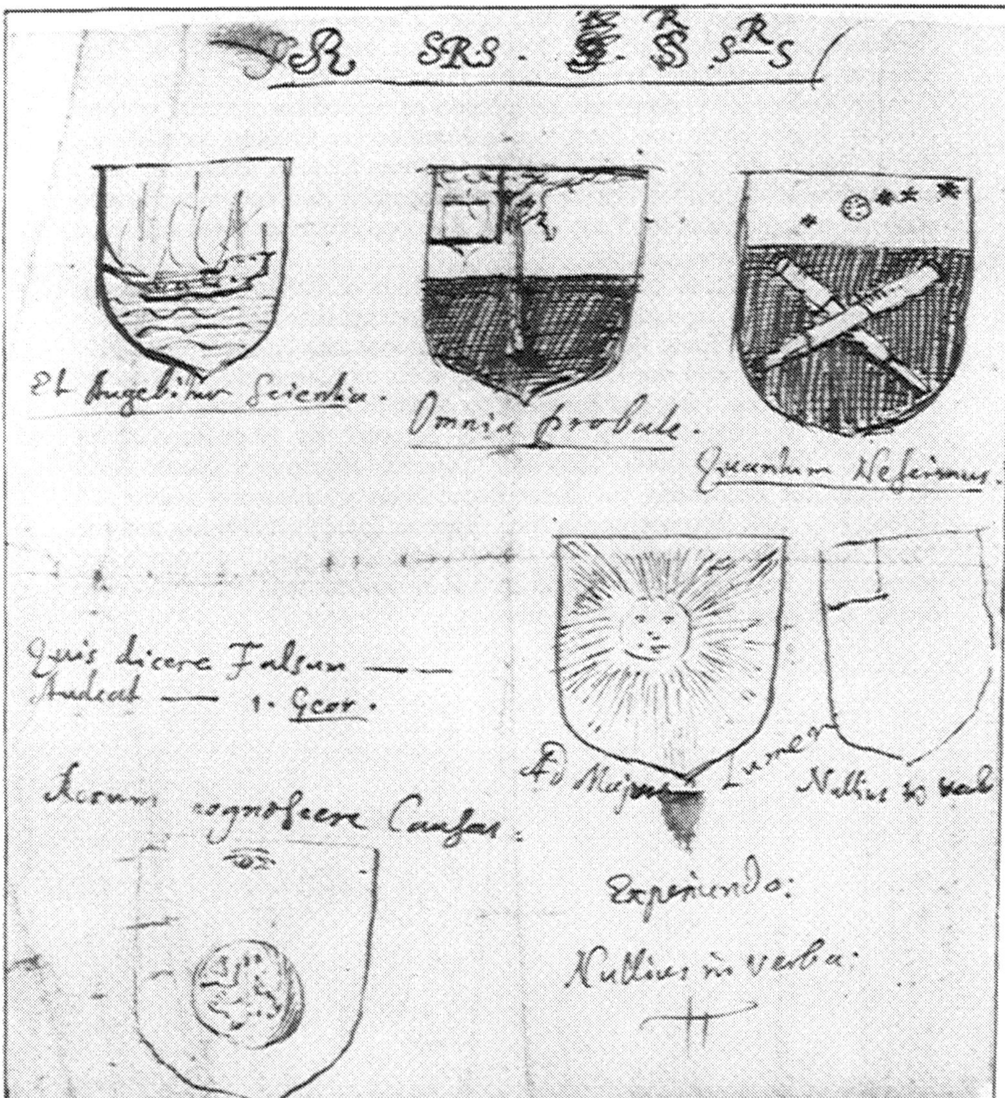

Plate 1. John Evelyn's drawings of alternative coats of arms and mottos suggested for the Royal Society, 1660. Reproduced from the original now preserved in the Evelyn Collection at Christ Church, Oxford, by courtesy of the Trustees of the Will of Major Peter George Evelyn. For elucidation of the designs shown, see the 'Note' following the Introduction.

1

INTRODUCTION

(a) The Importance of Being Institutionalized

The essays in this book are concerned with the institutionalization — or, to use a term with more contemporary resonance, the 'establishment' — of science in late seventeenth-century England. They deal exclusively with one institution, the Royal Society of London, but they raise broader issues about the impulse towards organization in the science of the day and the effects that this had, something which, in my view, has in the past been too much taken for granted. For one thing, as the essays in this volume illustrate, and as will be argued more fully later in this Introduction, the process of institutionalization proved much more complicated than had initially been expected: this had a significant effect both on the strategy that was adopted and on what the Royal Society as a corporate body was able to achieve, and it gives the lie to any simplistic presumption that the direction in which the fledgling Society would develop in its early years was somehow obvious. Equally crucial is another question which is easily treated as self-evident, due not least to the continuity and influence that the Royal Society has had ever since, namely of why institutionalization was perceived to be important in the pursuit of knowledge in England in the second half of the seventeenth century, and just what those who aspired to it believed that it would achieve.

This is the more important because of the novelty involved in the Society's foundation in 1660, which is itself often underestimated. For the Royal Society represented a new type of institution, a public body devoted to the corporate pursuit of scientific research, something unprecedented either in this country or elsewhere: though parallels may be found for facets of both its activity and its organization, the combination was unique, and it is worth spelling this out. First, the Society was concerned with experimental natural philosophy, or what we would call science. In the seventeenth century it is often difficult to disentangle such studies from other intellectual or applied pursuits, but the Royal Society was authentically devoted to a cluster of disciplines concerned with natural and mechanical phenomena to the exclusion of others, linked by common methods, and this is significant in itself. Indeed, this period saw a deliberate attempt to demarcate this as a separate sphere of activity, and in this process the Royal Society itself played an important role.[1]

[1] An example of such attempted demarcation is Robert Boyle's exchange with

The Society was also devoted to what we would call research, and this was again crucial in distinguishing it from other bodies at the time — from educational institutions like schools and universities which, even if incidentally encouraging scientific research, were primarily devoted to disseminating knowledge rather than producing it, and equally from bodies mainly concerned with upholding professional standards, such as the College of Physicians or the College of Arms, though obviously they had a certain degree of interest in research in their chosen fields. Suggestions were made at one point to give the Royal Society pedagogic functions, but only in order to raise revenue and hence fund its role in the pursuit and organization of research.[2] In addition, though the Society aspired to a role in accrediting intellectuals in its chosen field, which it achieved to a certain extent, this was marginal in comparison with the state of affairs in well-established and socially accepted fields like medicine and heraldry; such a role was also dependent, far more than that of the Physicians and the Heralds, on the performance, encouragement and promotion of research.

A third characteristic of the Royal Society was that it was a public institution. In this it was similar to the College of Physicians and the colleges of Oxford and Cambridge, and it differed from earlier centres of research in England and on the continent which had been wholly informal groupings or had been brought together by private patronage — such as the private academies of seventeenth-century France, or the Accademia del Cimento in Italy or (at home) the entourage of noblemen like the ninth Earl of Northumberland.[3] The Royal Society was a public body, a fact for which it sought recognition by public incorporation: and, though Charles II was parsimonious with funds, he did grant the Society a charter of incorporation, first in 1662 and then in revised form in 1663. He thus gave the Society a status comparable to that of bodies like the chartered companies of the City of London, and it is to corporations like these that the Royal Society is properly compared: it became one of the numerous legally-incorporated institutions with their discrete rights and privileges which went to make up the political and social fabric of England. As the virtuoso, Nathaniel Fairfax, rather aptly put it in writing to the Society's Secretary, Henry Oldenburg, a few years later, the Society was 'a body form'd to that end [the advancement of useful philosophy] as much as the Universityes ar

Henry More on the interpretation of experimental findings, of which a recent account appears in Steven Shapin and Simon Schaffer, *Leviathan and the Air-Pump* (Princeton, 1985), pp. 207f.

2 See below, 'Towards Solomon's House', p. 202. See also p. 293. On the role of the universities, cf. John Gascoigne, 'The Universities and the Scientific Revolution: the Case of Newton and Restoration Cambridge', *History of Science*, 23 (1985), 392, 424.

3 See Harcourt Brown, *Scientific Organizations in Seventeenth-century France* (Baltimore, 1934), W. E. K. Middleton, *The Experimenters: a Study of the Accademia del Cimento* (Baltimore, 1971), J. W. Shirley, *Thomas Harriot: A Biography* (Oxford, 1983), ch. 9.

for Scholars, the Inns of Court for Lawyers, Cityes & Towns for Merchants & Craftsmen'.[4]

The implication of this is equally striking, though equally often taken for granted, and this is the sheer elaborateness of the constitutional structure that the Society adopted in the charter that it acquired within eighteen months of its inauguration, a constitutional structure which might seem almost topheavy for an infant institution to whose actual informality attention has more than once been drawn.[5] This charter detailed the division of the Society into President, Council and Fellows, providing for annual re-election to office and making arrangements for filling vacancies during the year, laying down quorums for decision-making, insisting that the officers should take an oath on appointment, and so on.[6] It also provided for Fellows to be formally elected and for their election to be recorded in the Society's Register, while the accompanying statutes provided that all members should be obliged to pay an admission fee and an annual subscription: in the absence of the official endowment that was hoped for but was slow in coming, these corporate subscriptions formed the basis of the Society's early finance, in a manner similar to a professional body. In merging this formality of constitution, decision-making and finance with a devotion to research, the Royal Society set an example which was to prove deeply influential among other voluntary institutions founded to pursue scholarly ends from the eighteenth century onwards.

Equally important, the Society was endowed with a corporate personality. It was given the right to employ officers, to own property and to erect its own 'College or Colleges', to have a seal and a coat of arms, to correspond with foreigners and to sue and be sued in the courts. Perhaps more significantly, it was also given the right to appoint its own printer and to license its own books, a potentially worthwhile privilege when censorship was otherwise in the hands of the church and the government, and it was permitted the right to obtain the bodies of executed criminals for use in research, a right which (as the charter pointed out) the Society shared with the College of Physicians and the Company of Surgeons. There was also a rather vaguer provision for the Society to be assisted in its enterprise by Justices, Mayors, Aldermen, Sheriffs, Bailiffs, Constables, clergymen and other officers of the state. In all these provisions, the charter reads very like those of other corporate bodies such as those already referred to.

Considering how short a time the Society had been in existence when this charter was granted, it is striking both how confident its organizers were in

4 Fairfax to Oldenburg, 28 Sept. 1667, *Oldenburg*, iii, 492. On the legal status of an institution with similarities to the Royal Society, see H. J. Cook, '"Against Common Right and Reason": the College of Physicians versus Dr Thomas Bonham', *American Journal of Legal History*, 29 (1985), 301 – 22.

5 See Quentin Skinner, 'Thomas Hobbes and the Nature of the Early Royal Society', *Historical Journal*, 12 (1969), 238 and 217 – 39 passim; Hunter, *Science and Society*, pp. 33 – 4.

6 I have here conflated the content of the two charters. For their differences, and for documentation of them and the statutes, see below, pp. 18f.

the significance of what they were doing and how important they thought it properly to establish the Society as a public body. This sense of how much might thus be achieved is much in evidence in pronouncements from the Society's earliest years, such as a letter from the President, William, Viscount Brouncker, to the Earl of Ormonde in 1662 in which he attempted to secure an endowment for the Society from the fractions accruing from the Irish land settlement. For in it, Brouncker stressed how 'the interest of private persons, though very deserving', should properly be subordinated to 'the public concern of a society, whose designs, if protected and assisted by authority, may so much conduce to the greatness and honour of their prince, the real good of his dominions, and the universal benefit of mankind'. Thomas Sprat, in his *History of the Royal Society* of 1667 comparably emphasised how the design was 'public' and 'free from all suspicion of men, or private Interest'.[7]

Moreover, with this aspiration to a public, as against a private, role went a wish to build up an infrastructure of corporate resources, clearly in evidence in the establishment of the Society's 'repository' as recounted in a later chapter in this volume, but seen even earlier in the Society's concern for keeping records. This emerged at its third meeting, on 12 December 1660, when three books were ordered to be provided, 'one for the statutes of the society and names of its members; another for experiments, and the result of debates; and the third for occasional orders'.[8] Indeed, elaborate journal books and register books were preserved almost from the start, and the significance of this virtual obsession with record-keeping is again too easily taken for granted. As Oldenburg was to explain in a letter of 31 January 1668 to the Danzig natural philosopher, Johann Hevelius, all that he had sent to the Society had been deposited in its archives, 'and in this way the names of all of you who dedicated your efforts to improving the researches of the Royal Society will be handed down to posterity'.[9]

Great concern seems to have been felt to assure the permanence of the institution, to give it 'perpetual succession', as was stated a number of times in the Society's original charter and more often still in the revised version of 1663. Indeed, it is evidently this that explains why what to us may seem to be an inordinate amount of energy and expense was devoted to refining the constitutional structure of the Society in its first few years. Thus the elaborate first charter was found unsatisfactory within a year of being issued, so that it had to be replaced by a new version, changed in various respects which will be considered more fully below, a change which involved the Society in expenditure totalling £53.7.8, in addition to the £35.10s that had already been spent on the original that was now rendered

[7] Birch, *History*, i, 168; Sprat, *History*, p. 431.

[8] Birch, *History*, i, 7. On the repository, see below, 'Between Cabinet of Curiosities and Research Collection'.

[9] Oldenburg to Hevelius, 31 Jan. 1668, *Oldenburg*, iv, 137.

redundant.[10] There were also a succession of anxieties about details in the accompanying statutes that the Society formulated, although one lawyer who was consulted about them assured the Society 'that he found no exceptions at all against them, but must highly approve of them for the reason, convenience, and perspicuity therein; wishing, that all other laws might equal them in these qualifications'.[11]

Indeed, even more time may have been spent discussing such matters than is now apparent: John Evelyn records in his diary a series of Council meetings on Wednesday mornings from August 1662 onwards 'for the regulation of [the] Society, & making of such Laws & statutes as were conducible to its establishment & progresse', while a committee on this subject has left a couple of reports but no minutes of its meetings.[12] It was appropriate that the Society's charter and statutes were one of the components of the promotional *History* of the Society by Thomas Sprat, published in 1667, that were most closely supervised, the Secretary, Bishop John Wilkins, being requested to select 'the most material and least alterable particulars' of the statutes for insertion in it. Moreover, in commenting on these constitutional documents, Sprat wrote how the lawyers who had assisted in getting them ratified had thus 'compleated this foundation of the *Royal Society*', which it became 'the greatness of their Prudence to establish'.[13]

To us, all this may seem rather legalistic and vacuous. But it is interesting that in 1664 Hevelius requested information about the Society's 'constitution', and it would be wrong to underestimate the importance that was attached to such considerations at the time.[14] For beneath this concern for legal propriety lay a wish to obtain an organizational formula which would assure the Society's indefinite survival. The concern for permanence appears repeatedly: thus while Sprat expressed his conviction that the Society would be '*Immortal*', Oldenburg purveyed comparable sentiments to the overseas intellectuals to whom he wrote on behalf of the Society, 'the strength of whose design we expect to render it everlasting, God willing'.[15] Elsewhere, he speaks of the intention 'to establish their Institution to perpetuity', and it

[10] See Royal Society accounts s.v. 7 Aug. 1662 and a separate account attached to those for 1663 dealing with expenses from 21 Feb. to 30 May 1663 relating to the patent; it is possible that a further £6.10s paid to Moray on 6 Nov. 1661 is also connected with the first charter.

[11] Birch, *History*, i, 454. For references to concern on such points, see ibid., i, 127, 241, 249 – 50, 253, 259 – 60, 264 – 5, 268, 274, 287, 291, 293, 301, 305, 306, 309, 313, 316, 321, 323 – 4, 328 – 9, 340, 344, 346 – 7, 372 – 3, 385, 388 – 9, 394, 408, 410, 446; ii, 69, 95, 101, 131, 136, 141 – 2, 144, 145 – 6, 148, 149, 168. For the original in the hand of Sir John Hoskins of the paper printed in Birch, *History*, ii, 145 – 6, see D.M. 5.59; see also below, 'An Experiment in Corporate Enterprise', pp. 75, 104.

[12] Evelyn, *Diary*, iii, 332. On the committee, see last note.

[13] Sprat, *History*, p. 144; Birch, *History*, ii, 163.

[14] Birch, *History*, i, 414. Cf. Wallis to Hevelius, 5 April 1664, *Oldenburg*, ii, 166-8.

[15] Sprat, *History*, p. 79; Oldenburg to Sluse, 23 Oct. 1667, *Oldenburg*, iii, 537. Cf. ibid., pp. 595 – 6, 612.

is clear that it was widely presumed that 'a publick Establishment' was crucial to achieving the ends that the Society's founders had set themselves.[16]

This raises the question of *why* it should have been felt so important to institutionalize the new science in this way and at this particular juncture. One very general answer would invoke the influence of Francis Bacon. Certainly, Bacon had provided a generalized model for the Royal Society to follow, both in his view of 'Solomon's House' in his *New Atlantis* and in his exposition in *The Great Instauration* and other works how by collaborative effort, the reform of knowledge could be achieved and perpetuated in a way that had not been possible through the uncoordinated efforts of individuals.[17] But Bacon's influence explains little either about the exact means by which the Society's founders went about establishing an institution that was to be at once both public and permanent in the context of contemporary society, or about the time at which this took place. Indeed, the time is itself arguably crucial, for the context to the Society's foundation is provided by a more general urge to organize which is in evidence during the decades on both sides of 1660.

The Interregnum had seen a succession of schemes intended to bring organization to various facets of intellectual and public life, though few of these got beyond the drawing-board. There is Samuel Hartlib's much-promoted 'Office of Address', for instance, or William Rand's scheme for a College of Graduate Physicians, while even a body like the Council of Trade and Plantations should arguably be seen in the light of a similar feeling that organized bodies could achieve that which less formal ones could not. In addition, the latter years of the Interregnum saw various proposals for endowed institutions devoted to research which have a certain amount in common with the Royal Society as it was to materialize.[18]

It is also hardly coincidental that the first public museum in this country was set up in the late seventeenth century, the Ashmolean Museum at Oxford, its founder, Elias Ashmole, realizing plans which owed at least something to the aspirations of the Tradescants from whom he had bought the core of his collections. Ashmole was talking about the idea of a 'publique Musaeum' at least as early as 1674, and his subsequent negotiations with the University of Oxford took for granted the need for permanent establishment entailing the provision of a proper building and of statutes regulating its administration. Moreover the same preoccupation with permanence was displayed by one of the museum's principal early benefactors, John Aubrey, who eagerly embraced the opportunity that its foundation offered to secure

[16] *Phil. Trans.*, 2 (1667), 503. See also below, 'A "College" ', pp. 159, 161.
[17] The fullest statement of this argument is Margery Purver, *The Royal Society: Concept and Creation* (London, 1967), but it is there so simplistically put that the germ of truth that it contains is obscured. For an assessment, see Hunter, *Science and Society*, chs. 1 – 2.
[18] Charles Webster, *The Great Instauration* (London, 1975), pp. 67f., 97 – 8, 300f., 461 – 2 and passim.

the indefinite survival of his manuscripts by consigning them to a 'publick Repository'.[19]

Extrapolating still more widely, one may note an increasing sense during these decades of the need to move away from a dependence on the purely personal and hence potentially capricious towards more public and formal structures and procedures. In government, we see a shift towards more impersonal and bureaucratic institutions and more public-spirited ideals of public service, both in the Interregnum and in the later years of the century: moreover recent research has placed post-Restoration developments at an earlier date than has often been thought.[20] A still broader context is provided by a widespread discussion among intellectuals of the day of the 'public interest' and how it was to be defined in relation to the interests of individuals, which bears witness to a similar sense of potential conflict between public and private concerns.[21]

Why did all this occur, and in particular what is the explanation of the trend towards institutionalization of which the Royal Society's foundation forms part? Charles Webster, in *The Great Instauration*, has laid stress on the influence of 'puritan experiments in intellectual organisation', and it is undoubtedly true that a number of those associated with the Royal Society in its formative period had had links with the kind of proposals from the 1640s and 1650s already referred to, including Sir William Petty, John Wilkins, Robert Boyle and Henry Oldenburg.[22] But claims for a 'specifically puritan' influence are problematic: Webster has himself shown how mixed in their political affiliations the early activists of the Royal Society were, and by no means all of those who were instrumental in setting the Society up had puritan links (and neither, for that matter, had Ashmole). In fact, royalists, as well as puritans, were active in suggesting establishment as the proper way forward for the advancement of learning in the years around 1660. This is shown by the scheme for a 'college' that John Evelyn expounded in 1659, which he depicted in a manuscript view and plan which are reproduced for the first time later in this volume. It is further illustrated by Abraham Cowley's detailed *Proposition For the Advancement of Experimental Philosophy*, published in 1661, which visualized a comparable residential college for the pursuit of natural philosophy. There are hints of comparable notions in Sir Robert Moray, the man who was to submit to the King the petition for incorporation which was crucial to the Society's establishment

19 C. H. Josten, *Elias Ashmole* (Oxford, 1966), i, 143, 197, 205 – 6, 218, 248 – 9; Michael Hunter, *John Aubrey and the Realm of Learning* (London, 1975), p. 90.

20 See G. E. Aylmer, *The State's Servants: the Civil Service of the English Republic, 1649 – 60* (London, 1973); Howard Tomlinson, 'Financial and Administrative Developments in England, 1660 – 88', in J. R. Jones, ed., *The Restored Monarchy, 1660 – 88* (London, 1979), pp. 94 – 117.

21 J. A. W. Gunn, *Politics and the Public Interest in the Seventeenth Century* (London, 1969).

22 Webster, *Great Instauration* (n. 18), pp. 95 – 6. Cf. pp. 88, 99.

and which Moray probably wrote himself.[23] Hence it seems that — as so often in *The Great Instauration* — ideas have been identified as specifically 'puritan' which were in fact more widely shared, returning royalists having evidently come to see the need for permanence as much as their parliamentarian colleagues.

Another important potential influence on the nascent Society has recently been highlighted by the Italian scholar, Anna Strumia, after it had been consigned to a trivializing footnote by Christopher Hill: this concerns James Harrington's 'Rota' club, which met in 1659 – 60.[24] Certainly it is striking how many of those members of the Rota Club whose names we know subsequently became Fellows of the Royal Society, though not all were particularly early or prominent ones: according to my calculation, it is eleven out of twenty-seven, or 40 per cent.[25] Moreover among these were some who were clearly crucial in forming the Society's constitutional structure. Perhaps the most important of these was the Society's first 'Register', Dr William Croone; but also influential was Sir William Petty, while, from 1662 onwards, Sir John Hoskins was one of the lawyers who was most active in refining the Society's statutes.

Moreover, it is impossible not to be struck by the similarity between aspects of the Society's attention to organizational propriety, and that of the Rota club. The use of a secret ballot to reach decisions is a prominent feature of the Society in its earliest years (and thereafter), the rules for this being elaborately formulated at the Society's meeting on 12 December 1660.[26] Though balloting was not unknown among corporate bodies before this, it does not seem coincidental that the practice was made notorious by the Rota Club, of which it was a distinctive feature. Other practices common to the two groups include the imposition on a social grouping of a marked formality of proceedings, with regulations and minutes, contributions levied from members, and the delegation of matters to committees. A further practice in the Society's earliest period which conformed with the ideas of Harrington was the monthly rotation of chairmanship, though this

23 On these schemes, see below, 'A "College" ', pp. 157, 181 – 4, and Webster, *Great Instauration* (n. 18), pp. 97 – 8. On Moray, see Birch, *History*, i, 45, 50, 104; Alexander Robertson, *The Life of Sir Robert Moray* (London, 1922), ch. 8, esp. p. 152.

24 Anna M. Strumia, 'Vita Instituzionale della Royal Society Seicentesca in Alcuni Studi Recenti', *Rivista Storica Italiana*, 98 (1986), 520 – 3; Christopher Hill, *The Experience of Defeat* (London, 1984), p. 191 and n. 3.

25 See John Aubrey, *Brief Lives*, ed. Andrew Clark (Oxford, 1898), i, 289f.; Pepys, *Diary*, i, 14, 61. The men in question are: Petty (Hunter, *Royal Society*, catalogue entry 8), Croone (entry 20), Hoskins (entry 92), Aubrey (entry 127), Berkenhead (entry 138), Henry Ford (entry 145), Pepys (entry 187), Carteret (entry 188), Earl of Dorset (entry 196), Arderne (entry 265) and Robert Wood (entry 371). This includes 'auditors'.

26 Birch, *History*, i, 6. The purchase of a ballot box is recorded in the Royal Society's accounts s.v. 16 Jan. 1661; it cost £1.4.0, including gratuities.

was to be abolished by the first charter.[27] In addition, the Society's wish, classically expounded in Sprat's *History*, of achieving a mixed membership rather than 'a *Company* all of *one mind*', could be compared to the pursuit of 'balance' on Harrington's part which so much impressed supporters of his like John Aubrey, F.R.S.: 'nothing can be invented more faire and impartiall', Aubrey admiringly wrote of Harrington's constitutional schemes.[28]

What is more, underlying this in Harrington, as in the Royal Society, was a quest for permanence, an aspiration to a constitutional structure which would be so well-balanced that it would not succumb to the instability that Harrington saw as inevitable in the flawed constitutions of the past: in Harrington's case, this search was well-captured by the title of Charles Blitzer's study of his thought, *An Immortal Commonwealth*.[29] Harrington's response to this problem was unique, but his perception of the problem and its urgency was not, and one may argue that a broader background to the quest for permanent, public institutions of the kind that the Royal Society represents is a deep sense of instability that was widely shared in the aftermath of the Civil War, an event which crystallized an awareness of rapid and often disturbing change which characterizes mid-seventeenth century England as a whole.[30]

Nor did this widespread reaction to the instability of the Civil War evaporate after 1660: it seems to me that a misunderstanding of the actual state of affairs underlies Brian Vickers' recent statement that 'the social and political stability created by the Restoration . . . encouraged the institutionalization of science'.[31] For we easily overestimate the sense of permanence that was felt about the Restoration settlement in its early years: John Miller has cogently argued the extent to which Charles II's policy was based on the fear that he might be sent on his travels again, and it would not be surprising to find his subjects sharing similar anxieties, which were, indeed, to surface in the Exclusion Crisis.[32] Their relevance to people's institutionalizing

27 Birch, *History*, i, 6 and passim; Pepys, *Diary*, i, 13, 20, 24; J. G. A. Pocock, ed., *The Political Works of James Harrington* (Cambridge, 1977), pp. 69 – 70, 807f. On the Royal Society's committees, see below, 'An Experiment in Corporate Enterprise'. On balloting, see especially H. F. Russell Smith, *Harrington and his Oceana* (Cambridge, 1914), pp. 37 – 9 (see also ibid., pp. 101 – 8, on the Rota).

28 Sprat, *History*, p. 73; Aubrey, *Brief Lives* (n. 25), i, 291. See also Strumia, 'Vita Instituzionale' (n. 24), p. 523, and Sprat, *History*, p. 106, for 'Harringtonian' comments on Venice and Rome. It is perhaps of interest to note that Oldenburg's 'Liber Epistolarum' includes notes on government from Harrington's *Oceana*, with particular reference to the problem of 'erecting a new monarchy' (Royal Society MS 1, fols. 90v – 6).

29 Charles Blitzer, *An Immortal Commonwealth: the Political Thought of James Harrington* (Yale, 1960). Cf. Pocock, *Political Works* (n. 27), passim.

30 Hunter, *John Aubrey* (n. 19), p. 165. Cf. Paul Delany, *British Autobiography in the Seventeenth Century* (London, 1969), pp. 10 – 11.

31 Brian Vickers, *English Science, Bacon to Newton* (Cambridge, 1987), p. ix.

32 John Miller, 'The Later Stuart Monarchy', in Jones, ed., *Restored Monarchy* (n. 20), pp. 30 – 47.

aspirations is underlined by a reference to what might be done in the event of 'civil wars or like calamities' during Ashmole's negotiations with the University of Oxford for the establishment of his collections there.[33] Moreover concerns of this kind were arguably not even limited to England. As we shall see in a later essay, acclaim for the Royal Society as an institutional embodiment of the new science was to be found throughout Europe, and this owed at least something to a shared experience of turmoil, both England and other countries having been 'shaken recently by civil wars and now by foreign conflicts'.[34]

Hence the Royal Society was intended to be permanent, stable, safe, after the uncertainties of the mid-century, and to this extent, it was an authentic part of the ethos which brought about the Restoration. Yet, as its ability to take in a mixture of erstwhile royalists, puritans and others shows, it would be wrong to stress unduly its party-political line. Indeed, though it may seem to us paradoxical, the principal reaction to the experience of Civil War seems to have been less to try to enrol the new science to a specific viewpoint than to aspire to its 'establishment' as a worthwhile step in itself. As will be explained more fully below in my study of the Society's principal work of self-promotion, Sprat's *History*, its organizers seem to have sincerely believed that the enterprise to which the early Royal Society was dedicated *was* healing, that it would in some sense escape from politics by bringing together reasonable men from a wide range of ideological positions who could collaborate in gathering information which they hoped that all would be able to accept as undeniably true. What they aspired to was a structure within the parameters of which people could collaborate without risking the damaging divisions which the Civil War and Interregnum seemed to exemplify, and for this 'a form'd, and *Regular Assembly*' like the Royal Society seemed crucial.[35]

Moreover it may be argued that the kind of eirenic aspirations that underlay the founding of the institution had implications for the method which was to be pursued in its work. There was a strong incentive to find a healing, consensual philosophy, and for this the laborious empiricism advocated by Bacon was ideally suited, to an extent that would arguably not have been the case with Cartesianism, for instance. A philosophy based on the accumulation of empirical data might be 'impartial', 'balanced', appealing equally to the whole of a 'mixed' membership, by whom it was equally guaranteed, and this would end the divisiveness which traditional intellectual activity was often deemed to have fostered.[36] Hence the impulse

[33] Josten, *Elias Ashmole* (n. 19), iv, 1710 (this appears in the memorandum by Obadiah Walker on which Ashmole's statutes were based).

[34] Lubietzki to Oldenburg, 18 Aug. 1666, *Oldenburg*, iii, 212. See below, 'Promoting the New Science', pp. 251f.

[35] Sprat, *History*, p. 52.

[36] On this theme, see Hunter, *Science and Society*, pp. 27f.; P. B. Wood, 'Methodology and Apologetics: Thomas Sprat's *History of the Royal Society*', *British Journal for the History of Science*, 13 (1980), 1 – 26.

towards institutionalization itself entailed strong pressures towards the actual programme adopted.

Indeed, so strongly do its founders seem to have believed in the efficacy of the co-operative pursuit of empirical knowledge within the setting of a formal institution that in the first instance they simply devoted themselves to a frenetic round of activity in accumulating experimental or observational data about all sorts of natural phenomena without feeling the need for either a coherent plan of action or a public statement justifying what they were doing. It was almost as if it was deemed self-evident how the combination of empirical investigation and formal organization would produce the required results.

To us in retrospect, both the activity and its rationale may seem naive. It has thus been argued that Baconianism was the more assiduously espoused as a means of obscuring genuine differences of approach among natural philosophers, and it is certainly clear that many leading figures in the Society had a more overtly hypotheticalist position than would be gathered from certain early expositions of the Society's aims.[37] Moreover, in practice, the rather inchoate range of activity which characterized the Society in its earliest years also proved problematic, and the need for a more precise focus for the Society's work and a more articulate public exposition of its significance was to become apparent within a few years, as will be illustrated more fully in the second part of this Introduction.

There are other respects in which the Society arguably failed to live up to its promise. Thus for all its professions of a policy of openness, it was less comprehensive in its membership than it liked to imply. For one thing, it was in fact highly socially exclusive, and even such occasional figures of middling status as the government clerk, John Collins, or the entrepreneur and journalist, John Houghton, stand out among the membership: indeed, Houghton virtually earned himself a vote of censure when he tried to expand membership among men of his ilk in the 1680s.[38] Similarly, among medical practitioners, for instance, the Society recruited its members quite disproportionately from the educated élite represented by the Fellows of the London College of Physicians rather than surgeons or apothecaries, who remained rare in the Society's ranks throughout its early years.[39] Indeed, in some ways the Society arguably made its task unnecessarily difficult by cutting itself off from a natural clientele in terms of men with a professional interest in applied science whose role in the formative years of the scientific movement has recently been stressed.[40] It may be, however, that in terms of

37 Charles Webster, 'The Origins of the Royal Society', *History of Science*, 8 (1967), 114f.; Wood, 'Methodology and Apologetics' (n. 36).

38 See Hunter, *Royal Society*, pp. 41, 61. On Collins see Hunter, *Science and Society*, pp. 72 – 3.

39 See Hunter, *Royal Society*, p. 8 and passim. On the relationship of the different sectors of the medical profession, see H. J. Cook, *The Decline of the Old Medical Regime in Stuart London* (Ithaca, 1986), ch. 1.

40 See J. A. Bennett, 'The Mechanics' Philosophy and the Mechanical Philosophy', *History of Science*, 24 (1986), 1 – 28.

'establishment' its organizers thought it more important for science to be espoused by the 'powers that be', as evidenced by its impressive-sounding but rather misleading membership list, filled by the names of aristocrats, bishops and others who never attended meetings.[41]

Even the idea that science could somehow transcend politics may seem misconceived. As my essay on Sprat's *History* illustrates, in commissioning that book the Society's organizers greatly underestimated the difficulty of spelling out the broad implications of the new science in a manner that would not cause offence in an age alert to the slightest hint of heresy or subversiveness. Moreover, from our twentieth-century vantage point, it is easy to see how this attempt to create an apolitical scientific sphere was not 'neutral' in so far as it had the effect of investing a group of intellectuals with a privileged position. More insidious still were its 'statist' overtones, its tendency to induce enthusiasm for strong government either of a left or right-wing hue, which I have indicated at length elsewhere.[42] Moreover, in presenting itself as 'apolitical', the Society consciously rejected those whose religious views were deemed to represent a threat to the status quo, those 'Enthusiasts & Phanatiques' to whose ideas, significantly, John Aubrey saw the moderate Harrington's views as antithetic.[43] For similar reasons, the Society was also under pressure to drive a wedge between itself and Thomas Hobbes, that bogey-man of the Restoration establishment.[44]

Yet, in illustrating that the Society was not — indeed, could not have been — as 'apolitical' as its founders initially hoped, we must avoid going to the opposite extreme, of trying to align it more narrowly with political positions of its day than was actually the case and identifying a more overt programmatic implication in its scientific work than actually existed. Even in its relations with Hobbes, the Society managed to maintain a greater ambivalence towards him than was advisable in the circumstances or than might be gathered from the most recent account of this matter, even seeing no objection to having a picture of the philosopher hanging in its rooms.[45] Indeed, a number of recent writings on the early Royal Society referred to in the Bibliographical Essay seem to me guilty of underestimating the Society's pluralism and presenting it in an unduly sectarian light, often by an unjustified extrapolation from the attitudes of some individuals within it to those of others. Readers of J.R.Jacob's study of Henry Stubbe, for instance, might be forgiven for finding the idea of a strong Harringtonian influence on the Royal Society surprising in view of the clarity of the polarization between the two that Jacob depicts on the basis of the attack on Harrington by Matthew Wren (later F.R.S.), a member of the Interregnum Oxford group of natural philosophers whose members were to play so important a role in the Society's foundation. Moreover, this goes along with an unduly

41 See Hunter, *Royal Society*, p. 10.
42 Hunter, *Science and Society*, ch. 5.
43 Hunter, *John Aubrey* (n. 19), p. 56.
44 Hunter, *Science and Society*, pp. 178 – 9; Shapin and Schaffer, *Leviathan and the Air-Pump* (n. 1), pp. 129 – 39 and passim.
45 Aubrey, *Brief Lives* (n. 25), esp. i, 372, and see the references cited in n. 44.

'clerical' view of the whole natural philosophical community on Jacob's part, misleadingly generalized from the characteristics of individual members of it, which is also to be found in the recent work of Shapin and Schaffer.[46]

The partiality of the kind of politicized view of Stubbe's arguments with the Royal Society that we have recently been encouraged to adopt is illustrated by such reactions of contemporary outsiders as happen to be recorded. Take, for example, the Kendal schoolmaster Richard Jackson, who wrote on hearing of Stubbe's assault on the Society: 'Let others censure, let others laugh, I will not. Their method does promisse well to mankinde, and if in its infancy it meet with scoffes from the ignorant or envious world, it bootes not'. He compared the ridicule experienced by Columbus, 'or whatsoever checkes that Universall knowledge, that some mens Conceits have advanced them to'.[47] Jackson may have missed the point of Stubbe's attack — though it is equally possible that Stubbe's most recent commentator is at fault — but what is significant is who he expected to oppose the new science and why. Equally revealing is the attitude of the former critic of the universities, John Webster, who, according to certain accounts of the ideological polarization surrounding Restoration science, might have been expected to be Stubbe's ally: for in fact Webster was hostile to Stubbe's attack for reasons similar to Jackson's.[48]

In fact, what such commentators seem to be saying is not that the new science was a tool of a party, whether establishmentarian or otherwise, but that it was a new force which was widely unappreciated and, indeed, opposed, but which deserved encouragement for the hope for the future that it held out. Moreover this sense that the new philosophers were an embattled and unpopular minority brings us to a crucial point, significant both in its own right and for the Royal Society's institutional ambitions, which is currently in danger of being overlooked. For a number of recent studies have greatly overestimated the seriousness with which the new science was generally taken at the time.

Though a distaste for instability undoubtedly stimulated enthusiasm for the new philosophy, it is important to stress that the pursuit of science was only one of a number of possible responses to the events of the mid-century,

[46] J. R. Jacob, *Henry Stubbe, Radical Protestantism and the Early Enlightenment* (Cambridge, 1983), pp. 3 – 4, 14 – 17 and passim (see also his 'Restoration Ideologies and the Royal Society', *History of Science*, 18 (1980), 26 and n. 10); Shapin and Schaffer, *Leviathan and the Air Pump* (n. 1), ch. 7. Cf. also Christopher Hill, 'A New Kind of Clergy: Ideology and the Experimental Method', *Social Studies of Science*, 16 (1986), 726 – 34, esp. p. 732.

[47] Jackson to Fleming, 28 Oct. 1669, in J. R. Magrath, ed., *The Flemings in Oxford, vol. 1, 1650 – 1680* (Oxford Historical Society, 44 (1904)), 183 – 4. Cf. *Oldenburg*, vii, 54, 56, 106, 327.

[48] Webster, *The Displaying of Supposed Witchcraft* (London, 1677), p. 4. Cf. T. H. Jobe, 'The Devil in Restoration Science: the Glanvill-Webster Witchcraft Debate', *Isis*, 72 (1981), 343 – 56. See also Peter Elmer, *The Library of Dr John Webster: the Making of a Seventeenth-century Radical* (*Medical History, Supplement No. 6*, 1986), pp. 7 – 10.

and not a particularly common one at that.[49] Charles II and his ministers would, I suspect, have been amazed by a recent estimate of the intellectual activity which the Royal Society symbolized, that 'This experimental form of life achieved local success to the extent that the Restoration settlement was secured. *Indeed, it was one of the important elements in that security*'.[50] Charles himself was more interested in the new science than many of his subjects: but it is revealing that even he dismissively referred to the Royal Society as his 'fous', and few of those in positions of power around him — or in the country at large — were persuaded by the promises of either the practical or apologetic benefits of the new science to see such activity as anything more than a harmless curiosity.[51] They responded with according parsimony to calls for support and Richard Jackson was not alone in discerning a thorough-going apathy as the predominant contemporary reaction towards this intellectual enterprise. A further instance is provided by Richard Graham, Viscount Preston, who opined in his *Angliae Speculum Morale; the Moral State of England* (1670) that the achievement of contemporary science was 'never enough to be gratefully admired, though it hath the ill fortune to be little esteemed of now, but in ensuing ages, it will certainly be crowned with its due applause; for it is always seen that great Actions are deny'd their Bay's in that age in which they are born, because Envy and prejudicate Malice (the off-springs of the old Serpent) detract from them'.[52]

Up to a point, the new science could subsist by the efforts of dedicated individuals, who doggedly sought accurate knowledge about the natural world despite lack of support, hoping that this new and untried enterprise would somehow create a more efficient and equitable society, in which the old problems would be solved and the destructive divisions of the past would not recur. What is more, the sense of being pioneers of something unprecedented and important against the forces of obfuscation and inertia could itself have a significant galvanizing effect, actually binding together individuals who — apart from this common set of interests and sense of mission — may have had little in common. Indeed, by laying too much stress on the 'establishmentarianism' of such men, we fail to do justice to the sense of adventure that they felt about what they were doing.

But, in addition, we may here return to the role of the Royal Society, because it may be argued that it was precisely because of this sense that the new philosophy was undervalued that it was perceived to need the

[49] For alternatives, see, e.g., Hunter, *Science and Society*, p. 154 and the references there cited; John Gascoigne, 'Politics, Patronage and Newtonianism: the Cambridge Example', *Historical Journal*, 27 (1984), 3f.

[50] Shapin and Schaffer, *Leviathan and the Air Pump* (n. 1), p. 341 (my italics). For criticisms of political historians for their neglect of science on the basis of comparable overestimates, see Webster, *Great Instauration* (n. 18), p. 519, n. 41; Steven Shapin, 'Social Uses of Science', in G. S. Rousseau and Roy Porter, eds., *The Ferment of Knowledge* (Cambridge, 1980), p. 105 and n. 29.

[51] W. E. K. Middleton, 'What did Charles II call the Fellows of the Royal Society?', *Notes and Records*, 32 (1977), 14.

[52] Richard Graham, Viscount Preston, *Angliae Speculum Morale* (London, 1670), pp. 41 – 2. On the lack of support, see Hunter, *Science and Society*, pp. 38 – 40, 83 – 4.

'establishment' which that body represented. The point can be well-illustrated by two contrasting usages of the word 'institutionalization' which are otherwise liable to confuse the unwary (this is a further reason, apart from deference to contemporary usage, for preferring to employ the word 'establishment'). Whereas in 'commonsense' usage, 'institutionalization' is taken to mean the onset of formal organization ('the process whereby scientific activity is incorporated into formal organizations and socially recognized roles'), the sociological usage takes the word to refer to the establishment of collective norms ('the processes by which knowledge and associated behaviours are established as the basis for *standardized collective* reference and action').[53] In the formative years of the Royal Society, it was arguably precisely because the new science was not institutionalized in the sociological sense that it was seen as needing institutionalization in that word's commonsensical meaning.

To summarize, in setting up the first scientific institution in this country, the founders of the Royal Society were doing two things. In part, they were providing a focus through which the public acceptability of the new science could be enhanced. But they did not distinguish this from the actual scientific activity which they hoped would be intensified through the new forum which they had created. Indeed, the two went hand in hand, for it was partly because of the nature of the institution that they were establishing — a permanent body dedicated to the collaborative production of knowledge in a value-free atmosphere — that they thought that the enterprise would win the public acclaim which in their view it deserved. In the context that has been outlined here, 'establishment' has a real significance in itself which is easily underrated.

(b) Building an Infrastructure

Aspiring to scientific establishment is one thing; putting this into practice is quite another, and a number of authors have been guilty of confusing the two, sometimes even criticizing the Royal Society for the limitations of what it achieved by the standards of earlier schemes which had never been properly implemented, as Charles Webster does in *The Great Instauration*.[54] In fact, what seems to me so interesting about the episodes chronicled in this

53 The bracketed quotations are taken from Shapin, 'Social Uses of Science' (n. 50), p. 95 n. 3; on the differentiation of 'commonsense' and 'sociological' usage, see G. A. Abraham, 'Misunderstanding the Merton thesis: a Boundary Dispute between History and Sociology', *Isis*, 74 (1983), 374–5. See also Joseph Ben David, *The Scientist's Role in Society* (2nd ed., Chicago, 1984), ch. 5 and passim.

54 Webster, *Great Instauration* (n. 18), p. 97.

book is that they enable one to observe the actual process of institutionalization in what was effectively the first organization of its kind, where there were therefore no direct models for what was being attempted and precious few indirect ones. They thus help to restore a sense of historical contingency to a process which is of more interest in its own right than has hitherto generally been appreciated.

In so far as the Society's founders simply sensed that the new science required a symbol, they were clearly right, not only in this country but also among scattered enthusiasts abroad. This is demonstrated by the acclaim that the Royal Society received, most strikingly through Oldenburg's correspondence, as is expounded elsewhere in this volume: but it is in evidence even earlier, for instance in a letter from the Dutch scholar, Gronovius, to Theodore Haak of 18 August 1662.[55] Similarly, the Society became a symbol in this country for the kind of activity to which it was devoted, as is shown by the tendency of both its friends and enemies more or less consciously to conflate all of recent activity in these fields with what was achieved under the Society's direct auspices.[56]

But clearly the Society was supposed to provide more than a symbol: in addition, its founders intended that 'establishment' should achieve results in terms of the organization and execution of scientific research, its support, and its application to contemporary needs. Moreover, whereas a symbol was relatively easily created, a functioning institution was not, since a myriad of decisions had to be made, initiatives taken and structures set up and tested. Indeed, the whole process was to prove far more difficult than had been anticipated when the Society was first mooted, and what transpired was much more of a process of trial and error than is often realized. This section of the Introduction will illustrate this by examining in detail the Society's initial steps in this direction, which present a kind of microcosm of the Society's early development as a whole.

Even the Society's name may have been a matter for discussion. The title by which the Society has been known ever since was adopted in its first charter of 15 July 1662; it had earlier been employed by John Evelyn — who may have devised it — in describing the Society in a dedicatory epistle to a book published in 1661.[57] But, however used we may have become to it since, 'The Royal Society' was a slightly odd name for a body devoted to the study of natural philosophy, and it was clearly a surprise to some Fellows: Elias Ashmole, for instance, was evidently expecting the body to be called 'the Society of . . .' — perhaps 'the Society of Philosophers' — leaving a blank after 'Society' in his notes on the institution in its earliest months pending a decision on nomenclature which in the event was to necessitate

[55] Birch, *History*, i, 108 – 9; Royal Society Copy Letter Book, i, 49. See also below, 'Promoting the New Science'.

[56] Hunter, *Science and Society*, pp. 32, 48, 55.

[57] I.e., his translation of Naudé's *Instructions Concerning Erecting of a Library* (London, 1661), sigs. A3, A4v. Cf. Evelyn, *Diary*, iii, 306. It is unclear from the diary entry whether Evelyn invented the title or merely gave it a public airing.

his inserting 'Royal' above the line.[58] If the chosen formula reveals the Society's priorities in wishing to capitalize on its royal patronage, it is perhaps significant that, though the first charter referred to the body simply as 'The Royal Society', this was modified in the second to a more cumbersome but informative version, 'the Royal Society of London for promoting Natural Knowledge', possibly because the brief form was thought to be ambiguous and misleading.[59]

We also know something about the discussion which preceded the selection of the Society's symbolism in the form of the coat of arms and motto which were to be ratified on 17 September 1662 and officially confirmed by the King in the second charter, since a sheet of paper survives on which alternatives are noted in the hand of John Evelyn.[60] Probably, these comprise a mixture of Evelyn's own suggestions and those of other Fellows, jotted down at a meeting at which the matter was discussed: they show that, instead of the restrained and establishmentarian coat of arms which the King granted the Society — a shield blank except for a canton of the royal arms — it might have had something much more allegorical or mystical. One possibility was a Sun in splendour with the motto 'To The Greater Light'; another a celestial globe juxtaposed with an all-seeing eye; another a hand holding a plumb-line, possibly a Masonic symbol; while other designs had more directly scientific components, like a pair of crossed telescopes. On the other hand, the place of 'Nullius in Verba' (a paraphrase of a passage from Horace's *Epistles* affirming freedom from any master or school) might have been taken either by the Masonic-sounding formula already quoted, or by the more overtly Baconian 'Et Augebitur Scientia', a quotation from Daniel xii, 4, which had inspired much millenarian speculation earlier in the century.[61]

[58] Josten, *Elias Ashmole* (n. 19), iii, 810 – 11. John Graunt, in dedicating his *Natural and Political Observations . . . upon the Bills of Mortality* to the Society in 1662 calls it the 'Royal Society of *Philosophers* meeting at *Gresham-Colledg*' (Sir William Petty, *Economic Writings*, ed. C. H. Hull (Cambridge, 1899), ii, 322). The formula 'Royal Society' was apparently first used in the minutes in June 1662: Birch, *History*, i, 85.
[59] *The Record*, pp. 215, 237.
[60] See plate 1 and pp. 41 – 2, below. On the ratification (which is not noted in Birch) see Evelyn, *Diary*, iii, 336. For the authorization in the second charter, see *The Record*, p. 222. See also J. D. G. D., 'The Arms of the Society', *Notes and Records*, 1 (1938), 37 – 9.
[61] For the source of 'Nullius in Verba', see Horace, *Epistles*, I, i, 14. For the millenarian overtones of 'Et Augebitur Scientia', see Webster, *Great Instauration* (n. 18), pp. 9f. Of the mottos suggested, 'Omnia Probate' almost certainly originated with Evelyn, since these words are a variation of one of the mottos that he used on the bindings of his books, 'Omnia Explorate, Meliora Retinete'. For a design evidently by Ashmole, see below, p. 42, but it is not clear who suggested the remainder of either the designs or the mottos. Another Fellow with a documented interest in symbolic devices and mottos was Moray: see David Stevenson, 'Masonry, Symbolism and Ethics in the Life of Sir Robert Moray, F.R.S.', *Proceedings of the Society of Antiquaries of Scotland*, 114 (1984), 412f. and 420f. (esp. p. 423 for the use of the motif of the sun). For a discussion of possible patron saints for the Society, see Aubrey, *Brief Lives* (n. 25), ii, 144.

Perhaps more important are the constitutional developments that occurred in the Society's earliest years. For instance, though Thomas Sprat in his *History* of 1667 was to defend 'the largeness of their number', it was initially suggested that the membership should be limited. It was thus mooted at the Society's initial meeting 'that the stated number of the society be fifty five', though supernumeraries were to be admitted in the form of those of the rank of baron or above, professors at Oxford and Cambridge and perhaps Fellows of the College of Physicians; subsequently, on 23 January 1661, 'It was resolved, that no more be proposed' until those who had initially been invited to join definitely made up their mind. On 20 March 1661, however, 'It was resolved, that the number of the members of the society be enlarged', and thereafter the idea of a limited membership was sometimes specifically denied, as when John Wallis told Hevelius that 'the Society has no fixed number of members'.[62] The limitation to fifty-five was, however, picked up and exploited at the Society's expense by Thomas Hobbes in his *Dialogus Physicus de Natura Aeris* of 1661, while it is interesting to find that the idea of a finite membership was revived during debates on the Society's structure in the 1670s.[63]

There was also discussion as to the time of day at which meetings should be held, which is significant because it must have related to the extent to which participation in the Society was perceived to be a full-time or leisure activity for those involved: it was thus proposed in 1661 that the time be changed from 3 o'clock in the afternoon to 9 o'clock in the morning. Nothing, however, came of this: evidently 'some of the Doctors of Physick said diverse things against it', presumably because it would have impinged on their practice, and it is symptomatic that the principal change in the Society's arrangements for its meetings in its early years, which occurred later in the 1660s, was necessitated by the public duties of members taking priority over their philosophical pursuits.[64]

Other constitutional changes are revealed by the innovations introduced in the first charter and the modifications made in the second to the state of affairs ratified in that.[65] As originally constituted at its initial meetings in

[62] Sprat, *History*, pp. 71f.; Birch, *History*, i, 5 – 6, 13, 19. The status of Fellows of the College of Physicians was linked to discussions then in progress about the possibility that meetings might be held at the College. See also ibid., i, 7, 23; Wallis to Hevelius, 5 April 1664, *Oldenburg*, ii, 170.

[63] See Shapin and Schaffer, *Leviathan and the Air Pump* (n. 1), pp. 113, 350; and below, 'Towards Solomon's House', p. 196.

[64] Moray to Boyle, 11 Oct. 1661, Royal Society Boyle Letters vii, 18. Cf. Birch, *History*, i, 46, where a debate about altering the time is announced for the next meeting; it does not, however, appear in the minutes for it. On the later change, see Birch, *History*, ii, 144, and Pepys, *Diary*, viii, 242. See also Birch, *History*, i, 249 – 50, 260.

[65] For the charters and statutes of 1662 – 3, see *The Record*, pp. 215 – 63. Latin texts are also given in Birch, *History*, i, 88 – 96, 221 – 30, and an English translation of the second in Sir Henry Lyons, *The Royal Society, 1660 – 1940* (Cambridge, 1944), pp.

December 1660, the Society was served by a set of officers whose functions only partially overlapped with those to be laid down by the charters. There was already an annually elected Treasurer, but instead of a Secretary there was a 'Register', an office to which William Croone was nominated at the Society's inaugural meeting of 28 November 1660, who had a generalized responsibility for the Society's records but whose other tasks are unclear. In addition, as already noted, the nascent Royal Society had a rotating 'President' or 'Director' elected monthly, a position held by Sir Robert Moray, John Wilkins, Lord Brouncker, Robert Boyle and possibly the doctor, Sir George Ent. Administrative and other chores were dealt with by ad hoc committees, as is detailed more fully in a later essay in this book, while even the task of minute-taking was initially given to 'three or more of the society', who were to act as 'reporters'.[66]

It was the first charter which introduced the elaborate constitutional structure that has already been referred to, with a President, Treasurer and two Secretaries who were elected annually, and with the Society's major decision-making body taking the form of a Council of twenty-one, comprising these and seventeen other Fellows, ten of whom were replaced at an election held on St Andrew's Day each year.[67] Such arrangements were presumably thought to be more conducive to the permanence to which the Society aspired than the less formal arrangements initially adopted. In addition, the charter formalized the Society's membership, laying down that the Council should confirm the list of existing Fellows and insisting that all candidates should be duly elected.

The second charter, on the other hand, went into more detail on various points, mostly relating to the Society's constitutional structure, and, in particular, it introduced an elaborate system of quorums for decision-making and elections. Though such quorums had been mentioned at the Society's early meetings, the first charter had been surprisingly inexplicit on such topics, referring only to a quorum of seven Council members in a few specific connections, the figure evidently deriving from the similar number of Fellows who were expected to be present to witness the imposition of

329 – 40. Purver, *Royal Society* (n. 17), pp. 136 – 8, briefly discusses the differences between the charters, but dismisses all but a few of them as insignificant. For the granting of the second charter, see *Calendar of State Papers, Domestic, 1663 – 4*, pp. 15, 84. Sprat in his *History* (pp. 134 – 43) gives a shortened translation of the second charter, not the first, as wrongly stated by Cope and Jones in the 'Notes' to their edition, p. 14; Sprat also includes an abstract of the statutes, pp. 144 – 8.

66 Birch, *History*, i, 4, 6 – 7; Hunter, *Royal Society*, p. 77. See below, 'An Experiment in Corporate Enterprise', pp. 74 – 5.

67 It is perhaps worth noting that J. E. McClellan — whose understanding of the seventeenth-century history of the Society is limited — gives a misleading view of affairs in claiming the emergence of the Council's dominance as an eighteenth-century novelty: *Science Reorganized: Scientific Societies in the Eighteenth Century* (New York, 1985), p. 61.

oaths.[68] Now, however, a whole series of passages were inserted, laying down a different quorum for various types of decision. For the election on St Andrew's Day, thirty-one Fellows had to be present; for meetings at which new Fellows were elected, at least twenty-one; for other decisions, nine; while, as for Council meetings — at which most major matters of policy were dealt with — eleven had to be present for decisions on most topics, though nine or seven might be allowed on specific subjects. A quorum of nine was required for extraordinary meetings, which had to be properly summoned, while in each case the majority required for a valid decision was specified as two-thirds.[69]

Evidently these changes were the result of concern about proper decision-making and the need to guard against unconstitutional steps being taken and improper elections made. They evidently reflect the discussions from August 1662 onwards that have already been mentioned, in which a central role was played by one Fellow, Sir John Hoskins, a lawyer and erstwhile member of the Rota Club, who had not been very active before the granting of the first charter, for part of the time because he was travelling abroad.[70] In addition, disagreements may have been experienced, and, though none is specifically referred to in the minutes, it may be relevant that at the meeting at which the second charter was read 'It was ordered, that the debate concerning those, who were to be received and admitted into the society, be kept under secrecy', while the question of whether a secret or open vote should be used in Council elections had arisen in November 1662 while the new charter was pending.[71] An awareness of problems that might and indeed were to occur in later years is also suggested by the inclusion in the revised charter of a proviso allowing for the expulsion of Fellows, with a reference to the Society's statutes as providing the basis for this. Moreover the statutes which were prepared during 1663 and published in November that year did indeed go into detail concerning the reasons for which Fellows

[68] Birch, *History*, i, 5, 6; *The Record*, pp. 219, 221 – 2, 223 – 4.

[69] Ibid., pp. 237 – 50, passim. In addition, the words 'or deputy' were added to the requirement for the President to be present virtually throughout. Other more minor alterations are as follows: the document is described as 'literae patentes' rather than 'literae'; a clause is added to the effect that the Society's studies were to the glory of God and the improvement of the human race (p. 237); the proviso concerning the Statute on Alienation in Mortmain is added, as are the words 'tum realibus tum personalibus' concerning legal cases (p. 238); the period allowed for the election of Original Fellows is extended from one month to two (p. 239); the paragraph authorizing the holding of meetings is added (p. 241); the requirement to elect the clerk and the Sergeant at Mace annually is removed; provision is added in case elections were not carried out on St Andrew's Day (p. 246); what is said about the purpose of the statutes is more precise (p. 247); and the fact that correspondence was to be under the common seal of the Society was omitted. See also above, pp. 3 – 5.

[70] Cf. his letters to Aubrey at this time, Bodleian MS Aubrey 12, fols. 187 – 93. Cf. Webster, *Great Instauration* (n. 18), p. 92.

[71] Birch, *History*, i, 127, 236.

might be expelled and the means by which this should be carried out, together with numerous other organizational matters.[72]

Hence from a constitutional point of view, the charters and statutes show a significant process of evolution in the Society's earliest years. By 1663 a state of affairs had been reached which was to work well and, indeed, to provide a model for comparable organizations at a later date. Though the granting of a new charter in 1669 in connection with the royal grant of Chelsea College provided a further opportunity for emendation, only minor changes were made: the quorums required for certain decisions at Council meetings were reduced from seven to five, presumably due to business having been held up in the absence of the required number, while the number of Vice-Presidents was increased, and it was also provided that meetings could be held anywhere in England.[73] The last two changes were probably a response to the Society's experience during the Great Plague, when meetings of Society members at Oxford, though quorate, could not be recognised as more than 'a company of Virtuosi' because neither President nor Vice-President was present and because the charter of 1663 had specified that meetings could only take place within ten miles of London.[74] But it is symptomatic that such alterations were quite minor, and not all were ever invoked, as, for instance, that concerning meetings outside London.

Such changes as were made in subsequent years tended to be more in the nature of reassertions of original principles than fresh departures. For instance, on matters like expulsion, the issues involved seem to have been rehearsed every time the step was taken, though occasionally significant modifications were made to the Society's policy, perhaps most strikingly in 1682, when stricter entry requirements were imposed on candidates for election.[75] On other aspects of the Society's organization which had not been dealt with in the charters, a process of quiet evolution is in evidence, as for instance with foreign members of the Society, who had not even been mentioned in the charters or statutes: indeed in 1660 Oldenburg could even write of the newly founded Society, 'Whether foreigners will be admitted I doubt very much, though some say so'.[76] For whereas at first, most foreigners who were elected were men who happened to attend meetings of the Society while in London, gradually these were superceded by a class of

[72] *The Record*, pp. 239, 288, 300 – 1, and 287 – 301 passim. In addition, the extremely elaborate regulations that the statutes give concerning elections may be noted: ibid., pp. 291 – 3. On their publication, see Birch, *History*, i, 324.

[73] *The Record*, pp. 263 – 74. Cf. Birch, *History*, ii, 141 – 2. In addition, more detail was given about the role of the deputy, perhaps due to a concern that Brouncker might not be able to fulfill his duties owing to official business; it was also specifically stated that the President and Vice-President should take the oaths of allegiance and supremacy.

[74] Boyle to Oldenburg, 30 Sept. 1665, *Oldenburg*, ii, 537. For the restriction to ten miles, see *The Record*, pp. 246 – 7.

[75] See Hunter, *Royal Society*, p. 41 and ch. 3, passim. See also below, 'Towards Solomon's House'. pp. 219 – 20.

[76] Oldenburg to Boreel, 13 Dec. 1660, *Oldenburg*, i, 406.

overseas correspondents which looks forward to the network of overlapping membership of such societies to be found in the eighteenth century, and which was formalized from 1682 onwards, when such members were given a separate column in the Society's printed membership lists as 'Persons of Other Nations'.[77] Such alterations were minor, however, compared with the major steps taken in the early 1660s.

The charters and statutes are equally revealing about the evolution of the Society's scientific policy in its earliest years, showing a process of elaboration which evidently reflects the Society's experience during its formative period. As we have seen, the Society seems initially to have seen its experimental policy as almost self-evident, as if all that was necessary was to get on with the job of accumulating experimental and observational data. As a result, its proceedings in its first two years ranged over a whole gamut of topics, often in a rather unsystematic and unsustained way: various classic problems of natural philosophy were tackled, from ballistics to vegetation; oddities which seemed likely to elucidate the workings of nature were investigated; and a range of technological themes were pursued. The overall effect was highly miscellaneous, even allowing for the extent to which this impression is exaggerated by the nature of the minutes, which often give undue prominence to passing comments and sometimes fail to record what experiments were intended to prove even if this was clear in the minds of those who performed them. Indeed, from the outset one is made aware of the diffuseness and lack of coherence to which a rather disparate, voluntary body was always to remain prone.[78]

There are occasional passing hints in the Society's earliest months of the need for a more systematic programme, though nothing seems to have come of any of them. On 19 December 1660, for instance, each member was asked to suggest to a committee 'such experiments, as he should think fit for the advancement of the general design of the society'; while the idea of providing 'a book of theories' came up on 9 October 1661. In addition, an attempt was briefly made early in 1661 to instil more continuity between meetings.[79] By late 1662, however, some of the Society's leading figures had evidently decided that the Society's activities might be improved by the injection of a degree of professionalization. Whereas hitherto it seems to have been taken for granted that the reform of knowledge could be carried out entirely by the part-time efforts of volunteers, who were often

[77] See Hunter, *Royal Society*, appendix 10. For an earlier stage, with people being classified by their place of birth rather than of abode, see British Library Add. MS 4441, fol. 79 (reproduced here as plate 3). See also S.Gomes, A.Milian, C.Moreno and M.J.Pascual, 'Les Relaciones Internacionales de la Royal Society of London (1660 – 1700)', *Sylva Clius: Revista de Historia de la Ciencia* I (1987), 21 – 107.

[78] Birch, *History*, i, passim. For a slightly later example of the minutes failing to make clear the rationale of an experiment although this is known from other sources, compare ibid., i, 476, with the commentary in Hooke to Boyle, 21 Oct. 1664, Boyle, *Works*, vi, 494.

[79] Birch, *History*, i, 7, 20, 21, 29, 47. See also ibid., i, 80, 84, 104.

appointed 'Curators' for a specific experiment or topic, now it seems to have been felt that more could be achieved if the Society employed one or more people to act as 'Curators by Office' in a full-time capacity: such staff could systematically carry out experimental trials, follow up suggestions at meetings, and write up the Society's work, perhaps for publication. As a result, in November 1662, Robert Hooke was appointed to the post of Curator of Experiments to the Society, 'offering to furnish them every day, on which they met, with three or four considerable experiments'.[80]

The arrangements between Hooke and the Society concerning this post will be considered more fully below, but what is important here is the immediate effect of his appointment in giving the Society's programme greater coherence than hitherto. The minutes for the subsequent months show an unmistakeable contrast with those earlier, revealing a new degree of single-mindedness as a really sustained programme of investigation in a specific area of natural philosophy was inaugurated. Hooke carried out a rigorous series of experiments on the classic problems of pneumatics which unfolded from meeting to meeting, while on 23 February 1663 he brought in 'a scheme of inquiries concerning the air': this comprised a systematic experimental programme almost without precedent in the Society's proceedings, which he was to pursue over the next few months.[81] That this particular area of research was selected is not entirely surprising, for this was a highly topical area of investigation, the subject of extensive attention at this time by Robert Boyle, from whose employment Hooke had come to the Royal Society; it had also been the subject of Hooke's first book, published a year earlier, his *Attempt for the Explication of the Phaenomena, Observable in an Experiment Published by the Honourable Robert Boyle, Esq.* (1661).[82]

In addition, Hooke began to carry out sustained work on another topic, microscopy, reflecting another area of interest of the group of Oxford natural philosophers in which he had served his apprenticeship in the 1650s, though here his closest debt was not to Boyle but to Christopher Wren.[83] Now, Hooke was encouraged in a programme of presentations on this subject at meetings of the Society which, when written up, he was to publish in 1665 as *Micrographia*, one of the most successful books ever to appear under the Society's imprimatur.[84] He also served more miscellaneous functions in terms of investigating topics raised by other Fellows.[85]

80 Ibid., i, 123, 124. The concept of 'Curators by Office' is taken from the Society's statutes: *The Record*, p. 196.

81 Birch, *History*, i, 125f., esp. 202f. See also below, 'Science, Technology and Patronage', pp. 285 – 6, 288f.

82 On its topicality, see Shapin and Schaffer, *Leviathan and the Air Pump* (n. 1) and Simon Schaffer, 'Wallifaction: Thomas Hobbes on School Divinity and Experimental Pneumatics', *Studies in the History and Philosophy of Science*, 19 (1988), 275 – 98.

83 See J. A. Bennett, *The Mathematical Science of Christopher Wren* (Cambridge, 1982), pp. 73 – 4.

84 See John Harwood, 'Rhetoric and Graphics in *Micrographia*', in Michael Hunter and Simon Schaffer, eds., *Robert Hooke: New Studies* (Woodbridge, forthcoming).

85 Birch, *History*, i, 124 and passim.

This state of affairs was ratified by the second charter, in which provision was made which — significantly — had been lacking in the first, for the Society to appoint two or more 'Curators of Experiments'. It is perhaps also revealing of the Society's increasing realization of the limitations of what could be achieved by reliance on volunteers that the second charter also specified that one *or more* clerk might be employed. On the other hand, the provision for 'Sergeants-at-Mace' which was made in both charters was downgraded to 'praeterea', 'hereafter', in the second, and there is no evidence that such purely ceremonial officers were ever appointed.[86]

As far as curators were concerned, the statutes which were ratified in 1663 supplemented the charter by going into considerable detail as to the type of person who would be suitable for the office and the nature of their employment by the Society. The credentials of candidates for the post were to be examined to ascertain whether they were 'of good fame and virtuous conversation, knowing in philosophical and mathematical learning, addicted to and well versed in observations, inquiries, and experiments concerning natural and artificial things'. Their task was specified as to 'take care of the managing of all experiments and observations appointed by the Society or Council, and report the same; and to perform such tasks relating to the ends of the constitution of the Society, as the Society or Council shall from time to time appoint; such as the examining of sciences, arts, and inventions, bringing in histories of natural and artificial things'. It was also provided that they should be paid 'such a liberal allowance, as may be a sufficient maintenance for them, without seeking for or engaging in any other employment', and — subject to the Society acquiring sufficient revenue to support this — it was agreed that each curator might be paid an allowance of up to the equivalent of £200 at 1663 values.[87] Hence the role of paid staff was now clearly written into the Society's view of its work, and indeed one of the *leitmotifs* of the Society's later history was to be the employment of extra curators when funds permitted to supplement the assiduous work of Hooke: perhaps the best case of this is the episode studied in detail in chapter 8 below, when Nehemiah Grew was temporarily employed to do research under the Society's auspices in the 1670s, but further staff were also taken on more briefly in the early 1680s.[88]

The statutes are also of considerable interest because, in them, an attempt was for the first time made to lay down some methodological prescriptions for the Society's experimental work. Though brief, these have never received the attention that they deserve, as one of the few corporate policy statements by the Society in its very earliest years. Chapter 5, 'Of Experiments, and the Reports thereof', stated that 'In order to the propounding and making of Experiments for the Society, consideration shall be had of the importance of any Experiment, to the discovery of any

86 *The Record*, p. 245. My translation differs from that in *The Record*, p. 258, where the English word given is 'moreover'. See also Birch, *History*, i, 291.

87 *The Record*, pp. 296 – 8.

88 See below, 'Early Problems in Professionalizing Scientific Research'; Hunter, *Royal Society*, pp. 41 – 2.

truth or axiom in nature, or to the use and benefit of mankind'. It also made provision for the procedure to be adopted concerning experiments which it was not possible to carry out at a meeting of the Society: these were to be executed by two or more Curators, who were jointly to draw up a report. Moreover, of 'all Reports of Experiments to be brought into the Society', it was ordered: 'the matter of fact shall be barely stated, without any prefaces, apologies, or rhetorical flourishes; and entered so in the Register-book, by order of the Society. And if any Fellow shall think fit to suggest any conjecture, concerning the causes of the *phaenomena* in such Experiments, the same shall be done apart; and so entered into the Register-book, if the Society shall order the entry thereof'.[89] This sense that speculation about causes was acceptable so long as it was kept separate from factual observation is interesting, and it forms the background to the Society's careful attempt the following year to distinguish between 'certainties' and 'conjectures' in Hooke's *Micrographia*.[90]

Other developments occurred later in 1663, when an attempt was made to improve the Society's proceedings by reminding Fellows of their 'arrears', while three Fellows, Christopher Merrett, Daniel Whistler and John Hoskins, were requested to 'peruse the books of the Society, wherein the experiments and other philosophical matters, treated of at their meetings, are recorded; together with their journal-books; and consider, which relate to and depend upon one another; as also, wherein they may be defective, and how farther to be prosecuted'. They reported back to the Society early the following year, though a promised sequel failed to materialize.[91] Moreover that an evolutionary process continued as far as the Society's activity is concerned is shown perhaps most clearly by a further initiative that occurred in the following year, the appointment of a series of specialist committees covering various aspects of the Society's work. This attempt to improve the internal organization of the Society had not been forseen when its constitution was being drafted, since the only reference to committees in either the charters or the statutes had been a brief reference in the latter to the President calling for reports from them. This attempt at a division of labour arguably represented a further response to the difficulties of trying to reform the whole of knowledge about the natural world through the part-time activities of volunteers, and it will be considered further in a separate study below, though, as will be seen, the episode did not prove a great success.[92]

If the actual practice of natural philosophy was one area where a structure had to be built, another important part of the function that the Society was to serve concerned the promotion of the new science, the exposition to an ignorant and often hostile public of what the Society was doing and why.

89 *The Record*, pp. 289 – 90.

90 Birch, *History*, i, 491; Hooke, *Micrographia* (London, 1665), 'To The Royal Society'.

91 Birch, *History*, i, 341, 349, 366. For other references to the Society's method, see ibid., i, 377 and perhaps 499.

92 See *The Record*, p. 293, and below, 'An Experiment in Corporate Enterprise'.

The need to do anything in this connection seems to have dawned on the Society relatively slowly, since no mention is made of any projected account of 'the design of the society' till the summer of 1662.[93] Earlier, as already noted, it seems to have been presumed that all the Society needed to do was to get on with the job and to leave the results to speak for themselves. Indeed, it is symptomatic that in what is apparently the earliest publication of a specimen of the Royal Society's work, practically no help was given to the unenlightened reader as to what was significant about it. The item in question was Sir Robert Moray's account of the glass drops that Charles II asked the Society to investigate, which was included by Christopher Merrett as an appendix to his translation of Antonio Neri's *Art of Glass* (1662). For all that Merrett said about this by way of introduction was: 'The which I the rather desired; that this might be a pattern for experiments to be made in any kinde whatsoever, as being done with exceeding exactness'.[94]

From 1663, however, serious steps were taken to publish and expound the Society's activities. First, we have the project for an apologia for the Society which got under way with the election of Thomas Sprat as Fellow in 1663 and which was ultimately to culminate in 1667 as his *History*. The original idea may have been simply to print the statutes with a brief commentary, presumably because of the details about the Society's organization and mode of operation that they gave: thus William Brereton told Christiaan Huygens on 28 October 1663, 'As for the History of the Society, it should accompany the Statutes when they are printed, which will be, I believe, very soon'.[95] But gradually it must have become apparent that a fuller apology was required, and by the time that Sprat had finished his task, the epitome of the statutes that was included represented only a small part of a much larger work.

Meanwhile, 1664 and 1665 did indeed see two books brought out under the Society's auspices and using the imprimatur that it had been granted in its charter. The first was John Evelyn's *Sylva, or a Discourse of Forest-Trees* (1664), a 'Publique Fruit of your Royal Society', as Evelyn described it in his dedication to the King, which exemplified the Society's hopes to assist in practical matters through the collaborative efforts of its members.[96] Then in 1665 appeared Robert Hooke's *Micrographia*, based on the presentations which Hooke had made at meetings of the Royal Society which have

[93] Birch, *History*, i, 84 – 5.

[94] Christopher Merrett trans., Antonio Neri, *The Art of Glass* (London, 1662), pp. 353 – 4. Cf. Birch, *History*, i, 37 – 41. For the background, see Laurel Brodsley, Sir Charles Frank and J. W. Steeds, 'Prince Rupert's Drops', *Notes and Records*, 41 (1986), 1 – 26.

[95] Christiaan Huygens, *Ouvres Complètes*, v (The Hague, 1893), 4. For the origins and progress of Sprat's volume, see below, 'Latitudinarianism and the "Ideology" of the Early Royal Society'.

[96] John Evelyn, *Sylva* (London, 1664), sig. A3; see also Hunter, *Science and Society*, p. 93. It is interesting that the second, third and fourth editions of John Graunt's *Observations . . . upon the Bills of Mortality* were also issued under the Society's imprimatur: C. A. Rivington, 'Early Printers to the Royal Society, 1663 – 1708', *Notes and Records*, 39 (1984), 22.

already been referred to. As has recently been expounded in detail, this work was closely supervised by members of the Society, and it was intended to exemplify the potential of the new philosophy in giving a novel view of the natural world: in this case, literally a 'view', since it is clear that care was taken over the elaborate plates that the book contained.[97]

Sprat's book, which was intended to serve an apologetic function *par excellence*, finally materialized after various delays in 1667. That book, however, proved far more problematic than many commentators have realized, illustrating difficulties in doing justice to the Society's activities and their implications which do not seem to have occurred to those who commissioned the work and who rather cursorily supervised it, and showing how much more difficult the task they had set themselves was than had initially been anticipated. The effect was to instil a degree of caution in such activity on the Society's part, and no further such general apologia was to appear. Instead, it was rather the example of *Micrographia* and *Sylva* that the Society was to follow, in terms of promoting the new science through offering piecemeal accounts of its members' findings, often in volumes published with the Society's imprimatur.[98] Equally important was the highly successful *Philosophical Transactions* which Henry Oldenburg was to bring out with the Society's blessing from 1665 onwards. The latter, with its ability to disseminate findings 'which, not enough for a Book, would else be lost', was to prove highly effective as an agency for publicizing the new science and the Royal Society, overlapping with and extending the function that Oldenburg's correspondence also served, which is assessed more fully below.[99] Indeed, the degree of success in promoting the new science which Oldenburg had while more 'official' ventures like Sprat's *History* flagged is symptomatic in itself, and it was to suggest directions of development for the Society which take us far beyond the initial stages in organization to which this section has been devoted.

(c) Conclusion: the Fruits of Experience

The account given in the previous section of the formative stages of the Society's organization and activity is a natural prelude to the detailed

97 See Harwood, 'Rhetoric and Graphics in *Micrographia*' (n. 84).

98 For a list of such books, see Rivington, 'Early Printers' (n. 96), Appendix II.

99 Hunter, *Science and Society*, p. 52, and ibid., pp. 51 – 4, passim. On Oldenburg's correspondence, see below, 'Promoting the New Science'. For an instance of the usefulness of *Philosophical Transactions*, see Michael Hunter, 'Science and Astrology in Seventeenth-century England: an Unpublished Polemic by John Flamsteed', in Patrick Curry, ed., *Astrology, Science and Society* (Woodbridge, 1987), esp. pp. 269 – 70. I understand that a full study of *Philosophical Transactions* is forthcoming shortly from Charles Bazerman.

scrutiny of further crucial episodes in the Society's early years to be found in the chapters that follow. It seems worth ending this Introduction, however, by attempting to draw some general conclusions about how the Society developed in its first decades, and hence about the complex initial stages of the process of institutionalization itself.

At the outset it is worth pointing out that we will misunderstand the early Royal Society if we presume undue unanimity among its members. That this is the case with their views on political and religious issues is underlined by the disparate affiliations of the Society's founders and their stress on how their enterprise might transcend such differences, as outlined in the first part of this Introduction; it is further illustrated below in my account of the context to Sprat's *History*. This is no less true, however, of views of the proper ends of natural philosophy and of the best ways of achieving these through formal organization, since it is clear that greater diversity of opinion on such matters existed within the Society than has often been presumed. On magical phenomena, for instance, members of the Society are to be found voicing widely disparate attitudes, and this is hardly less true of views on the proper method of studying nature. For the Society's commitment to Baconianism disguised a considerable range of degrees of sophistication, from some who were happy with almost random accumulation of data to others who insisted on the need for a clear theoretical orientation to their work.[100] A typical example of the way in which some had moved further from Bacon's views than others is provided in chapter 3 below, by the contrast between the classification of natural things formulated by the committee for collecting the phenomena of nature set up in 1664 and those outlined at much the same time by Robert Boyle; in addition, a whole range of evidence of this kind is documented and discussed in chapter 6, 'Towards Solomon's House'.

Moreover, as is there illustrated, this had clear implications for views about how the Society should be organized and run, since a range of options was considered concerning almost all aspects of the Society's activities, perhaps particularly at times of crisis in its institutional life. Even within the formative period documented in the second part of this Introduction, disagreements must clearly have been voiced in the course of formulating the organizational measures dealt with there, though most of these can now only be divined. Later, however, it is possible to reconstruct a succession of disagreements, not only between what might be described as 'virtuoso' and 'scientific' values, due to the dependance of the Society on a rank and file which was prone to dilettantish attitudes, but even among those with a serious commitment to the Society's enterprise. For some of these were more tolerant of the role of 'plain, diligent, and laborious observers' than others who, because of their view that theoretical knowledge was indispensable to research, argued that the Society's membership should be more exclusive than their colleagues did. Moreover some attached more

[100] See Hunter, *John Aubrey* (n. 19), esp. pp. 132f.; K. T. Hoppen, 'The Nature of the Early Royal Society', *British Journal for the History of Science*, 9 (1976), 1 – 24, 243 – 73; below, 'Towards Solomon's House'.

significance than others to a coherent programme of experimentation, preferably serviced by paid professionals, of the kind inaugurated by Hooke's appointment as Curator in 1662.[101]

Disagreement is also seen in connection with institutional initiatives, emerging in 1668, for instance, on the question of whether the Society should have a purpose-built 'college' as provided for in the charter, which might have provided a home for the Society's meetings, its possessions and various experimental facilities. For one of the reasons why this scheme was to prove abortive — as we shall see in chapter 5 — was that there was a considerable lack of unanimity in the Society as to the desirability of such a premises. While Oldenburg and others saw this as crucial in galvanizing the Society's efforts, providing the facilities and accommodation required for a serious research institution, others seem to have considered it an unnecessary administrative burden for a body which they thought could flourish perfectly well in less elaborate surroundings.

That this reflected rather different views of the proper function of a body like the Royal Society is suggested by the fact that those who expressed scepticism of this kind are known to have included the Oxford Professor of Geometry, John Wallis: for he may have been inclined to regard the Society's proper role as more modest than did some of its eager protagonists, since, like other Oxford dons, his sense of the worth of what the universities were doing was perhaps offended by the rival claims of a novel institution devoted to intellectual pursuits.[102] It is also interesting to juxtapose Wallis' later opinion on the responsibility of his fellow Oxonians to pay subscriptions to the Society: 'it seemes a little hard that having been scarce there this six year, to pay as those that be there constantly'.[103] Such a stress on the benefits received in return for membership of the Society was understandable, but it was almost entirely antithetic to the view of many that the Society needed support in its own right for its institutional role in advancing the studies to which it was devoted. Moreover others took such attitudes even further, like one figure who failed even to join since 'he could not discover what advantage of knowledge could come to him that way which he could not arrive at otherwise'.[104] Here was a challenge to the very concept of a scientific institution which the Royal Society represented, for, while that had been founded on the presumption that public enterprise was preferable to private, at least some contemporaries continued to consider that private facilities might be equally effective.

Yet another area where disagreement is in evidence concerns the Society's attitude towards technology. Quite apart from the fact that some members were clearly more fastidious than others, views on this topic, too, might be complicated by Fellows' external affiliations. An example of this is provided

101 See below, 'Towards Solomon's House' and 'Early Problems'. For the quotation, see Sprat, *History*, p. 72.

102 See below, 'A "College" ', esp. p. 180; on university hostility, see Hunter, *Science and Society*, pp. 145 – 7.

103 Hunter, *Royal Society*, p. 103.

104 Hunter, *Science and Society*, p. 52.

by attitudes towards experiments on navigation, and particularly Petty's trials with his 'double-bottomed' boat. For in the first instance these were supported by the Society: we learn from the minutes that John Wilkins wrote Petty a letter on the Society's behalf 'encouraging him to prosecute his invention', which evidently comprises an enthusiastic communication which still survives, in which Petty was urged to proceed with 'an Invention of such publike & Generall benefit', 'one of the first attempts to render their institution usefull'.[105] But on a later occasion, Petty was told that navigation, 'being a state concern was not proper to be managed by the society', though Petty 'for his private satisfaction' might have the sense of 'particular members of the society'. This evidently reflected the views on the subject of the President, the Navy Commissioner Lord Brouncker, who seems to have had reservations about the merits of Petty's vessel on primarily strategic grounds: perhaps since he was President, however, this was what the Society went along with.[106] The way in which, in practice, attitudes might split over the desirability of technological improvement was also shown in January 1663, when a 'thrifty kind of fewel [fuel]' was discussed and Hooke was ordered to consider possible improvements: for Sir Paul Neile suggested 'that the practice of this thriftiness would prove prejudicial to shipping'.[107]

It is clear, therefore, that one factor that has to be taken into account in reconstructing the Society's early development is a struggle between different priorities and interests. But, important though such disagreement may have been, in itself it provides only one component of the situation, and arguably more important was a dynamic process by which the Fellows as a whole discovered what was and what was not practicable, and modified their aspirations accordingly. For in fact, as we have already seen, there was a significant degree of trial and error in the Society's attempts to translate the ideal of organized science into practice, and many facets of the Society's early development can only be understood in these terms.

This is, in fact, particularly well illustrated by the history of the Society's concern with technology. For, even if the attitudes of Fellows towards this part of the Society's work may have differed, what is unmistakable is a general shift in the Society's approach to such activity — a trend to which attention has more than once been drawn in the past, and of which further evidence is provided in the course of this book.[108] The Society's initial hopes to take initiatives of immediate practical benefit are especially well illustrated by my study of its early committees in chapter 3. John Buckland's public-spirited proposal for planting potatoes, to which consideration was given by a special committee in 1663, was an immediate response to the

105 Birch, *History*, i, 141, 155. Cf. ibid., i, 124, 131. E.L. O.1.1. For degrees of fastidiousness, see Hunter, *Science and Society*, ch. 4.

106 Birch, *History*, i, 249. Cf. *Oldenburg*, ii, 51 n. 6; Aubrey, *Brief Lives* (n. 25), ii, 147; Hunter, *Science and Society*, p. 94.

107 Birch, *History*, i, 177.

108 See especially Margaret 'Espinasse, 'The Decline and Fall of Restoration Science', *Past and Present*, 14 (1958), 71 – 89. See also Hunter, *Science and Society*, ch. 4.

threat of famine. The Agricultural Committee's attention to Ralph Austen's suggestions for afforestation was motivated by a comparable concern for the public good. The Mechanical Committee's attention to schemes of potential public benefit such as the improvement of London's water supply and its hope to gain the support of those with an 'interest' in them bears witness to an equally clear wish to exploit the technological spin-offs of the new science. Moreover this ambition is more broadly in evidence in the Society's earliest years, not least in its hope that it might become an official arbiter of the validity of technological proposals to see if they were 'new, true, and useful', a proposal put to Lord Ashley, Chancellor of the Exchequer, which was said to have gained the approval of the King.[109]

Yet this was a classic case of aspiration proving more difficult to realize than had been anticipated. Even within the lifespan of the committees, it is notable how the kind of hopeful initiatives which historians have often rather tokenistically fastened on as evidence of the Society's commitment to utility were allowed simply to lapse inconclusively. Moreover, this was a foretaste of a broader shift which was already well-advanced during the 1660s, in which the Society's priorities were redefined away from technology and towards pure science. In part, this was because it became clear that the Society's original belief that it could do everything at once had been overambitious, and that it needed to scale down its terms of reference. But within this, it also became apparent that worthwhile results were much more difficult to achieve in the technological sphere than in the realm of pure science, partly because of the difficulty of conveying 'the practick part', partly because of the sheer diffuseness of the audience that needed to be reached to disseminate technical change on a worthwhile scale. Meanwhile the agencies of international communication that were being established by men like Henry Oldenburg meant that findings in pure science were rapidly and effectively made known to a select international public, with the consequent kudos that accrued to their originators. It is hence hardly surprising that it was to work of this kind that the Royal Society came increasingly to devote itself.[110]

The retreat from technology is arguably only one facet of a process by which, within a relatively few years, grandiose aspirations were severely modified in the light of experience. The most striking instance of disillusionment — which had major implications for what the Society was able to achieve — concerned the support to which it aspired in contemporary society. The Society's original high hopes are illustrated clearly by the catalogue of the classes and interest-groups from whom support for the new science might be expected given by Thomas Sprat in a passage in his *History of the Royal Society*, which, though not published till 1667, bears the stamp of the early 1660s, and was arguably already somewhat out of date by the time it appeared.[111] Sprat reached his climax

109 Birch, *History*, i, 116, 391, 397. On the committees, see below, 'An Experiment in Corporate Enterprise'.
110 Hunter, *Science and Society*, pp. 104, 110–11 and ch. 4, passim.
111 Sprat, *History*, pp. 129f.

with the King himself, and considerable ingenuity had initially been applied to trying to think up sources of royal revenue which might appropriately be given to a new institution: a grant of lands left by the sea was one possibility that was considered, while another was that 'if any thing should be found, that at present yielded no revenue to the king, the grant of it might fall to the Society'.[112] In addition, all the promotional efforts already referred to were dedicated to Charles II, though aimed at Restoration society more generally in an attempt 'to raise us up benefactors'.[113] But the response of both the King and his subjects was to prove deeply disappointing, as has already been indicated in the first section of this Introduction. As a result, the Society's organizers are often to be found complaining how the lack of such endowment stunted the institution's growth and restricted the initiatives that were feasible.[114]

In its first decade, the Society was the beneficiary of only two major gifts, both of which are dealt with in the pages that follow, and even with these there were revealing complications. One was Sir John Cutler's establishment of a Lectureship for Robert Hooke under the Society's auspices, which the Society clearly valued especially highly because it came from one of the 'men of *Trafick*' to whom it was hoped that the new science would appeal.[115] But it is symptomatic that this involved complex negotiations — and, ultimately, litigation — due to the Society's wish at the same time to capitalize on Cutler's gift and to adapt it to its own purposes: the topic on which the lectures were to be given was altered, and attempts made to use the benefaction to subsidize Hooke's work as Curator of Experiments in a slightly shady transaction dealt with in full in chapter 9. Moreover, though there are hints that the Society hoped that Cutler might establish a permanent fund for scientific research and exposition, in fact the endowment died with Hooke, and it would have predeceased him but for the assiduity with which he pursued his arrears from the 1670s onwards.[116]

The other endowment of the 1660s came from a Fellow of the Society, Daniel Colwall, who provided the finance which enabled the Society to purchase a large collection of rarities as the basis for a substantial institutional museum early in 1666. This was a gesture which was deliberately intended to demonstrate the institutional vitality of the Society after the intermission of activities caused by the Great Plague, and it certainly enabled the Society to acquire a resource which was quite highly regarded at the time. But there were again complications, as is indicated in chapter 4. For one thing, it is revealing that the Society acquired a ready-made museum in preference to other facilities whose provision was

[112] Birch, *History*, i, 377, 379. See also D.M. 5.37.

[113] See Evelyn, *Sylva* (n. 96), sig. A3; Hooke, *Micrographia* (n. 90), sigs. A1 – 2; Sprat, *History*, 'Epistle Dedicatory'. For the quotation, see below, 'Latitudinarianism', p. 49. Cf. also Birch, *History*, i, 84.

[114] Hunter, *Science and Society*, p. 40.

[115] Sprat, *History*, pp. 129 – 30. See also below, 'Science, Technology and Patronage'.

[116] See ibid., esp. p. 288.

considered at this time, such as a purpose-built laboratory or observatory: this may have resulted from a sense that they were getting a bargain or, more likely, from a belief that — whatever its scientific value — this was a facility which would be widely esteemed among members of the leisured class from which the Society's expectations of support were highest. More important, shortage of funds meant that in subsequent years the Society was to encounter serious difficulties in maintaining and administering the collection: for the limited staff on whom this task fell were also burdened by other responsibilities.[117]

Moreover, though the repository may have materialized in 1666, the next such initiative to occur, the proposal to erect a premises for the Society in 1667 – 8, proved an ignominious failure. As we have already seen, matters were not helped by disagreement within the Society as to whether such a building was desirable at all — though this may have owed something to apprehensions that the support might not be forthcoming to maintain it even if it was built. But it is equally striking that, despite initial hopes that benefactors might be forthcoming both from outside the Society and from among its peripheral, high-ranking membership, in fact the response from either quarter was disappointing in the extreme. Perhaps the Society should have taken Wren's advice of starting the building before funds were available to complete it in the hope that this gesture might attract contributions — a strategy successfully deployed by other institutions such as the College of Arms, though it is interesting that the Royal Society's caution was matched by that of the University of Cambridge over plans to erect an 'Academic Theatre' in the 1670s.[118] As it was, however, the project was simply abandoned, and the north bank of the Thames was left bereft of the purpose-built Baroque research institute which had been so grandiosely planned.

What is particularly revealing about this episode is where such support as materialized came from. Clearly the Society began with high expectations of the traditional patrons of intellectual activity, the aristocracy, to whom it made a special point of applying for funds for the building. Yet the Society was almost pathetically disillusioned, and it is perhaps significant that though, in the 1670s, the Society considered introducing a sliding scale of membership fees according to social rank, with a subscription of as much as £30 per year for Dukes and Marquesses, this was not pursued.[119] Instead, those who offered to help pay for the abortive college were a more varied band of enthusiasts, mostly of lower status, and the same is true of a further specific project of the following decade dealt with in chapter 8, the

[117] See below, 'Between Cabinet of Curiosities and Research Collection', esp. pp. 127 – 8, 139f., 153f.

[118] See W. H. Godfrey, Sir Anthony Wagner and H. Stanford London, *The College of Arms, Queen Victoria Street* (London, 1963), ch. 2; Robert Willis and J. W. Clark, *The Architectural History of the University of Cambridge* (Cambridge, 1886), iii, 403; below, 'A "College" ', p. 173 and passim.

[119] See below, 'A "College" ', pp. 162 – 3; 'Towards Solomon's House', pp. 192 – 3.

provision of a research post for Nehemiah Grew through £50 a year being raised by extra subscriptions from Fellows. Moreover this was paradigmatic of support for the Society's ordinary activities, as is illustrated at length in my study *The Royal Society and its Fellows, 1660 – 1700*: throughout this period, the Society was kept afloat financially by the piecemeal contributions of individual Fellows.[120]

This was something to which the Society had to become accustomed, and it is something of which at times a virtue was made. Sprat, for instance, in one passage of his *History*, argued how the Society could hardly expect money from the state before it had proved its worth on the basis of purely voluntary support, lest by claiming 'that their Enterprise requir'd the *Exchequer of a Kingdom*; they would onely have been contemn'd, as vain *Projectors*'.[121] On the other hand, it certainly affected the Society's structure, making it difficult for it to be intolerant of members who came to meetings 'only as to a Play to amuse themselves for an hour or so', as long as they supported the Society by paying their subscriptions. It is also clear that activities were sometimes geared to an attempt to maximize attendance: the employment of Grew was related to such a wish as well as to the desire to promote research, while the provision of facilities — as with the repository in 1666 — could also be seen as a means of enticing involvement.[122]

As in so many facets of the Society's early institutional life, this system of seeking financial support largely from piecemeal subscriptions was novel, looking forward to the mode of support of scholarly institutions at a later date, as also to the use of subscription-publishing which became common in the eighteenth century: indeed, it is interesting to observe Grew himself being propelled towards the latter via the episode in which his research was paid for by voluntary contributions in the early 1670s.[123] But there were problems about it, since the Society had to learn the hard way about the dependability of such support and hence just what was feasible on its basis. In the case of Grew, for instance, the initial idea of raising a salary from subscriptions proved not to be sustainable: instead the Society was driven to such strategems as making payments for 'presentations' at meetings and using more traditional sources of patronage in the form of the perquisites available from lecturing at Gresham College. The same was true of the Society's finances in general, since repeated crises were encountered as arrears mounted and a campaign of retrenchment had to be set in train, and it is not surprising that, at times like these, there was envy for the public funding that scientists enjoyed in contemporary France. On the other hand, ironically, when a certain amount of public funding for the Royal Society *did* finally materialize in the 1680s, this simply had the effect of encouraging

120 See below, 'A "College" ', pp. 182 – 3; 'Early Problems', pp. 264 – 5. See also Hunter, *Royal Society*, esp. ch. 1.

121 Sprat, *History*, p. 77.

122 Hunter, *Science and Society*, p. 34; see below, 'Early Problems', p. 265; 'Towards Solomon's House', p. 126.

123 See below, 'Early Problems', p. 277. In addition, see Hunter, *Science and Society*, p. 69; *Royal Society*, ch. 1.

individual Fellows to fall behind with their subscriptions, and was thus to some extent counterproductive.[124]

Nor was it only for finance that the Society depended on the efforts of volunteers. It was also these who, in the absence of funds for more than a minimal paid staff, largely sustained its activities. Hence circumstances almost forced the Society to acquire the characteristics of a voluntary institution. When one looks at the Society's actual activity as recorded in its journal books, one is reminded how — apart from the role played virtually throughout this period by Hooke as Curator of Experiments — it was on a core of such volunteers that the Society depended. The channels in which it developed tended to reflect the enthusiasms of these activists, while the extent and nature of its corporate life oscillated according to their loyalty and commitment. One thing that is clear from my study of the committees of 1664–5 is the extraordinary appetite for activity of a handful of members, whose names constantly recur, even at the meetings of committees they were not supposed to be on, and these same men sustained the proceedings of the Society's main meetings. The names of some of these are familiar — such as Boyle, Oldenburg, Petty and Wilkins — but others are not, including the President, Lord Brouncker, and the indefatigable Sir Robert Moray, 'so active a Member', as John Wallis put it in a letter of 1668, that he expressed his confidence that Moray's return from a visit to Scotland would 'add some vigour to the proceedings of the Royall Society'.[125] Moreover the dependence of the Society on activists like these was underlined at times like the early 1670s, when Moray and Wilkins died within a year of one another and crisis ensued.

The result was that the Society had to learn what all organizers of voluntary institutions have learnt since — the sheer unpredictability of such volunteers, and the extent to which the Society had to limit itself to what they were prepared to do. The experiment with committees is in this sense paradigmatic of the Society's general experience: an attempt to involve a wide range of Fellows and to maximize their output by imposing specialization on them foundered on the inescapable rock of their dogged refusal to do more than they liked. Moreover, the dependence on volunteers also led to a tendency for activities to degenerate into miscellaneity and aimlessness, which was in evidence not only in the Society's very earliest years, as already noted, but repeatedly thereafter. Time and again one sees an attempt to impose a clearer programme on the Society, for instance by introducing a programme of research on the laws of motion in 1668 or by commissioning a series of 'discourses' from active members in 1674, as is indicated below in chapter 6.[126]

[124] Hunter, *Royal Society*, pp. 44, 50 and ch. 3, passim.

[125] Wallis to Moray, 14 July 1668, E.L. W.1.52. See also below, 'An Experiment in Corporate Enterprise'.

[126] See below, 'Towards Solomon's House', pp. 190–1, 209; see also Hunter, *Royal Society*, ch. 3.

But it was not just a question of the Society being let down in its expectations of support in contemporary society and of its having to adapt the level of its activities to the resources and manpower that it had at its disposal. In addition, it may be argued that the Society had to learn important lessons about what a public institution could usefully achieve in terms of the pursuit of knowledge, and to be disabused of certain presumptions on this score. Indeed, it could be argued that, in certain respects, institutionalization was a rather mixed blessing, since it appears to have encouraged a presumption that formal structures and procedures were almost by definition better than informal ones, and corporate initiatives to be preferred to personal ones. The result was that the Society's aspirations for the organization of research are littered with false trails, particularly in its earliest years.

A clear instance of this was the Society's aspiration to comprehensiveness in collecting information, an activity which, though stimulated by its Baconian legacy, seems to have been seen as particularly appropriate to an institutional body. This is shown clearly by the history of the committees set up in 1664 which have already more than once been mentioned, their inauguration itself representing a grandiose organizational gesture whose scale is perhaps encapsulated by the title of one of them, which was dedicated to 'collecting the phaenomena of nature and all philosophicall experiments hitherto observed, made and recorded'. For all of these bodies seem to have seen it as their task to scour books to find out what was already known about their chosen subject and hence what needed to be further improved: they thus displayed a perhaps surprising bookishness for a Society with the motto, 'Nullius in Verba', though an interesting distinction was made by one committee between 'what is delivered in Books, and practised' and 'what is practised and not found in Books'.[127]

The equivalent of this in terms of applied science was the Society's ambition to compile an encyclopaedic 'History of Trades' in which technical processes were to be described with a view to discovering and disseminating improved practices.[128] Moreover, the extent to which this kind of project was associated with the ideal of institutionalization is illustrated by the history of the Oxford group of natural philosophers during the Interregnum, the Royal Society's closest precursor in terms of formal organization: for here too organization was associated with grandiose plans 'to gather together such things as are already discovered and to make a booke with a generall index of them, then to have a collection of those which are still inquirenda and according to our opportunityes to make inquisitive experiments', and such plans represent an arguably neglected component of philosophical endeavour in the period.[129]

Perhaps more surprisingly, the Society's 'repository' was associated with

127 See below, 'An Experiment in Corporate Enterprise', p. 116.

128 On the History of Trades, see below, p. 285, and the references there cited.

129 H. W. Robinson, 'An Unpublished Letter of Dr Seth Ward Relating to the Early Meetings of the Oxford Philosophical Society', *Notes and Records*, 7 (1949), 69.

a comparable ambition to make this a 'complete' collection, comprising common objects as well as the rarities that such accumulations had traditionally comprised. This was well expressed by Sprat in his *History*, who saw 'a General Collection of all the Effects of *Arts*, and the Common, or Monstrous *Works* of *Nature*' as 'one of the Principal Intentions' of the Society. Moreover he overoptimistically claimed that by the time he wrote they had 'already drawn together into one Room, the greatest part of all the several kinds of things, that are scatter'd throughout the *Universe*'.[130]

It was part of the process of education to which I have already referred, however, that in practice all of these encyclopaedic projects flagged, desirable as they may have seemed in theory. Quite apart from the fact that they stretched the limited resources of the infant Society, it became apparent to some that the information thus amassed was not particularly useable, and that a more selective approach — often one informed more fully by hypotheses — was required. Such a more thematic approach to research was in evidence in the programme adopted by Hooke in his early work as Curator of Experiments, as outlined in the second section of this Introduction, and it was thereafter frequently championed in exchanges within the Society. There is a parallel in relation to technology, for — quite apart from the general retreat from activity in this area that has already been noted — an equivalent change is in evidence, which is documented in the case of Hooke in chapter 9, below. For, as is there revealed, Hooke actually began work on the History of Trades in the 1660s, compiling two essays of this kind which survive among his papers: but these are without parallel among his extant remains, and he was otherwise to concentrate on the more precise focus for a fruitful interchange between science and practice provided by the world of instrumentation and gadgetry.[131]

As for the repository, practical and intellectual factors coalesced to inspire a concentration on exotic phenomena — ironally, the class of objects preferred all along by the virtuosi of whom Sprat and others were so disdainful: for a realization gradually dawned that common objects were better studied in their natural habitat than as desiccated specimens in a museum, and that a book rather than a room was the best place to bring the two together in an overall classification of nature.[132]

To an extent, a similar process is even in evidence in connection with the Society's extensive records, the keeping of which, as we saw in the first part of this Introduction, was a symptom of the Society's institutional ambitions. For, though these archives have retrospectively proved invaluable to historians, they do not seem to have been very heavily used at the time. Some contemporary copies survive of material from the Society's Register Book, but such transcription seems to have occurred on a rather haphazard

130 Sprat, *History*, p. 251. See also below, 'Between Cabinet of Curiosities and Research Collection'.

131 See below, 'Science, Technology and Patronage'. On exchanges within the Society, see below, 'Towards Solomon's House'.

132 See below, 'Between Cabinet of Curiosities and Research Collection'.

basis, and relatively little use was otherwise made of such documentation.[133] It is symptomatic that in 1678 William Croone suggested that 'an universal index' to the Society's Register Book was needed, in which he was seconded by Sir Christopher Wren, and the idea of 'a general index, like a concordance, to the journals and register-books of the Society' recurred again more than once thereafter.[134] But, here again, a rather different approach from that initially adopted was to come to the rescue, in the form of Oldenburg's *Philosophical Transactions*: for this made selected papers available in a relatively accessible (indeed, indexed) form, in a way that was not feasible with unpublished records held at the Society's premises. Indeed, this was shown particularly clearly in the aftermath of Oldenburg's death, when the Society's earlier records were raided to provide material for the *Transactions*, the papers published thus entering the public realm after hitherto being virtually lost to sight.[135]

The mention of Oldenburg brings us to the most crucial theme of all, and this is the extent to which the Society was forced to learn how much depended on personal initiative even in a public institution. It is symptomatic that the Society's correspondence had itself been deputed to a committee before Oldenburg came to the fore as correspondence secretary, and Oldenburg's career is the clearest example of a general tendency for collaborative projects to succumb to work done by individuals.[136] The demise of the committee system was prototypical from this point of view, as was the way in which a venture that began as a corporate one, Evelyn's *Sylva*, only came to fruition through a degree of editorial supervision on the Evelyn's part which meant that the work was increasingly associated with him alone. What is crucial about the case of Oldenburg — as is explained at length in chapter 7 — is the fruitfulness of the interrelationship between the personal and the institutional that he exemplifies. Both the *Philosophical Transactions* and the extensive and effective correspondence which he indefatigably carried on for over a decade depended on his personal efforts, as was underlined by the fact that both were virtually to collapse after his death. Yet Oldenburg's epistolary success was largely dependent on the authority that he acquired as Secretary of the Society, since the ideal of institutionalization provided a stimulus to the mutual exchange of scientific findings that would otherwise not have existed, while the Society undoubtedly gained immeasurable prestige from Oldenburg's efforts on its behalf. Even if not quite the kind of impersonal institutional initiative

133 For examples, see Royal Society Boyle Papers 20; Royal Society MS 215; British Library MSS Sloane 243, 698. It is perhaps worth noting here a subsidiary function of the Society's records, as a means of establishing priority of invention: see Birch, *History*, ii, 24 – 5, 212, 247.

134 Birch, *History*, iii, 395 – 6; iv, 195, 399, 402. Such a work was in fact compiled by Richard Waller in 1689: see his 'General Index to the Journalls, Registers and Letter-Books of the Royal Society', shelved with the original Journal Books at the Royal Society.

135 *Phil. Trans.*, 12 (1677 – 79), 923 – 1074, passim.

136 See below, 'Promoting the New Science', esp. p. 247.

originally intended, therefore, Oldenburg achieved a highly effective institutionally-based agency of scientific dissemination and cooperation, which went far to vindicate the hopes of the Society's founders as to the kind of positive effect that institutionalization might have.

Oldenburg's is not the only case which illustrates the counterpoint between the aspiration to formal structure on the one hand and, on the other, the activity of individuals on which the Society's achievement actually depended. This is further borne out by two case-studies of individual scientists in the pages that follow, Nehemiah Grew and Robert Hooke. There is no question that the Society's provision of funding stimulated Grew to research that he would otherwise not have attempted. It is no less clear that the credit for this accrued as much to the institution under whose auspices it was done as to its author, not least since Grew's work materialized as a series of publications under the Society's imprimatur. Yet the institutional needs of the Society also had a less positive effect: for one thing, as already noted, a subsidiary incentive for employing Grew on the Society's part was to provide 'entertainment' for Fellows at its meetings, and the subsequent replacement of Grew's stipend by piecemeal payments for such displays almost certainly had a trivializing effect on his work. A further impact of the needs of the institution is in evidence later in the 1670s when Grew was deflected from his pioneering work on plant and animal anatomy to the cataloguing of the Society's repository, a task that was important from the point of view of the Society's infrastructure, but was of rather less scientific significance, as is more fully elucidated below.[137]

With Hooke, one has an equally complicated picture, related to his role both as Curator of Experiments and as Cutlerian Lecturer. On the one hand, Hooke's status as an employee of the Society was always much in evidence in the general tone of the Society's instructions to him and its expectation that the servicing of the Society's meetings should be his chief priority.[138] This general picture is borne out by specific episodes documented here. One case in point would be Hooke's being forced to subsume his individual application for a patent for inventions that he had devised into a corporate one made for the benefit of the Society as a whole: this may well have rankled with Hooke, though we have no evidence of his views on the subject.[139] Another involves the Society's treatment of him in the negotiations over the Cutlerian Lectureship, when Hooke was not even consulted about the final wording of the formal agreement of the terms of the foundation, evidently because it was felt more appropriate for such a benefaction to be settled with a corporate body rather than with an individual. It might also be felt that Hooke was helped less by the Society than he should have been in the litigation which later ensued, for which the

137 See 'Early Problems' and 'Between Cabinet of Curiosities and Research Collection'.

138 On this theme, see Steven Shapin, 'Who was Robert Hooke?', in Hunter and Schaffer, *Robert Hooke* (n. 84).

139 See below, 'An Experiment in Corporate Enterprise', pp. 89 – 90.

Society was partially responsible.[140] Yet, for all this, Hooke's relationship with the Society was highly fruitful from both a personal and an institutional point of view. The significance of his Curatorship of Experiments has been indicated in the second part of this Introduction, while it was effectively under the Society's auspices that Hooke prepared and wrote the book which overnight established his reputation as one of the leading natural philosophers in Europe in his day, *Micrographia*, in which he not only startled contemporaries by his adept display of the finds of the microscope but also divulged findings and speculations on a much wider range of scientific topics.

Equally important was the impact on him of the Cutlerian Lectureship, as is indicated in chapter 9. For — though altering its proposed subjectmatter and attempting to make the post overlap with Hooke's Curatorship — the Society *did* accept Cutler's suggestion that the appointment should be a 'Lectureship', with its overtones of public exposition to an audience, an activity which the Society had not previously seen as falling within its remit. Moreover, this certainly had an effect on Hooke's output. For, in contrast to the accounts of experiments which he produced as Curator, which tended to represent simple expositions of data with limited conclusions, the lectures — which comprise a high proportion of Hooke's writings — were more reflective and analytical, as befitted the more formal context in which they were delivered.[141] In addition, the subjectmatter of his lectures was certainly influenced by the formula agreed between the Society and Cutler, 'the History of Nature and Art'. In part, Hooke followed this rubric by illustrating how observation of natural phenomena might be aided by the use of instruments, as in his published *Lecturae Cutlerianae* of the 1670s. Equally significant, however, are a series of writings by Hooke on the proper aims and methods of scientific research which either originated as or were adapted from Cutlerian Lectures, including both his seminal '*General Scheme, or Idea* of the Present State of Natural Philosophy, And How its Defects may be Remedied' of 1668, and further, as yet unpublished, presentations made later in his life. Moreover, in stimulating Hooke thus to expound methodological views at unusual length, the role of the Royal Society as an institution was clearly important, since Hooke evidently saw such programmatic statements as especially appropriate to the audience he was there addressing.

This complex but highly fruitful relationship between institutional pressure and individual initiative is typical of the way in which the Royal Society developed in its early years. Its context is provided by the themes that have been explored in this Introduction: the impulse towards institutionalization; the efforts made to set the Society up in an effective form; the disagreements which surfaced as to what such a body should be doing and how; and the lessons that had to be learnt about what support it could expect, and what kind of functions it could best serve. It is the highly

140 See below, 'Science, Technology and Patronage'.

141 For a sense of distinction between discourses/lectures, and experiments/papers, see Hooke, *Diary*, passim.

intricate connection between these different factors which has to be taken into account in order to understand how the Society actually developed, and this is exemplified by the remaining chapters of this book. Their *raison d'être* is the fullness of the detail that they are able to offer about the way in which events unfolded in a variety of significant episodes. This itself gives a sense of the interplay of institutional, social, intellectual and other factors, for which the kind of summary attempted here is no adequate substitute. They offer a story not simply of high ideals being thwarted by harsh realities, but of a process with a momentum of its own: only if this is taken into consideration will we understand why the Royal Society developed in the way that it did, and what it was able to achieve. Before we start extrapolating to broader conclusions about the rise of modern science, it is important that we know what actually happened. Moreover, when we rediscover the intrinsic interest of the process by which the new science became 'established', we may find therein implications hitherto overlooked, not only concerning the role of the early Royal Society, but concerning the impact of institutionalization on science more generally.

NOTES ON ILLUSTRATIONS

(1) *Alternative Coats of Arms and Mottos suggested for the Royal Society* (Plate 1)

The original of the document reproduced as plate 1, formerly owned by William Upcott, is now preserved among the Evelyn Papers at Christ Church, Oxford. A facsimile of it was published by C. J. Smith is his *Historical and Literary Curiosities*, which appeared in collected form in 1840, having previously been issued in parts during the 1830s. The original is preserved loosely inserted in Upcott's copy of the fasciscule in which this item was reproduced, no. 5: this comprises a plate (dated 1837) with a page of accompanying text, bound in a blue manila wrapper on which the title of the item has been printed. Smith has endorsed this in manuscript to the effect that only twenty-five copies were printed, while Upcott has noted that Smith died in 1839 and has added a presentation inscription 'to his much esteemed Friend' Mrs Mary Jane Evelyn, dated 24 November 1844; this gift presumably included the original.

The document itself has been placed in a mount, presumably of nineteenth-century date, on which has been written in pencil: 'Arms & Mottos proposed for the Royal Society 1660. Original design by John Evelyn'. On the back of the sheet are the following endorsements in Evelyn's hand: 'Armes for R. Society 1660', and, the other way up, 'Armes & Mottos proposd for the R: Society 1660', followed by the mark ⧺ (as reproduced by Smith at the top of his engraved facsimile) and the words 'Cono [?] W: N:'. There are also some scribbles, including the device '≡⊙'

and the words 'Chapmans obligation'. As well as giving Evelyn's endorsement as a heading, Smith also added to his published version copies of Evelyn's signature and of the actual arms granted to the Society, the latter taken from the official record preserved at the College of Arms.

The possible crests and mottos are as follows:

First, various possible variations on the monogram 'S.R.S.' [i.e. Socius Regalis Societatis]

Then (1) a shield with a ship under sail with the motto 'Et Augebitur Scientia', 'and knowledge shall be increased': cf. Daniel, xii, 4.

(2) a shield divided in two horizontally, the lower part hatched, with a hand holding a plumb-line issuing from a cloud at the top. In the left-hand corner is a blank canton, probably added after the design had been drawn, within which a smaller shield has been drawn. The motto is 'Omnia probate', 'try all things': cf. 1 Thessalonians v, 21.

For another version of this coat of arms, see C. H. Josten, *Elias Ashmole* (Oxford, 1965), i, plate 12, and iii, 924 – 5n., where it is suggested that Ashmole was responsible for it. In Ashmole's version, the motto attached to this coat of arms is 'Rerum Cognoscere Causas', 'to know the causes of things': cf. Virgil, *Georgics*, ii, 490 (i.e., as attached in Evelyn's paper to a different design: see below, no. 6).

(3) a shield with the lower part hatched, on which are picked out two telescopes, object-glasses upwards, crossed to form a St Andrew's Cross; the upper part shows the earth and planets on a white ground. The motto is 'Quantum Nescimus', 'How much we do not know'. This does not appear to be a specific quotation but a commonplace sentiment: cf. the use of a comparable formula as a motto by the Dutch scholar, Daniel Heinsius (Th. H. Lunsingh Schleurleer and G. H. M. Posthumus Meyjes, eds., *Leiden University in the Seventeenth Century* (Leiden, 1975), p. 424).

(4) a shield with the sun in splendour, with the motto: 'Ad Majorem Lumen', 'To the Greater Light'. To the left of this are the words: '[Solem] Quis dicere Falsum —/ Audeat — 1. *Geor.*', 'Who will dare to call [the sun] untrue': cf. Virgil, *Georgics*, i, 463 – 4.

(5) a shield with canton only and the motto 'Nullius in [?] Verb[a]', as adopted.

(6) a shield with a terrestrial globe at the centre and a human eye at the top, with the motto above, 'Rerum cognoscere Causas', 'To know the causes of things': cf. *Georgics*, ii, 490. Adjacent to this is the word 'Experiundo:' 'by experience', while 'Nullius in Verba' is repeated, with a further flourish by Evelyn below it.

(2) *Charles II's copy of Sprat's* History of the Royal Society? (Frontispiece and Plate 2)

The frontispiece and plate 2 are reproduced from a copy of Thomas Sprat's *History of the Royal Society* (1667) now in the British Library, shelfmark 90.d.18. This is a large-paper copy with gilt edges. It is bound in red morocco with a motif of intertwined C's under crowns on the spine and at the corners of the patterned boards, in the centre of which the coat of arms of George III has been superimposed.

The book is evidently from the library of Charles II and the binding was probably executed for the King by Samuel Mearne: for comparable specimens, see H. M. Nixon, *Royal English Bookbindings in the British Museum* (London, 1957), plates 9, 15a (and, on the descent of such volumes through George III to the British Museum, ibid., p. 3). When Evelyn saw Charles II's library in 1680, he was much disappointed by it, describing it as 'consisting chiefely of such books as had from time to time been dedicated, or presented him' (Evelyn, *Diary*, iv, 215). In the light of this, it seems quite possible that 90.d.18 was the copy of Sprat that was presented to the King by the Society.

CAROLVS
II
REGALIS

2

LATITUDINARIANISM AND THE 'IDEOLOGY' OF THE EARLY ROYAL SOCIETY: THOMAS SPRAT'S *HISTORY OF THE ROYAL SOCIETY* (1667) RECONSIDERED*

In this chapter I intend to take a fresh look at one of the most famous texts of the scientific movement in seventeenth-century England, which, it has been said, 'should also be read as a latitudinarian manifesto'.[1] In part I want to consider what light is thrown on Thomas Sprat's *History of the Royal Society* by some new manuscript evidence concerning its genesis and reception which I have found, including a draft by Henry Oldenburg for the passage in which Sprat attempted to deal with problems about the repeatability of experiments noted by no less a figure than Robert Boyle,[2] and some gleanings from the manuscript remains of another prominent Fellow of the Society, John Evelyn. Of the latter, the most significant are some hints to be derived from Evelyn's correspondence with the Somerset virtuoso, John Beale, now preserved in the Evelyn Collection at Christ Church, Oxford; but the most poignant are perhaps my own notes

* I am grateful to the following for their comments on a draft of this chapter: Mark Goldie, John Henry, J. E. McGuire, Steven Shapin and Paul B. Wood. I also benefitted from the discussion that followed my presentation of a shortened version of it at the conference, 'Latitudinarianism, Science and Society' held at the William Andrews Clark Memorial Library, Los Angeles, in April 1987. Material in the Evelyn Collection at Christ Church, Oxford, is quoted by kind permission of the Trustees of the Will of Major Peter George Evelyn.

1 J. R. Jacob, *Robert Boyle and the English Revolution* (New York, 1977), p. 155.

2 This text is printed below, pp. 68 – 9. On its background, see Paul B. Wood, 'Methodology and Apologetics: Thomas Sprat's *History of the Royal Society*', *British Journal for the History of Science*, 13 (1980), 6 – 7. My general indebtedness to Wood's important article will become apparent in the course of this paper.

Plate 2. The engraved frontispiece to Sprat's History of the Royal Society *(1667); engraved by Wenceslaus Hollar from a design by John Evelyn. On the history of this plate, see Hunter,* Science and Society, *pp. 194 – 7 and below, p. 52. Reproduced by permission of the British Library from British Library shelfmark 90.d.18. For the likelihood that this was Charles II's copy, see above, pp. 42 – 3.*

on Evelyn's markings in his copy of Sprat's *History*, made when the book was on show at a London auction house in 1978. For the volume was stolen before the sale, and my notes may now constitute the only record of these annotations. I have therefore appended a record of them to this essay, together with a transcript of the paper by Oldenburg.

More important, however, I want to use this and other evidence to reconsider the context of Sprat's book, and especially the relationship of the views that it expresses to the opinions of the members of the Society on whose behalf it was published. A number of authors, notably Margery Purver and J. R. Jacob, have acclaimed the book as the Society's 'official manifesto', representing, if not the Society's 'official ideology', then at least its 'public stance'. They have seen the book as written 'under the watchful care of the Society' and Miss Purver, in particular, has reacted strongly against the alternative view that Sprat's book was 'an expression of his own or John Wilkins's private opinion'.[3]

I am not going to argue that either of these positions is unreservedly true. In fact, as will become apparent in the course of this paper, I am going to suggest that the truth lies somewhere between them: that the book was commissioned by the Society, but that it was not very closely supervised. It is my contention, however, that this ambivalence has a significance in itself which has hitherto been overlooked, accounting for many of the difficulties which the Society experienced about the book at the time and which historians studying it have encountered since.

Moreover this raises a still broader issue, and this is that, in so far as the book has been taken to be — and, indeed, may have been intended to be — a statement of the Society's 'ideology', it raises the question of the extent to which the Society actually *had* a corporate 'ideology', and, if it did, just how coherent this was, how accurately it represented the views of the Society's members, and how indispensable it was to their view of their scientific work and its implications. Indeed, I would argue that the men who founded the Royal Society had a rather hazier view of the broad public role of science than has sometimes been thought, and that, though they tried to formulate a kind of 'ideology', they found this a much more difficult task than they expected or than many modern scholars have presumed.[4]

[3] Margery Purver, *The Royal Society: Concept and Creation* (London, 1967), p. 9 and passim; J. R. Jacob, 'Restoration, Reformation and the Origins of the Royal Society', *History of Science*, 13 (1975), 169; J. R. Jacob, 'Restoration Ideologies and the Royal Society', *History of Science*, 18 (1980), 25; J. R. and M. C. Jacob, 'The Anglican Origins of Modern Science: the Metaphysical Foundations of the Whig Constitution', *Isis*, 71 (1980), 258. For an important critique of the views of Purver, see Charles Webster, 'The Origins of the Royal Society', *History of Science*, 6 (1967), 106 – 28.

[4] I have deliberately placed the word 'ideology' in inverted commas throughout this paper, since it is not clear to me whether the public stance that I wish to argue that

Not the least of the implications of this concern the debate between scholars as to what kind of position the Society adopted. Was it, as J. R. Jacob has argued in an influential and widely distributed paper, an essentially positive rather than negative stance, one which not only 'forestalled those who said at the time science is injurious to religion' but also gave science 'an ideological edge'? Is it true that 'the leading Fellows who preached a social message in the Society's name did so with a view to building a particular kind of kingdom that would serve particular ends', consciously promulgating a stance with fairly precise resonances in the political and religious thought of the day?[5]

Alternatively, is there truth in the position which Jacob has sought to modify if not to refute — namely the view classically expounded by Barbara Shapiro that the Society's position was intended as one of altruism and withdrawal, that 'latitudinarian science and moderation, far from being ideological, represented the end of ideology'?[6] To us in retrospect this position may seem a naive one, but that is no reason why it should not have been genuinely held at the time. Moreover, if this was indeed the stance of those associated with the new science it would give a different and more diffuse meaning to the 'Latitudinarianism' of the Society than would the view adopted by Jacob, and this, in turn, may assist us in understanding the true place of the new science in contemporary society.

The Royal Society was founded in 1660 with the aim of reforming knowledge by experiment, and it is wrong to underestimate the single-minded devotion to this intellectual enterprise as the motive force behind the Society's early activities. 'It aimes at the improvement of all usefull Sciences and Arts, not by meer speculations, but by exact and faithfull Observations and Experiments', as Oldenburg explained in a letter to an overseas correspondent in 1668, thereby echoing the sentiments of the Society's charter, which stated that its 'studies are to be imployed for the

---

the Royal Society adopted is properly so described. The scholar who has been mainly responsible for the use of the term in this context is J. R. Jacob, but I have been unable to find a definition of it in his writings. For further discussion of this concept and its meanings see, in a historical context, Morris Berman, *Social Change and Scientific Organization* (London, 1978), pp. xxiif., and D. R. Kelley, *The Beginning of Ideology* (Cambridge, 1981), pp. 4 – 5 and passim, or, more generally, Raymond Williams, *Keywords* (revised ed., London, 1983), pp. 153f.

5 Jacob, 'Restoration' (n. 3), pp. 155, 169 and passim (this article has achieved wide circulation by being included in the Open University reader, *Seventeenth-century England: A Changing Culture. Volume 2: Modern Studies*, ed. W. R. Owens (London, 1980), pp. 241 – 52). See also Jacob, *Robert Boyle* (n. 1), pp. 153f.; 'Restoration Ideologies' (n. 3), pp. 25 – 8; 'Anglican Origins' (n. 3), pp. 258 – 9.

6 J. R. and M. C. Jacob, 'Scientists and Society: the Saints Preserved', *Journal of European Studies*, 1 (1971), 88, and see Barbara Shapiro, 'Latitudinarianism and Science in Seventeenth-century England', *Past and Present*, 40 (1968), 16 – 41 (reprinted in Charles Webster, ed., *The Intellectual Revolution of the Seventeenth Century* (London, 1974), pp. 286 – 316); *John Wilkins, 1614 – 72: an Intellectual Biography* (Berkeley and Los Angeles, 1969), ch. 8.

promoting of the knowledge of natural things, and useful Arts by Experiments. To the glory of God, and the good of mankind.'[7]

Although this was always the centre of the Society's perception of itself, however, there was also a sense at the time that the Society needed what might be called an 'ideology' or public stance. In other words, there was a feeling — not least because of the sheer grandeur of the ultimate outcome that the new scientists predicted for their endeavours — that they needed to spell out the non-scientific implications of what they were doing.

For this, I think that there were three reasons. One was connected with the granting of the charter that I have already referred to, which gave the Society a constitutional structure and a permanence which differentiated it from earlier, less formal groups devoted to natural philosophy. By acquiring a charter and joining the ranks of the legally incorporated institutions, the Society gained a degree of public visibility which raised questions about natural philosophy's public import which otherwise might never have surfaced. Much in evidence in Sprat's *History* is a feeling that as public body as the Society had thus become *needed* some kind of position statement as to its role.

Related to this is another factor, which comes across strongly from the pages of Sprat and other sources concerning the Society in its early years, and this is that the infant institution had numerous critics. These attacked the Society's enterprise from a number of viewpoints and in a number of ways — from the sermons of High Churchmen like Robert South to the lampoons of the wits. It was in response to the 'Objections and Cavils' of 'the Detractors of so noble an Institution' that Sprat's book was explicitly written.[8] It is thus hard to accept Barbara Shapiro's claim that 'there was no serious opposition to the new philosophy' except for 'a tiny group of critics' — even if apathy may have been as great an enemy as outright hostility — for there is no mistaking the defensiveness that the new philosophers felt about their enterprise.[9]

Thirdly, I believe that the Society's organisers were anxious to spell out the broader implications of what they were doing because of a genuine — if naive — belief that the enterprise of the Royal Society *was* innocuous and reconciling, and that, if this were expounded, the suspicions that people harboured towards it could be swiftly overcome and many others enrolled in support of a worthwhile activity in which the Society's members often felt themselves to be embattled pioneers. We may be in danger of making the early Royal Society seem too Machiavellian in its motives, except that this did go with a hope to attract potential patrons for the design. In the words of John Evelyn — urging the poet Cowley to complete his verses for

7 Oldenburg to Norwood, 10 Feb. 1668, *Oldenburg*, iv, 168; Sprat, *History*, p. 134 (I have preferred Sprat's translation of the second charter to more modern ones).

8 Sprat, *History*, sig. B4v and passim. Cf. Wood, 'Methodology' (n. 2), p. 2, and Hunter, *Science and Society*, esp. chs. 6 – 7.

9 Barbara Shapiro, *Probability and Certainty in Seventeenth-century England* (Princeton, 1983), p. 73. On apathy, see Hunter, *Science and Society*, pp. 83 – 4, 196 – 7.

Sprat's book in 1667 — what was needed was a way to 'bespeak it some veneration in the world' and 'to raise us up benefactors'.[10]

The most rudimentary attempt to 'place' the Society in this way is to be seen in the deliberate enrolment of the socially eminent to its ranks, and the advertisement of this fact through the publication of printed lists of the Society's members, of which the first appeared in 1663, to be followed by annual publication thereafter. Though in part intended as ballot papers for the annual elections of officers to the Society, these were clearly widely distributed, doing much to spread esteem for the Society, not least due to the distinction of many of the rather nominal Fellows whose names appeared on them.[11]

Indeed, in a status-conscious age, it is striking how often the *eminence* of the membership was advanced as an argument for the respectability and innocuousness of the Society's enterprise. Reviewing Sprat's *History* in the *Philosophical Transactions*, Henry Oldenburg went to the trouble to compute that the Society included in its ranks not only 'three of the Greatest Princes of *Europe*', but two archbishops, four bishops, twenty-nine dukes, marquesses, earls, viscounts and barons, thirty-five knights, fourteen D.D.s and B.D.s, and so on. This had been still more graphically expressed by a manuscript membership list dated 20 November 1663 which survives in a volume of Royal Society papers in the British Library and which actually divides up the membership by social rank, each being given a separate section.[12]

Clearly, however, there was a limit to the potential of lists like these as a defence of a novel intellectual enterprise. It was for this reason that, in June 1662, we hear of the appointment of a committee of various leading Fellows 'to draw up a paper concerning the design of the society', while, writing to Christiaan Huygens in May 1663, Sir Robert Moray spoke of plans to publish 'a little treatise, by which you can learn all that concerns the Society'.[13] Possibly Moray was already referring to a putative work by Sprat, who had been elected to the Society in April that year, presumably with the intention that he should write such a book, since his composition of the *History* represents virtually his only known participation in the Society's affairs. Sometime in 1663, he evidently embarked on the work we have.[14]

I need not say much about Sprat, who, as is well-known, was a young graduate of Wadham College, Oxford, and a protégé of its former Warden, John Wilkins, but I wish to stress one point to which in my view inadequate

[10] Evelyn to Cowley, 12 March 1667, in Evelyn, *Diary and Correspondence*, iii, 194 – 5. Cf. Birch, *History*, i, 84.

[11] See Hunter, *Royal Society*, pp. 9 – 10, 55.

[12] *Phil. Trans.*, 2 (1667), 508; British Library Additional MS 4441, fol. 79, reproduced here as plate 3.

[13] Birch, *History*, i, 84 – 5; Sprat, *History*, p. xiii.

[14] Loc. cit.; cf. Hunter, *Royal Society*, catalogue entry 132. For a later episode in which Sprat was involved, see below, 'Science, Technology and Patronage', p. 326.

attention has been paid, and this is that Sprat was a self-conscious writer, brought in to serve a specific literary function. As J. I. Cope and H. W. Jones put it, Sprat was part of 'the circle of Buckingham, Rochester, Clifford, Cowley, Dryden — the Court Wits and witlings who set the tone for restoration literature'.[15] An awareness that Sprat was a polished author is much in evidence in the comments of the Society's supporters concerning his book when it appeared. The virtuoso, Nathaniel Fairfax, described Sprat as 'a Gent. who has enough of the knack of fair speaking'; others spoke highly of his 'elegant prose'; while many of the passages which John Evelyn marked in his copy of the work appear to be *obiter dicta* which he intended to copy out into his commonplace book for reuse at a later date.[16]

It is interesting that a 'professional' like Sprat was chosen in preference to Fellows with literary pretensions such as Evelyn, who had already promoted the Society in print, and whose *Public Employment and an Active Life prefer'd to Solitude*, published in 1667, suggests that he could perfectly well have purveyed some of the sentiments that Sprat was to express.[17] Clearly stylistic excellence was at a premium in compiling the Society's apologia, and there is almost an element of the *prima donna* in Sprat's relationship with the Society. In deference to Sprat's skill as an author, he was evidently given considerable initiative as to exactly what he wrote and how he expressed it, though within agreed parameters and subject to close supervision concerning the book's account of the Society's constitutional status and its scientific work. There is also evidence of a certain remoteness in Sprat's relations with the Society, about which anxious remarks are to be found among the Society's organizers at different stages during the progress of the project.[18]

At this point it is worth briefly expounding what is known about the making of Sprat's book, not least since this provides such hard evidence as

[15] Sprat, *History*, p. xxv. On Sprat as a stylist, see Robert Cluett, 'Style, Precept, Personality: A Test Case (Thomas Sprat, 1635 – 1713)', *Computers and the Humanities*, 5 (1971), 257 – 77.

[16] Fairfax to Oldenburg, 28 Sept. 1667, *Oldenburg*, iii, 491 – 2; Beale to Evelyn, 11 Sept. 1667, Evelyn Collection, Christ Church, Oxford, Correspondence (bound volumes), no. 64. See also *Phil. Trans.*, 2 (1667), 503; Oldenburg to Boyle, 24 Nov. 1664, *Oldenburg*, ii, 321; Webster, 'Origins' (n. 3), p. 115. For Evelyn's markings, see below.

[17] John Evelyn, trans. G.Naudé, *Instructions Concerning Erecting of a Library* (London, 1661), sigs. A2 – 7: cf. also *A Panegyric to Charles the Second* (London, 1661), p. 14. Evelyn's *Public Employment* has recently been reprinted in Brian Vickers, ed., *Public and Private Life in the Seventeenth Century: the MacKenzie-Evelyn Debate* (New York, 1986). For an interesting insight into the relationship between Sprat and Evelyn, see a letter from Evelyn to Sprat of 12 Nov. 1667 in Evelyn's Letterbook, Evelyn Collection MS 39, no. 297: this accompanied verses which Evelyn had written on the death of Cowley, which he deprecatingly considered would seem a 'crude paper' to 'great Witts' like Sprat and other *litterateurs*.

[18] For a full account, see Wood 'Methodology' (n. 2), p. 4.

there is on the issue of its status as an official statement by the Society.[19] Having begun the work in 1663, Sprat had made considerable progress with it by the end of the following year, since in a letter to Boyle of 24 November 1664, Oldenburg wrote: 'Mr Spratt intends to begin next week to print the History of our Institution'. He added that the book had been perused by the President, William, Viscount Brouncker, and by such prominent members as Sir Robert Moray, John Wilkins, John Evelyn 'and others', though he regretted that Boyle himself was not going to be able to see it before publication. He then gave an exposition of the work's structure and content which tallies almost exactly with what Sprat says about his book at the start of the published version.[20]

Whether Charles Webster is right to presume from this that the book as we now have it was already written, however, is dubious.[21] Oldenburg's correspondence in the mid 1660s is sprinkled with over-optimistic hopes for the impending appearance of the work, and against his implication that the entire book was complete late in 1664 we have to balance Sprat's own statement on page 120 of the published version that 'thus far I was come in my intended *work*, when my *hand* was stop'd and my *mind* disturb'd' by, first, the Great Plague of 1665 and then the Fire of London in 1666. In his 'Advertisement to the Reader' he states 'That much of this Discours was Written and Printed above two years before the rest'. This suggests that by late 1664 he had completed Part I — which gives 'a short view of the *Ancient*, and *Modern* Philosophy; and of the most Famous Attempts, that have been made for its *Advancement*' — together with at least the first seventy pages or so of Part II, which comprises a narrative of the Society's origins, foundation, aims, methods and composition.[22]

But Sprat's apology that, because of the intermission during the writing of the book, the reader 'may chance to find some Expressions that by reason of the difference of time may seem not well to agree with the last' in its first two parts makes it clear that, even if already drafted, Part III as we have it was substantially composed at a later date. Moreover, that work continued even on Part II after the date of Oldenburg's letter to Boyle is shown by the fact that in December and January 1664 – 5 references in the Society's minutes show that efforts were being made to strengthen up what Oldenburg had singled out as the weakest part of the book: as he put it, 'I know not whether there be enough said of particulars, or, to speake more truly, whether there are performances enough, for a R. Society, that hath been at work so considerable a time'. A further committee to 'review'

[19] For earlier accounts, see Sprat, *History*, pp. xiiif.; Purver, *Royal Society* (n. 3), ch. 1; Hans Aarsleff, 'Thomas Sprat', in C.C.Gillispie, ed., *Dictionary of Scientific Biography* (New York, 1970 – 80), xii, 582, 584; Webster, 'Origins' (n. 3), pp. 111f.; and, above all, Wood, 'Methodology' (n. 2), pp. 3f.

[20] *Oldenburg*, ii, 320 – 1; Sprat, *History*, p. 4.

[21] Webster, 'Origins' (n. 3), pp. 112 – 3.

[22] *Oldenburg*, ii, 288, 301, 399, 632; iii, 73. Sprat, *History*, sig. B4, pp. 4, 120. Cf. Aarsleff, 'Sprat' (n. 19), p. 582.

Sprat's efforts was appointed in May 1665, whose work was presumably still in progress when the plague struck in the summer of that year.[23]

There is also evidence that Sprat was increasingly dilatory in his task even before the suspension of the work due to the catastrophes that struck London in 1665 and 1666, evidently due to other preoccupations, including the attack on Samuel Sorbière which he composed at this time.[24] Indeed, John Beale in Somerset seems to have presumed that Sprat's book would never appear, thereupon preparing a substitute of his own, and the fact that he was encouraged by prominent members of the Society suggests that they may have had their doubts too. One of them, John Evelyn, even designed a plate for Beale, which was later deemed appropriate for redeployment as a frontispiece to Sprat's apologia.[25]

It may have been in connection with his preparations for this abortive work that Beale was apparently shown the section of Sprat's text that had been set in type in 1664. For, retrospectively, Beale was to mention an opinion he had formed about the book 'before many sheets were printed off', which must therefore have been at this stage. He claimed to have had reservations about the book's apologetic, as against its purely descriptive, functions, and he took particular exception to a specific passage on page 34, in which Sprat half-implied that God was likelier to bestow an 'extraordinary Light in Nature' on virtuosi than on others. Though Beale 'did at first sight desire' that this be suppressed, in fact nothing was done about it, which suggests that at this stage Beale was unusual in his sensitivity about the precise message that Sprat was putting across in the more discursive passages of his book: this is a point to the significance of which we will return.[26]

After the gap caused by the Plague and Fire, by 1667 Sprat had returned to work on the project. Moreover it is clear that the bulk of Part II as well as Part III of his book was substantially remodelled at this stage, even if drafted earlier, for references in the minutes show that the question of the 'instances' of the Society's scientific work that Sprat should include was again taken up, as also of what he should extract for his work from the Society's journal book and its statutes. At this stage, the person who was authorized to take the initiative on these specific matters was John Wilkins, who already before the Plague had had more to do with supervising Sprat's work than any other Fellow: indeed, Sprat had flatteringly referred to Wilkins' role in a passage of the book which was already in print by this time. But — since considerable stress has been laid on the part played by Wilkins — it is worth pointing out that it was specified that Wilkins' decisions should be ratified by the Council and by the President, Lord Brouncker, while on at least one

23 *Oldenburg*, ii, 321; Birch, *History*, i, 507; ii, 3, 6-7, 47. Cf. ibid., ii, 51, for the wish to vet Sprat's passage concerning the Society's imprimatur.

24 For accounts of this episode see Sprat, *History*, pp. xvi – xvii; Aarsleff, 'Sprat' (n. 19), pp. 584 – 5. See also Evelyn to Sprat, 31 Oct. 1664, in Evelyn, *Diary and Correspondence*, iii, 144 – 7.

25 Hunter, *Science and Society*, pp. 194 – 7. See plate 2.

26 Beale to Evelyn, 11 June 1670, Evelyn Collection, Correspondence, no. 96.

point (concerning the statutes) Wilkins' supervision was supplemented by that of a committee.[27]

This is virtually all that we know of the book's preparation in its later stages: the evidence relates exclusively to the extracts from the Society's statutes and its scientific work, and nothing is said about the supervision of Part III of the *History* by John Wilkins or anyone else. The only other clue is the hitherto unpublished paper by Oldenburg printed here, which, though never before noted, is clearly the source for pages 243 to 245 of Sprat's text. This demonstrates that on one key issue Sprat was working from a draft prepared by a leading figure in the Society: whether this was unique is unclear, but it is quite likely that it was, in view of the delicacy of the topic that the paper deals with — the sceptical critique of experiment made a few years earlier by one of the Society's most eminent members, Robert Boyle.[28] Interestingly, this, like the scientific papers and statutes, was a section of Part II.

Collation of Sprat's text with this draft and with the originals of the scientific papers and constitutional documents that he included shows that various alterations were made, particularly to the former. Sprat virtually rewrote Oldenburg's notes on the Society's experimental policy, mainly making stylistic embellishments, but in one case slightly obscuring the point of Oldenburg's argument. The papers illustrating the Society's work that he included were subjected to more minor emendations, while his version of the statutes comprises a workmanlike epitome.[29] The latter almost certainly *was* supervised, in view of the solicitousness over legal matters that the Society showed throughout its early years, and it is possible that the other changes were also made by Wilkins or other prominent Fellows rather than by Sprat. But the fact that the 'instances' had still not been finally selected by 23 May 1667, whereas the book was published later that same summer, makes it equally likely that Sprat was given a free hand in modifying such passages, and highly probable that the supervision that the book received once written was cursory in the extreme. The controversial Part III, especially, in which Sprat defended the new science at length, may well not

[27] Birch, *History*, ii, 138, 161, 163, 169, 171, 176; see also the passages cited in nn. 13 and 23; Sprat, *History*, p. 94. In addition, Robert Hooke, the Society's Curator of Experiments, was ordered to assist Wilkins. It seems to me that, in reacting against the views of Purver, Wood, 'Methodology' (n. 2), pp. 2 – 4, downplays the extent to which others shared the responsibility with Wilkins at this point.

[28] See above, n. 2.

[29] See the text below and compare Sprat, *History*, pp. 243 – 5, particularly his treatment of the second part of the penultimate paragraph of the draft; note also his omission of the passage about 'more or lesse skill, care, patience' in the first paragraph. On the alterations made to the papers from the Society's archives, see ibid., pp. x – xi. For the statutes, see *The Record*, pp. 287 – 301. I am not inclined to attach significance to any of Sprat's omissions, but one might draw attention to his interpolation on p. 146 of a note to the effect that the Vice-President should take the President's powers in his absence, which had not appeared in the statutes, but which evidently reflects a concern on the part of the Society at this time which was to be taken up in its 1669 charter: see above, Introduction, p. 21.

have been read by anyone, and certainly not by a committee, before it went to press, though the fact that Sprat was working from an agenda that he had been given is suggested by the accuracy of the brief summary of the burden of this part of the work that Oldenburg could already give in 1664.[30]

The special solicitude that was shown regarding the exact content of the scientific and constitutional sections of Sprat's book suggests that it was about these that the Society felt most anxiety. On the other hand, the relative lack of evidence of supervision of the part of the work in which Sprat dealt with the broader implications of the new science suggests that only a rather generalized concern was felt about this — as if all that was required were a few soothing platitudes on a range of agreed themes, which Sprat could be left to express in his superlative prose. Indeed, it may well be that — aside from their concern to demonstrate the viability of the Society's constitution — what the Society's organizers visualized was something more similar to the *Saggi di Naturali Esperienze* published by the Accademia del Cimento in the same year as Sprat's book, which virtually eschews broader material and concentrates on the scientific work of the group.[31] Moreover it may be noted in passing that, as far as the Society's international readership was concerned, it was precisely these scientific sections which were eagerly read and absorbed, while the rest of the content was apparently virtually ignored.[32]

This brings us to the nature of the message that Sprat was putting across in the course of his book, and the relationship of this to the collective views of the members of the Society. The way in which Sprat's presentation even of the Society's scientific work was affected by his apologetic intention has been ably indicated by Paul Wood, and I do not need to go into it here.[33] As Wood has shown, Sprat misrepresented the real nature of the Society's activity in the interests of proffering an uncontentious and undemanding range of activities to potential supporters. In particular, he failed to do justice to the role of hypothesis in the work of the Society's leading researchers, and upgraded the role of the pure accumulation of data.

In addition, it is clear that Sprat was inspired by ulterior motives to set up a somewhat artificial polarization between science and magic, between the clarity of the aims and method of the new empirical science and the obscurity of alchemists and others.[34] Insofar as, in the course of this, Sprat

[30] See above, n. 20. On the date of publication, see Sprat, *History*, p. ix, and Beale to Evelyn, 11 Sept. 1667, Evelyn Collection, Correspondence, no. 64: Beale, living in Somerset, had received his copy by that date.

[31] See W. E. K. Middleton, *The Experimenters: A Study of the Accademia del Cimento* (Baltimore, 1971). For another possible model which is almost equally descriptive, Paul Péllison-Fontanier's *Relation Contenant l'Histoire de l'Académie Françoise* (1653), see Sprat, *History*, p. xv.

[32] *Oldenburg*, e.g. iv, 63, v, 126, vi, 487.

[33] Wood, 'Methodology' (n. 2), passim. See also Webster, 'Origins' (n. 3), pp. 114f.

[34] Wood, 'Methodology' (n. 2), pp. 16f.; Brian Vickers, 'The Royal Society and English Prose Style: A Reassessment', in Brian Vickers and N. S. Struever, *Rhetoric and the Pursuit of Truth: Language Change in the Seventeenth and Eighteenth Centuries* (Los Angeles, 1985), pp. 41f.

gave the impression that members of the Royal Society wholly eschewed a concern with magical phenomena like apparitions or with alchemical investigations, he again dispensed a misleading view of the actual state of affairs, in which a more mixed position was the norm. After all, one member, Joseph Glanvill, had in the previous year published a 'Philosophical Endeavour towards the Defence' of witches and apparitions in which he was prominently identified on the title-page as 'A Member of the Royal Society'.[35] Evidently on such matters Sprat adopted a line which he saw as publicly prudent, something which says more about his own views than about the preoccupations of the heterogeneous body of virtuosi which the Society comprised, and such passages have to be treated with the greatest of caution as a 'public statement' by the Society.

What, however, about Sprat's more general comments about the place of the new science in contemporary society, particularly those in the lengthy Part III of his work? These are widely read and cited concerning the aspirations and affiliations of the new philosophy, but what significance should be attached to them and to the message that they purvey?

The passages in question may be briefly summarized. Already, in Part I, Sprat had included a certain amount of such material, including some forty-five pages on the history of learning, from the ancients to the '*Modern Dogmatists*', and an exposition of the Society's position in relation to this. In Part II, there is also some material which goes beyond an account of the Society's constitutional structure and scientific method, as in Sprat's narrative of the Society's origins in the aftermath of the Civil War and his description of the width of its support in religious, social and national terms. But in Part III such material burgeons, with a twenty-page defence of the experimental philosophy in educational terms, both in its own right and in relation to 'the usual wayes of *Education*' and their content, together with a three-and-a-half page annexe on the value of science '*for the cure of mens minds*'. This is followed by what Sprat describes as 'the weightiest, and most solemn part of my whole *undertaking*', the justification of the Royal Society and the new science in religious terms, which occupies thirty-three pages. There are then twenty-five pages on the advantageousness of experiments to the manual arts, followed by nine and a half pages justifying experiments as '*a proper Study for the Gentlemen of our Nation*', six pages recommending them to the wits, and twelve arguing for the advantageousness of science to the national interest more generally, before a List of the Fellows of the Society and a concluding peroration.[36]

If, as I have stated elsewhere, Sprat's book is to be seen as 'a summation of polemical attempts to define the role of science in the atmosphere of the mid seventeenth century',[37] what can be said about its antecedents and context? One response to the problem of dealing with science's relationship with

35 Sprat, *History*, pp. x, 37–8, 97, 340. See Charles Webster, *From Paracelsus to Newton* (Cambridge, 1982), pp. 93–6; K. T. Hoppen, 'The Nature of the Early Royal Society', *British Journal for the History of Science*, 9 (1976), 1–24, 243–73.
36 Sprat, *History*, pp. 28, 323, 342, 345, 403 and passim.
37 Hunter, *Science and Society*, p. 29.

broader issues which went back at least to the mid 1640s in England and earlier still on the continent was to deny that there was one. Thus the Accademia dei Lincei in Rome in the early years of the seventeenth century had explicitly avoided political discussion, and this was echoed by John Wallis in his famous account of the 1645 London group: 'Our business was (precluding matters of Theology and State Affairs) to discours and consider of *Philosophical Enquiries*, and such as related thereunto'.[38]

A similar attitude is in evidence in the early years of the Royal Society in an anonymous memorandum, usually, but I believe erroneously, attributed to Hooke on the basis of an unsubstantiated claim by C. R. Weld, who first published the text in the early nineteenth century (all that can in fact be said about it is that it is certainly of late seventeenth-century date). This defines the 'business and design' of the Royal Society as 'To improve the knowledge of naturall things, and all useful Arts, Manufactures, Mechanick practises, Engynes and Inventions by Experiments — (not meddling with Divinity, Metaphysics, Moralls, Politicks, Grammar, Rhetorick, or Logick)'.[39] The sense that the Royal Society should eschew religious matters is found later, as, for instance, when it refused to license a book for publication in 1696 on the grounds that its subject was 'Theological' and hence 'not within the Cognizance of the Society'.[40]

At some points in his book, Sprat echoes this kind of line, arguing that the Fellows of the Royal Society 'meddle no otherwise with *Divine things*, than onely as the *Power*, and *Wisdom*, and *Goodness* of the *Creator*, is display'd in the admirable order, and workman-ship of the Creatures', seeing experiments as '*calm*, and *indifferent* things', above the hubbub of politics.[41] Elsewhere, however, he juxtaposes this with a more positive view of science's role not only as appealing to all men, but as a useful form of intellectual exercise which, through its methods and attitudes, could form the intellectual underpinning for the revived Anglican Church of his day. This juxtaposition might seem uneasy, but it is arguably symptomatic of the rather naive — even inept — presumption of the protagonists of the new science that what they were doing was not only harmful to no-one, but somehow might at the same time be positively appealing.[42]

For the same reason, there was also a proclivity to try to align the new science with the status quo, a fairly obvious strategy for a society which was ill-established and the subject of hostility. Thus Sprat was fullsome in his profession of loyalty to the King and he stressed again and again how well

38 Stillman Drake, trans., *Discoveries and Opinions of Galileo* (New York, 1957), p. 78; C. J. Scriba, 'The Autobiography of John Wallis', *Notes and Records of the Royal Society*, 25 (1970), 40.

39 C. R. Weld, *A History of the Royal Society* (London, 1848), i, 146. On the attribution see below, 'Towards Solomon's House', n. 211.

40 Royal Society Copy Council Minutes, ii, 126. Cf. also Glanvill to Oldenburg, 17 Dec.1669, *Oldenburg*, vi, 372.

41 Sprat, *History*, pp. 82, 426.

42 Ibid., Part III, passim. For a different reading of this juxtaposition, see Steven Shapin and Simon Schaffer, *Leviathan and the Air-pump: Hobbes, Boyle, and the Experimental Life* (Princeton, 1985), pp. 306 – 7.

the new science was suited to the 'present prevailing *Genius* of the *English* Nation'.[43] In this, it seems to me, Sprat's aim was simply to align the new science with as many consensus values as possible, his specifics as to what was excluded mainly comprising elaborations of the enmity towards 'fanaticism' whose wide support among divers sections of the political nation was what had made the Restoration of the Stuarts possible a few years earlier.

Can one, however, be more precise, and, in particular, is it convincing to see Sprat's book, as J. R. Jacob has, as a statement of a fairly specific ideological position, echoing views expressed by Robert Boyle and his circle at the time of the Restoration? According to this view, Sprat was doing more than voicing bland commonplaces: his message included not only 'religious moderation . . . civil obedience, private enterprise, and profit', but also 'ecclesiastical comprehension'. He thus took up ideas advocating a broad church, liberty of conscience and perhaps even limited episcopacy which Boyle and his friends had outlined and which are central to the Latitudinarianism which Professor Jacob sees the Royal Society as exemplifying.[44]

In assessing such claims, it seems to me crucial to try to distinguish between sentiments which could have been expected to appeal to a very wide range of people in Restoration England, and ones which had a more limited cachet. Until we have evidence that non-Latitudinarians *did not* believe in overseas trade and empire, it is not very helpful to describe these as Latitudinarian, as against commonplace, views, and the indications do not look promising on this front: even if some contemporary commentators were more enthusiastic about commerce than others, a growing acceptance is to be found among all.[45] But I think that there is one, crucial part of Jacob's case which *did* have precise resonances at the time and which, if proven, would therefore show the Royal Society to have been Latitudinarian in a positive and precise sense rather than a negative and broad one, and this relates to the issue of comprehension. For ideas of comprehension were notoriously controversial in Restoration England, and hence were party-political in a way that many of Boyle's and his friends' notions were not.[46] Is it true, then, that the Royal Society, and Sprat, can be claimed to have adhered to this more exactly defined viewpoint?

43 Sprat, *History*, Ep. Ded., p. 78 and passim.

44 Jacob, *Robert Boyle* (n. 1), pp. 155 – 6 and ch. 4, passim; 'Restoration' (n. 3), p. 169 and passim.

45 See R. B. Schlatter, *The Social Ideas of Religious Leaders, 1660 – 88* (London, 1940), esp. ch. 5.

46 See Norman Sykes, *From Sheldon to Secker: Aspects of English Church History, 1660 – 1768* (Cambridge, 1959), ch. 3; G. F. Nuttall and Owen Chadwick, eds., *From Uniformity to Unity, 1662 – 1962* (London, 1962); W. G. Simon, 'Comprehension in the Age of Charles II', *Church History*, 31 (1962), 440-8; D. R. Lacey, *Dissent and Parliamentary Politics in England, 1661 –* 89 (New Brunswick, 1969); R. A. Beddard, 'The Restoration Church' in J. R. Jones, ed., *The Restored Monarchy, 1660 – 88* (London, 1979), pp. 155 – 75.

Now it is true that on the face of it there appears to be some support in Sprat for such a view, since he repeatedly trumpets the virtue of sober, collaborative research involving people from a wide range of backgrounds. In the most generalized sense, Sprat was perhaps thus claiming that intellectual enterprise offered 'a model for the nation as a whole'. But is Jacob right to paraphrase Sprat as meaning that 'the Royal Society might serve then as the basis of *ecclesiastical comprehension*', and, by way of background to this, what evidence is there that such a party-political view — even if expressed by Sprat — was generally shared within the Royal Society?[47]

Bishop Wilkins, as is well-known, was a keen advocate of comprehension, who was associated with attempts to introduce legislation along these lines at much the time that Sprat's book appeared.[48] But, perhaps because he suits perfectly certain views about the affiliations of the new science in this period, we have arguably heard too much about Wilkins in connection with the Society's public stance in its early years. If one looks more broadly, one comes up with a more varied pattern which makes it hard to credit even the Society's central core with anything approaching unanimity on this crucial issue.

Of those who took a leading role in directing the Society's affairs in its early years, one, Sir Robert Moray, was certainly a protagonist of moderate episcopacy, as was shown by his activities both at the time of the Restoration and in Scottish affairs thereafter.[49] But what about another figure who had a role matching Moray's in the Society's early administration, Seth Ward, Bishop of Salisbury? It was Ward whose open opposition proved fatal to Wilkins' comprehension scheme in 1668, and throughout these years Ward was active in the 'Sheldonian' wing which opposed moves towards broadening the basis of the church championed by those of Wilkins' ilk.[50]

There are hints of a similar attitude on the part of the figure who was arguably more important than anyone else in directing the Society's affairs during these years, the President, Lord Brouncker, despite the fact that Brouncker's political views are on the whole rather shadowy, like those of many prominent members of the Society. For in his diary, Brouncker's colleague at the Navy Office, Samuel Pepys, recorded that on 21 August 1665 he heard Brouncker 'speak of some presbyter-people that he caused to be apprehended yesterday at a private meeting in Covent garden, which he

47 Jacob, *Robert Boyle* (n. 1), p. 156; Jacob, 'Restoration' (n. 3), pp. 169, 170 (my italics).

48 Shapiro, *John Wilkins* (n. 6), ch. 6. See also the works cited in n. 46.

49 Alexander Robertson, *The Life of Sir Robert Moray* (London, 1922), pp. 105f.; see also pp. 170f. and passim, and David Stevenson, 'Masonry, Symbolism and Ethics in the Life of Sir Robert Moray, F.R.S.', *Proceedings of the Society of Antiquaries of Scotland*, 114 (1984), 405 – 31, esp. pp. 426f.

50 Simon, 'Comprehension' (n. 46), pp. 443 – 5 and 447 n. 27; Sykes, *From Sheldon to Secker* (n. 46), pp. 69, 80. See also Walter Pope, *The Life of Seth*, ed. J.M.Bamborough (Oxford, 1961), pp. 71 – 4. On the significant role of Ward (and Moray) in the Society, see especially below, 'Reconstructing Restoration Science', p. 344.

would have released upon paying £5 per man to the poor; but it was answered they would not pay anything, so he ordered them to another prison from the guard [i.e., the Guard-House at Whitehall].' One cannot draw absolutely definite conclusions about Brouncker's churchmanship from such persecution, since Latitudinarians as well as High Churchmen were to be found advocating that those who refused to join the state church should suffer for it: but the fact that Brouncker was a layman, and that he carried out this policy during the Great Plague, when others in London suspended the persecution of nonconformists, suggests a zealous High Church attitude on his part.[51]

Looking slightly more widely at the Society's active membership, we find at least two regular attenders of Council meetings in the late 1660s, who, as Members of Parliament, were committed defenders of the established church who were actively involved in promoting the new Conventicles Act passed in 1670, Wilkins' implacable opposition to which earned him the King's disfavour. Thus the former royalist exile, Sir Gilbert Talbot, assisted the passing of the measure by reading out in the Commons a letter denouncing nonconformist insolence, while Sir John Lowther, the great improver of Whitehaven in Cumberland, was another 'firm Anglican' who played an active role in the making of the Act.[52] If such men belonged to the active core on which the Society's work depended, it is also worth pointing out that the Society's more nominal supporters included churchmen who were certainly not Latitudinarians, such as Gilbert Sheldon and John Pearson in the 1660s and Henry Paman in the 1670s; moreover, among lay members on the periphery of the Society were some extreme supporters of the Anglican reaction, including Sir John Berkenhead, Sir Winston Churchill and Sir Nicholas Stewart.[53]

If, in the light of this, one goes back to Sprat's text to see if he advocates comprehension in any sense that has resonances in the political atmosphere in Restoration England, the answer is that he does not. In other words, he was saying something about intellectual activity which was only intended to be extrapolated in a negative rather than a positive sense to the political realm. This is seen especially clearly in what he says — and, indeed, in what the early Royal Society exemplified — on the question of attitudes to Roman Catholics. Thus Boyle and his friends would have had no truck with

[51] Pepys, *Diary*, vi, 199. For Latitudinarian advocacy of intolerance, see, e.g., J. I. Cope, *Joseph Glanvill, Anglican Apologist* (St Louis, 1956), pp. 79f., and, on the suspension of persecution during the plague, Shapiro, *John Wilkins* (n. 6), p. 170. On Brouncker's importance, see particularly below, 'Reconstructing Restoration Science', pp. 341 – 2.

[52] B. D. Henning, *The House of Commons, 1660 –* 90 (London, 1983), ii, 770, iii, 524, and Hunter, *Royal Society*, catalogue entries 71, 157. On Wilkins and this Act, see Shapiro, *John Wilkins* (n. 6), pp. 188 – 9.

[53] Hunter, *Royal Society*, catalogue entries 190, 222, 354 (for churchmen), 138, 178, 238 (for laymen). Cf. Henning, *House of Commons* (n. 52), i, 635 – 6, ii, 72, iii, 485. On the Society's active core, see Hunter, *Royal Society*, pp. 30, 119 – 20.

tolerance for Catholics in their state church.[54] But Sprat states categorically that the Royal Society numbered Catholics among its Fellows, and this was indeed the case: Catholic Fellows in the Society's active core included Sir Kenelm Digby as well as Charles Howard of Deepdene and his brother, Henry Howard, later sixth Duke of Norfolk.[55]

Moreover, lest one suspect that the Society was made up only of figures inhabiting a spectrum from Latitudinarian Anglicans towards the 'right', it is worth pointing out that these Catholics can be matched by a practicing nonconformist, Nehemiah Grew, elevated to the Secretaryship in 1677 after researching under the Society's auspices in the 1670s, while there was no objection to electing the Quaker, William Penn, to the Society in 1681. The dangers of trying to discern an undue degree of consensus in the Society is also shown by the presence among its activists of figures tainted with religious heterodoxy, including the highly Erastian Sir William Petty: even Moray is perhaps to be placed in this category in the light of Charles II's opinion that 'I believe he is head of his own church'.[56]

This is not, of course, to say that the Society was moving towards a position of 'complete indifference to religion', as Lotte Mulligan has claimed:[57] on the contrary, its members would have been shocked by this, as they repeatedly and quite genuinely exclaimed. But they *were* moving towards a religious pluralism, reflecting the fact that all the evidence now suggests that the motivation to scientific activity in this period was convergent rather than specific — in other words, it was not inspired by any single religious position but by a range of religious or secular stimuli, which brought people together in pursuit of shared goals which were largely defined by the internal dynamic of scientific enterprise.

This is also borne out by what we know of contemporary impressions of the Royal Society's political and religious stance. For in these the Society often seems to have taken on a veritably chameleon-like character which tells us more about the sentiments of the commentators than about the actual position of the Society. Thus while the University orator at Oxford, Robert South, indulged in 'Satyrical Invectives, against Cromwell, Fanaticks, the Royal Society, & New Philosophy', John Beale reported a completely different attitude among certain of his gentry neighbours in Somerset, who 'were fallen out with God & the King, & could not endure to heare well of any Royall Act, or Royall Society'.[58] Such accusations arguably cancel one another out, their chief significance being to illustrate the range of parties from which hostility might be expected. Moreover, as already

[54] Jacob, 'Restoration' (n. 3), p. 157; *Robert Boyle* (n. 1), p. 135.

[55] Sprat, *History*, p. 427. Cf. ibid., pp. 62-3. For these Fellows, see Hunter, *Royal Society*, catalogue entries 24, 123, 218.

[56] Stevenson, 'Sir Robert Moray' (n. 49), p. 426. See also Hunter, *Royal Society*, catalogue entries 8, 289, 387.

[57] 'Debate: Science, Politics and Religion', *Past and Present*, 66 (1975), 142.

[58] Wallis to Oldenburg, 16 July 1669, *Oldenburg*, vi, 129; Beale to Evelyn, 24 May 1666, Evelyn Collection, Correspondence, no. 52.

noted, this was precisely the reason why Sprat's book had seemed necessary in the first place.

It is in this context — as an apologia for a heterogeneous body widely regarded with suspicion — that Sprat's message makes sense. On the one hand, he *was* trying to say that science was above religious and political distinctions. This was not, however, exclusive of a complementary attempt to invest the Royal Society with a generalized, uncontroversial role in Restoration society, taking advantage of areas where he thought a positive point could be made of the Society's aspirations without entailing divisive overtones, but avoiding any very precise statement of position, especially of a party-political kind. This was, after all, essential for a society whose active core contained people from a range of ecclesiastical positions and which managed to attract and retain at least the nominal support of churchmen and others with a wide range of opinions. Indeed, one might go further and argue that, when allowance is made for the rather unfavourable environment in which the institution subsisted in its early years, it would have been difficult for the Society to have been as party-political as commentators such as Jacob have implied. The whole object of the exercise was to placate potential critics rather than to risk making new enemies.

Turning to the reception of Sprat's book, such evidence as we have suggests that it was Sprat's generalized attempt to appeal to everybody and antagonize nobody that impressed supporters of the new science when they read the work in the immediate aftermath of its appearance. John Evelyn's annotations to his copy, for instance, seem predominantly to show an approval of Sprat's well-turned phrases in praise of the new philosophy and of the eirenic commonplaces that he purveyed.[59] John Beale forgot his earlier anxieties and itemized the parties whom Sprat had courted, 'besides what he hath done in the whole for the Soc[iety]' — academics, lawyers, parliamentarians, merchants and so on — concluding: 'I make no doubt but it will prove an Invincible Light to all the world, & for all ages'. Nathaniel Fairfax expostulated: 'the Hist[ory] of the R.S. I have read, admird & lov'd. & I am hugely pleased to se a busines of that importance, intrusted with a penman, so well fitted for the work', thereupon going on to praise Sprat for capturing 'the life & soul' of the Society and its potential in terms which Oldenburg subsequently echoed himself when expatiating on the book to John Winthrop in Connecticut.[60]

Those who believed that Sprat had written a winning book, however, were quickly outnumbered by those with qualms about it. A consideration of these reservations will return us to the theme of the general difficulty of writing a statement of the 'ideology' of a corporate body like the Royal Society to which allusion was made at the start of this paper, and the specific

[59] See below. In addition, some evidently note factual references.

[60] Beale to Evelyn, 11 Sept. 1667, Evelyn Collection, Correspondence, no. 64; Fairfax to Oldenburg, *Oldenburg*, iii, 491 – 2. Cf. Oldenburg to Winthrop, 13 Oct. 1667, ibid., iii, 525 – 6.

problems that stemmed from the lack of supervision that Sprat had evidently received in writing a book to which the Society officially put its name.

In spite of the efforts made to improve the scientific content of the book while Sprat was writing it, Oldenburg remained concerned about this facet of the work, encouraging Joseph Glanvill to publish his *Plus Ultra* (1668) as a supplement to Sprat in illustrating the solid achievements of the new philosophy.[61] Indeed, Oldenburg's dissatisfaction with the *History* from this point of view is clearly in evidence in the review of it that he wrote for *Philosophical Transactions*. For he extrapolated from Sprat's text to give actual *numbers* of the Society's 'most remarkable particulars', which Sprat had omitted, to show how 'The *Particulars* upon which *Heads* are more numerous, and of greater moment and variety, than perhaps Detractors and Cavillers imagine or expect: they exceed indeed the number of 700; of which the *Experiments* and *Observations* both together amount to above 350; the *Relations*, to about 150; the *Queries*, *Directions*, *Recommendations*, and *Proposals*, to above 80; the *Instruments*, to about 60; the *Histories* of *Nature* and *Art*, to above 50; and the *Theories* and *Discourses*, to as many'.[62]

Despite his lack of satisfaction on this point, it is worth observing that there is no evidence that Oldenburg dissented from the general view of the proper method of science that Sprat presented, however little it represented the actual practice of scientists: to this extent the special degree of supervision that this part of the book had received arguably paid off. Yet even here the book had its deficiencies. One of these was Sprat's defence of experiments against reservations 'rais'd by the *Experimenters* themselves' — in the person of Boyle — which, as we have already seen, Sprat based on a draft by Oldenburg. Paul Wood, who did not know this, describes Sprat's handling of the matter as 'inept'. So it is, but the ineptness was largely that of Oldenburg, whose argument Sprat closely followed. It thus cannot be put down to lack of scientific expertise on Sprat's part, but evidently to the difficulty of constructing a readily comprehensible justification of the new science which tried to do justice to its complexities. The fact that Oldenburg's manuscript draft, and especially his 'Answer' to the 'Objection', is rather hesitant and repetitious — traits which Sprat's version tended to iron out — is perhaps symptomatic of the trouble he had in trying to devise an appropriate public stance.[63]

If this was true of the scientific component of the book, it was much truer of its religious and political facets. By 1670, John Beale had reverted to his earlier dissatisfaction with the book from this point of view, observing how

61 Oldenburg to Boyle, 1 Oct. 1667, ibid., iii, 503.

62 *Phil. Trans.*, 2 (1667), 505. Cf. Sprat, *History*, p. 155. It is interesting that Evelyn marked this paragraph in his copy of the journal, now British Library Eve.a.149: the only other point he marked was that at which Oldenburg referred to the book's dedication to the King (p. 503).

63 Wood, 'Methodology' (n. 2), p. 7. See also below and Sprat, *History*, pp. 243 – 5.

'I was alwayes of [opin]ion, that the Hist[ory] of R.S. should be purely narrative & [suitable] alike for all Climates, of Rome, Spaine, New England'. He now reiterated how he had felt all along 'that the Defensive & Apologetical part should be quite sever'd from the History; & confin'd to a peculiar Treatise', largely because of the difficulty of doing justice to the religious policy of an institution 'which admitted of all partyes into their Membership'.[64]

Comparable anxieties may help to explain what might otherwise seem an anomaly in the coverage that Oldenburg gave to Part III of the book when he reviewed it in *Philosophical Transactions.* Oldenburg devoted almost a whole page to Sprat's arguments concerning the innocuousness and potential of the new science in relation to education and nearly as much to its value to the manual arts. On the other hand, he dismissed the religious arguments which took up more space than any other in this section of Sprat's work in a mere five lines, summing up Sprat's verdict on the new science as 'not at all dangerous', and thus suggesting that he was either embarrassed or not particularly impressed by this section.[65]

In the wider public arena, it may be argued that Sprat's book, so far from being 'immediately successful', as some authors have claimed,[66] was little short of a disaster, at least as far as its domestic reception was concerned, though (as already noted) it may have fared better overseas. In 1668, the Cambridge don, John Worthington told Henry More that he considered Sprat 'perfect Hylobares' — an allusion to the figure of the materialist in More's *Divine Dialogues* — criticizing him as one of those devoted to 'nothing but what gratifies externall sense, or what sense doth reach'.[67] Worse was to follow, for over the next three years Sprat's *History* — together with Glanvill's *Plus Ultra* — was to stimulate the two most serious attacks on the new science to be published during the period, those of Meric Casaubon and Henry Stubbe. Casaubon's *Letter . . . to Peter du Moulin* (1669) was a searching critique of *Natural experimental Philosophie, and some books lately set out about it,* while one of Stubbe's books was actually entitled *A Censure upon Certaine Passages Contained in the History of the Royal Society, As being Destructive to the Established Religion and Church of England* (1670). Moreover it comprised just that, taking a series of what he saw as questionable and dangerous statements in Sprat's book and subjecting them

64 Beale to Evelyn, 14 March 1670, preserved among a folder of loose letters in a box of 'Evelyn Period' papers in the Evelyn Collection (it is severely damaged by damp, and I have supplied the bracketed words); Beale to Evelyn, 11 June 1670, Evelyn Collection, Correspondence, no. 96.

65 *Phil. Trans.*, 2 (1667), 506 – 8.

66 Webster, 'Origins' (n. 3), p. 115. Cf. p. 118, and, e.g., Purver, *Royal Society* (n. 3), passim. For an estimate more similar to that given here, see Aarsleff, 'Sprat' (n. 19), p. 581.

67 Worthington to More, 5 Feb. 1668, in John Worthington, *Diary and Correspondence*, ed. R. C. Christie, ii, part 2 (Chetham Society, 1886), 265.

to detailed and effective scrutiny: significantly, all but one of these came from Part III of the work.[68]

The effect was devastating, especially since Sprat's book *had* been officially commissioned by the Society, even if, as we have seen, it had apparently not been closely supervised. The dilemma in which this placed the Society is shown by the reaction of Sir Robert Moray to Stubbe's threatened assault, which Stubbe reports in one of his later books. 'When the first *brute* of my designing to write against the R.S. did reach *London*, Sir *R.M.* writ to the Lady *E.P.* to inform them of my intentions; adding, That there was nothing in which the R.S. as a Body, could be concern'd, excepting this *History*: and if I would civilly represent unto them any defaults therein, they would take it kindly, and amend them'. Stubbe therefore wrote to him concerning both Sprat's *History* and Glanvill's *Plus Ultra*, 'and DEMANDED that the R.S. should disclaim *both of them by some authentick Declaration*'. Revealingly, however, the Society felt obliged to stick by the *History* as an 'official' statement, although embarrassed by the reaction it had inspired, while finding it easy to disown Glanvill on the grounds that 'He was a Private Person, and that the *sense of the R.S.* was not to be collected from the Writings of every *single Member*'.[69] The problem was that, while it could not be denied that the *History was* an authorized statement, its blunt printed statements of policy provided hostages to fortune which need not otherwise have existed.

Clearly the Society's supporters were particularly shocked precisely because they had not thought that they were saying anything controversial. Indeed, they would have been relieved to learn of the subversively radical motives which we have recently been encouraged to see in Stubbe, which might have reassured them that they were on the side of the angels: as it was, they were worried by Stubbe precisely because he seemed to be appealing to commonplace anxieties of the day.[70]

Telling evidence in this respect comes from a letter written by John Evelyn to John Beale in the aftermath of Stubbe's attack. After a lengthy tirade against the 'petulant Scribbles' of 'this angry monster', which bears evident testimony to his shock at Stubbe's onslaught, Evelyn explained how: 'For my part I have read & seriously ponder'd both Dr *Sprat* and

68 A number of extracts from Stubbe's attacks are printed in the 'Notes' to Sprat, *History*. See also ibid., Appendix B. For modern accounts of these attacks, see esp. Hunter, *Science and Society*, ch. 6 and the references cited on pp. 212 – 3. Two works there referred to as forthcoming have now appeared: M. R. G. Spiller, *'Concerning Experimental Natural Philosophie': Meric Casaubon and the Royal Society* (The Hague, 1980) (see also Michael Hunter, 'Ancients, Moderns, Philologists and Scientists', *Annals of Science*, 39 (1982), 187 – 92) and J. R. Jacob, *Henry Stubbe, Radical Protestantism and the Early Enlightenment* (Cambridge, 1983).

69 Stubbe, *A Reply unto the Letter written to Mr Henry Stubbe* (Oxford, 1671), pp. 19 – 20. Lady E.P. was evidently Lady Puckering: cf. Stubbe to Moray, 27 May 1667, E.L. S.1.89. Cf. ibid., 91. For a slightly different view on the issue of the Society's authorization, see *A Letter to Mr Henry Stubs* (London, 1670), p. 3.

70 See Jacob, *Stubbe* (n. 68), passim. For a caveat, see Hunter, *Science and Society*, pp. 139 – 40.

Glanvil, and can find nothing which a sober person can take just Exceptions at'. He went on to assert the kind of 'safe', 'apolitical' role for science which, as I have suggested, had always been at the back of the minds of the virtuosi, even when they had tried to argue that it was possible to be above politics and yet solve the problems of politics at the same time. Sprat and Glanvill, he claimed, 'have said nothing but what is justifiable for the bespeaking the favour and Civilities of honest and learned men' to the Society, whose activities he proceeded to characterize in a revealingly low-key manner, consoling himself by noting with satisfaction how even the hitherto apathetic universities were beginning to take an interest in the studies which the Society had pioneered, in addition to reverting to further bitter recriminations against Stubbe.[71]

What is clear is that, by trying to give an account of science's broader role which they hoped would offend no-one, the Society's organizers found that they had stirred up a hornet's nest of hostility, illustrating that some of the religious, intellectual and political implications of the new science *were* potentially more controversial than they seem to have appreciated, especially in the context of commonplace attitudes of the late seventeenth century. Moreover the problem was exacerbated by the choice of author, for Sprat was almost *too* glib in his ability to make the best showing of every facet of the Society's activities, and the notions of the brash young graduate were less mainstream in the context of the 1660s than he perhaps thought, or than many of his modern readers have realized. As it was, there was much to shock susceptible contemporaries, including Sprat's rather 'Hobbist' sentiments on various points, his iconoclastic attitude towards the existing educational curriculum, his disdain for the ancients, and perhaps above all his rationalistic and anti-intellectualist religious views.[72]

The fact that this difficulty had apparently not even been foreseen by the Society's organizers, who would otherwise surely have ordered that the text should be carefully read before it was sent to press, suggests to me that the reaction to Sprat awakened them from a considerable state of naivety concerning the difficulties involved in composing a public statement of this kind. Indeed, in these circumstances, the apparent lack of supervision of the book's controversial later sections is astonishing, though it would not have surprised Stubbe: he thought it highly likely that the Fellows 'had not patience to read it, or *any Book* of *that bulk*; but, as in other cases, gave their *assent* and *applauds* upon *trust*'. He reiterated, however, 'that the R.S. did own it, any man knows that was in *London* at its publication, not to mention the character which Mr. *Glanvill* and the Transactor fix on it'.[73]

The result was that the effort to define science's broader role represented by Sprat's book was quietly abandoned. In 1668, there was talk of a

[71] Evelyn to Beale, 27 July 1670, Evelyn Collection MS 39, no. 329. Evelyn's characterization of the Society is quoted in Hunter, *Science and Society*, p. 44, and his comment on Oxford in ibid., pp. 146 – 7.

[72] See Hunter, *Science and Society*, ch. 6. For 'Hobbist' sentiments, see Sprat, *History*, e.g. pp. 33, 379.

[73] Stubbe, *A Reply* (n. 69), p. 19.

proposed second edition with additions which might have helped to rectify the book's shortcomings from the scientific point of view: but, although the book soon became scarce, no revision or even reprint occurred during the rest of the seventeenth century.[74] There are hints that a fresh defence of the new science was planned early in 1678, this time by Thomas Smith, a divine who had just been elected to the Society and may have been recruited specifically for this purpose, while further such discussions may have occurred in 1693.[75] But in neither case did anything materialize, and the nearest approach to a follow-up to Sprat's apologia for the Society comprised some remarks 'to *vindicate* that *Assembly*' that John Evelyn included in the preface to the third edition of his *Sylva: or a Discourse of Forest-Trees* in 1679.

Moreover these are interesting precisely because their burden is so internalist: it was for its contribution to the task of improving knowledge about the natural world, not its religious, political or social potential, that Evelyn defended the Royal Society. His comments on the broad implications of the new science were defensive, and were virtually limited to saying that, so far from the King and kingdom being diminished in reputation by such activities, the adulation of foreigners had accrued to national esteem.[76] And, though claims have been made for an 'ideology' for the Royal Society in the period of the Exclusion Crisis, the evidence is very limited by comparison with the open attempt at a public statement on major issues made by Sprat.[77]

Only in the eighteenth century was Sprat's *History* reprinted, first in 1702, and then again in 1722 and 1734, and I think that this may itself be significant. For one thing, shifts in attitudes meant that religious and intellectual positions which had seemed controversial in the 1660s were by

[74] Birch, *History*, ii, 266; Oldenburg to Beckman, 30 March 1668, *Oldenburg*, iv, 278; Beale to Evelyn, 11 June 1670, Evelyn Collection, Correspondence, no. 96. It is perhaps worth noting here a defence of the Royal Society published in 1670 as *A Letter to Mr Henry Stubs* (see above, n. 69): but this is almost wholly devoted to the refutation of specific accusations by Stubbe (see Oldenburg to Williamson, 14 March 1675, *Oldenburg*, xi, 226, for what is apparently a later reference to it). For a reference in 1681 to the 'right of copies' if Sprat and other works were to be reprinted, see Birch, *History*, iv, 82. See also Hooke's diary entry for 23 Feb. 1681 in City of London Guildhall Library MS 1758, fol. 74: 'I proposed subscriptions for Soc: history at Jonathans'.

[75] Hooke, *Diary*, pp. 342, 345; R. T. Gunther, ed., *Early Science in Oxford*, x (Oxford, 1935), 264 – 5. On Smith, see Hunter, *Royal Society*, catalogue entry 333.

[76] John Evelyn, *Sylva* (3rd edn., London, 1679), sigs. A1-4, a1. On the background to this material see Hunter, *Science and Society*, pp. 177 – 8. Quotations from this preface are included in the notes to Sprat, *History*: see especially that to p. 78, l. 31 (though it is there wrongly stated that the material first appeared in the 1670 edition).

[77] Jacob, 'Restoration Ideologies' (n. 3). It should also be noted that it is far from clear that any of the writings of this period cited in this article were intended to represent the corporate views of the Royal Society.

this time becoming increasingly common.[78] In addition, the new science was by now much more 'established' than it had been in the pioneering days of the early Restoration, due to the increase of fashionable participation in it, to the rise of public lectures and demonstrations, and to the inclusion of experimental philosophy in the university curriculum.[79] By this time, assisted by the acclaim that Newton received for his *Principia*, the new science had 'arrived' in a way that it had not half a century earlier, acquiring a superior prestige to which appeal was frequently made, and Sprat now seemed almost prophetic in his claim as to how science represented the country's 'present prevailing *Genius*'.[80]

This, however, has important negative implications for the late seventeenth century, when it is easy to overrate the popularity that the new science enjoyed. In fact, at this stage the new science was a marginal and widely unappreciated activity, which was susceptible to hostility because of the extent to which it could be deemed to offend against crucial religious and learned preoccupations of the day.[81] It was for this reason that the new scientists were so often on the defensive in these formative years. Not only does this explain why they had to work so hard to distinguish themselves from thinkers with overlapping concerns who were commonly perceived as pernicious, such as Hobbes.[82] It also accounts for the fact that Sprat's apologia seemed necessary in the first place, and, if this is forgotten, the message that Sprat attempted to purvey in his book is easily misunderstood.

But the aftermath to the publication of the *History* illustrates the difficulty of constructing a satisfactory apologia in such circumstances and shows that, even if a statement of the religious and political implications of the new science seemed in theory desirable, it was in practice dispensable. Moreover the investigation of the context of Sprat's *History* given here should indicate how unwise it is to take the views expressed in that book as reflecting a 'party line' on the Society's part. In my view, even the book's broad message about science's compatibility with rational religion and with national prosperity and harmony — in a word, its 'Latitudinarianism' — is to be read negatively rather than positively. In his *History*, Sprat was not aligning the Royal Society with the ecclesiastical policy of Bishop Wilkins. The object was to find a generalized creed which would enjoy at least the tacit support of people from a wide range of positions, which would disarm criticism and might also enlist support. This is the true, if facile, message of

78 Hunter, *Science and Society*, esp. p. 159.

79 Ibid., esp. pp. 84 – 5, 191 – 2; A. E. Musson and Eric Robinson, *Science and Technology in the Industrial Revolution* (Manchester, 1969), ch. 1; John Gascoigne, 'Politics, Patronage and Newtonianism: the Cambridge Example', *Historical Journal*, 27 (1984), 1 – 24, esp. pp. 14f.

80 Sprat, *History*, p. 78.

81 See above, nn. 8 – 9.

82 I would thus place a somewhat different construction from Shapin and Schaffer on the evidence they adduce in *Leviathan and the Air-pump* (n. 42): it seems to me that they underestimate the *defensiveness* of the posture of Boyle and his colleagues. Cf. Hunter, *Science and Society*, pp. 172 – 3, 178 – 9.

an apologia which in fact proved much more problematic that either its author or his sponsors expected.

DOCUMENTS

(1) Oldenburg's draft for pp. 243 – 5 of Sprat's History

This item survives as Royal Society Classified Papers xxiv.42. It is in the hand of Henry Oldenburg throughout, but it is clear that it was written in two stages: after the first two paragraphs, which set out the problem, the concluding two paragraphs which give an 'Answ[er]' to it are written in a different, darker ink, as if added later, perhaps after consultation with others. There are numerous deletions and insertions reflecting Oldenburg's hesitations and second thoughts in drafting the paper: these are recorded in footnotes. A few editorial insertions have been made to clarify the sense; contractions have been expanded throughout.

Objection
Against the Experimental way
of Philosophy;
Thought necessary to be answered in M. Sprats history.

There is a great nicety and difficulty in the making of many Experiments: The succes of very many is[1] various[,] inconstant, and differing, not only in the hands of different, but even[2] the self same Experimenters, upon the account 1. Of the more or lesse skill, care, patience; 2. Of the diversity of the materials, whereof some are genuine, some sophisticated; some simple, some mixed; some fresh and vigorous, some old and effete: And particularly in Chymical Trials, many dissolvents are well, many ill prepared; some Spirits[3] too litle, some too much purified, etc. 3. Of some circumstances, which are either constantly unobvious, or at least scarce discernable, till the Tryall be past: which makes the[4] Experiments very obnoxious to contingencies.

Upon these and the like considerations, it is to be feared that, though the R[oyal] Society should take due care of having their Experiments and

[1] After 'is', 'so' has been deleted. In the next line, 'differing' has been altered from 'different'.

[2] 'even' replaces 'also', deleted.

[3] After 'Spirits', 'etc' has been deleted.

[4] 'the' replaces 'such', deleted.

Observations made[5] skilfully,[6] and with care, and recorded with[7] faithfulnesse, yet other Companies both of this present Age, and of future times, will meet with successes much differing from theirs, and yet pretend to as much skill and caution, as possibly they can doe. And if there be no agreement in Experiments, at least in many such, as are important to determine truth by /verso/ [8]there appears small hopes of ever raising a Systeme of solid Philosophy, etc.

Answ[er]. All circumstances concurring in the making of the Experiments, [9]will render[10] the effects to be constantly the same.[11] If the effects differ, it will put men upon investigating[12] the causes of that difference, which must needs discover a great many considerable and luciferous things.

It is presumed that the present[13] R[oyal] Society, which is to lay the foundation of this work, and to give an Example to their successors[14] will[15] make use of very expert, carefull,[16] diligent and circumspect men in the making of Experiments, and in the recording[17] the proces and all the circumstances of it with all possible exactnes.

(2) John Evelyn's annotations to his copy of Thomas Sprat's History of the Royal Society *(1667)*

The information that follows was derived from Evelyn's copy of this work, which was stolen from Christies when sent for auction there in 1978. Since the present whereabouts of this volume is unknown — if it survives at all — it seems worth preserving here what may constitute a unique record. Details of the book are given in the Christies sale catalogue, *The Evelyn Library, Part 3, M-S*, 15-16 March 1978, lot 1405 and plate 24. Line numbers have been taken from the modern edition of Sprat's text by J. I. Cope and H. W. Jones. All notes were in the margin unless otherwise stated. 'l.' denotes a vertical stroke beside the lines of text specified. If more than one line is given (e.g.

5 After 'made', 'and recorded carefull' has been deleted.
6 After 'skilfully,' 'car' has been deleted.
7 'recorded with' is inserted above the line.
8 Before 'there', 'it seems' has been deleted.
9 Before 'will', 'though by severa' has been deleted.
10 replacing 'make', deleted,
11 After 'same', a comma has been replaced by a full-stop and 'though i' has been deleted.
12 After 'investigating', 'and discovering' has been deleted.
13 'present' is inserted above the line, as is 'to' in the next line.
14 'their successors' replaces 'posterity', deleted.
15 After 'will', 'use' has been deleted.
16 After 'carefull', 'and ci' has been deleted.
17 This was replaced by a word which was then deleted and is now illegible.

21/2), this is because it is not clear whether the marking was supposed to be adjacent to the upper or the lower of the two.

p. 4, l. 32: 'new?'
p. 8, l. 31: '-'
p. 12, ll. 25-6: 'strongster' altered to 'stronger' in text
p. 26, l. 24: 'of' underlined in text and 'or?' inserted in margin
p. 27, l. 29: 'ich' deleted in 'which' in text, and 'le/' added in margin
p. 61, ll. 10-11: l.
p. 67, l. 10: 'Author' in text underlined, and the following written in margin: 'Mr *Graunt* a Bodice-maker in Birchin-lane, a man of extraordinary natural parts & solid judgement'
p. 69, ll. 6-8: l.
p. 75, ll. 19-20, 25-6: l.
p. 89, in margin at top of page: 'Lipsius/ Scaliger/ Salmosius/ De Cartes'
p. 90, ll. 3/4-5: l.
p. 91, ll. 5-6, 7-8: l.
p. 93, ll. 11-13: l.
p. 94, l. 16/17: 'Dr Wilkins' [i.e., refering to ll. 14-15]
p. 97, ll. 27-8: l.
p. 98, ll. 28-30: l.
p. 100, ll. 13-14: l.
p. 101, ll. 34-5: l.
p. 107, ll. 5-6: l.
p. 110, ll. 18-end: l.
p. 111, ll. 1-9: l.
p. 112, ll. 18-20: l.
p. 127, ll. 4-6: diagonal stroke
p. 163, l. 18: 'sheds' in text altered to 'opens'
p. 164, ll. 27-30: l.
p. 165, ll. 3-4, 24-5, 31-2: l.
p. 200, title: after 'Receiv'd', the words 'by Mr Evelyn' have been inserted in the text, not in Evelyn's hand
p. 208, l. 29: cross above line in text between 'spungy' and 'Stone', and note in adjacent margin: 'of which I have one'
p. 213, ll. 17-8, 19-20: l.
p. 277, l. 20: horizontal line '——'
p. 313, l. 32: 'r' in 'more' deleted in text and 'v/' added in margin
p. 314, l. 20: 's' deleted in 'diamets' in text and 'er/' added in margin
p. 350, ll. 6-8, 12-16: l.
p. 352, ll. 17-18: l.
p. 355, ll. 9-11, 18-22: l.
p. 360, ll. 10/11-15: l.
p. 364, ll. 10-12: l.
p. 367, ll. 12-16: l., with an extra l. to ll. 12-13.
p. 370: above page number: 'mispag. 362
ll. 20-29: l.
p. 371: above page number: 'mispag. 363'

p. 382, ll. 15-18, 19-21/2, 25-8: l., with 'nb' beside latter
p. 386, ll. 14-16: l.
p. 388, l. 16: 'East' deleted in text and 'so' [sic] in margin
p. 391, ll. 14-16: l.
p. 393, ll. 2-5, 32-5: l.
p. 397, ll. 5-7, 29-31: l.
p. 398, ll. 28/9-31: l.
p. 400, l. 3: horizontal line '——'
p. 401, ll. 16-18: l.
p. 402, ll. 17-19, 21-2: l.
p. 404, ll. 1-3, 8-9, 31-2: l.
p. 407, ll. 7-8: l., and 'nb' [?]
p. 412, ll. 4-5, 15-16, 22-4: l.
p. 429, ll. 9-14: l.
p. 438: at end of text: 'perlegi' and mark ╫
p. [439]: after the title 'ERRATA', 'correcta — ' inserted in scribal hand
Back fly-leaf: 'Art of shooting flying # p. 277'

November 20. 1663. The Royall Society consisted of 131 Fellowes, 21 of them being of the Council, have their names heere written in

The Duke of Buckingham. George Villiers
The Marquis of Dorchester. Henry Pierrepont

7 The Earle
of Argile ..... Archibald
of Craford & Lindesay. John
of Devonshire .. William Cavendysh
of Kincardine. Alexander
of Northampton. James
of Peterburgh. Henry
of Sandwich .... Edward Montagu

2 Viscount .... [illegible]
Viscount Massarene ..... John Clotworthy.

7 The Lord
Annesley ..... [illegible] James
Ashley ...... Antony Ashley Cooper
Berkeley ...... George
Bruce ...... Robert
Cavendish ..... William
Hatton ...... Christopher
Lucas ...... John

4 Knights of ye Bath
Sir John Denham.
Sir Francis Fane
Sir Edward Harley
Sir Richard Powle

1 Knight & Baronet. Sir Robert Paston

17 Knights
S. John Birkenhead.
S. Edward Bysshe.
[illegible]
S. John Finch
S. Robert Harley
S. Andrew King
[illegible]
S. Anthony Morgan
S. [illegible]
S. Thomas Nott
S. William Persall
S. Peter Pett
S. William Petty
S. James Shaen.
S. [illegible]
S. Peter Wyche
S. Cyril Wyche.

47 Esquires
Elias Ashmole
John Austen
John Awbrey
[illegible]
Richard Boyle
[illegible]
[illegible]
John Brooke
John Clayton
Daniel Colwall
Thomas Coxe
John Driden
Andrew Ellis
[illegible]
Henry Ford
William Gonelton
John Graunt
William Hammond
James Hayes
Thomas Henshaw
[illegible]
John Haskins
Charles Howard
Richard Jones
James Long
Antony Lowther
William Neile
Philip Packer
[illegible]
Peter Pett
Thomas Povey
Henry Powle
Henry Proby
[illegible]
Robert Southwell
Alexander Stanhope
Thomas Stanley
Samuel Tuke
Edmund Waller
Edward Waterhouse
Joseph Williamson.
Francis Willughby
William Winde
John Winthrop
[illegible]
Edmund Wylde

8 Strangers borne in
France
Peter Beaufort Valves de Frejou
Nicaise .... le Febvre
Samuel Sorbiere ....
Germany
Theodore Haak
[illegible]
William Schroter
Holland
Christian Huygens de Zulichem
Cornelius Vermuyden

32 Doctors

6. Of Divinity
SETH WARD Bishop of Exeter
Edward Cotton
William Holder
John Pell
John Wallis
John Wilkins Deane of Rippon

1 Of Law
Christopher Wren, Savilian professor of Astronomy.

25 Of Physick & Medicine
Thomas Bayns
Peter Bale
George Bate
Ralph Bathurst
David Bruce
Walter Charleton
Thomas Cox
William Croon
Timothy Clark
George Ent
Francis Glisson
[illegible]
Nathaniel Henshaw
William Hoare
Christopher Merret
Jasper Needham
Walter Pope
Henry Power
William Quatremain
Charles Scarburgh
George Smith
Christopher Terne
Daniel Whistler
Thomas Willis
Thomas Wren.

2 Batchellars of Divinity
Isaac Barrow
Francis Potter
2. Masters of Arts
Robert Hook
Thomas Spratt

18	Lords
22	Knights
47	Esquires
32	Doctors
2	Batch: Divs
2	Mast: Arts
8	Strangers
131	in all.

3

AN EXPERIMENT IN CORPORATE ENTERPRISE: THE ROYAL SOCIETY'S COMMITTEES OF 1663-5, WITH A TRANSCRIPT OF THE SURVIVING MINUTES OF THEIR MEETINGS*

The minutes preserved in the Royal Society's journal book are one of the best-known sources for the Society's history in its early years, itemizing in detail the experiments that were carried out, the discussions that ensued, and the reports and letters that were presented at meetings. That the minutes are so frequently cited is thanks largely to the edition of those for the period up to 1687 produced by Thomas Birch in the eighteenth century under the title, *The History of the Royal Society* (1756 – 7). Birch's edition is not perfect, but it was good for its time, and it is a tribute to its completeness that it has been worth publishing such lacunae in its coverage as have come to light, such as a record of two meetings in 1674 which were omitted from the journal book although noted in the draft minutes.[1]

The minutes on which this essay is based, of which a text is appended to it, are therefore interesting not least because they constitute the largest addition to our knowledge of the collective activities of the Society in its formative years to be published since the eighteenth century. This further

* I am indebted to Christine MacLeod for her advice on patents and her comments on a draft of this chapter. Jim Bennett has also kindly answered my queries, while Keith Moore helped to facilitate the transcription of the committee minutes by providing print-outs of microfilm copies of them, and James Robertson assisted me in elucidating the books referred to by the various committees.

[1] See I. Kaye, 'Unrecorded Early Meetings of the Royal Society', *Notes and Records*, 8 (1951), 149 – 52. See also Hunter, *Royal Society*, p. 81, and, on the way in which Birch slightly abbreviated the text, *Oldenburg*, viii, xxvi – vii.

Plate 3. Manuscript List of Fellows of the Royal Society divided up according to social rank and professional status, dated 20 November 1663, from British Library Additional Manuscript 4441, fol. 79. The names of the Council members were distinguished from those of other Fellows by the use of red ink, which has faded so badly that these names are now barely visible. Reproduced by permission of the British Library. At a Council meeting on 9 November 1663 it had been ordered that an alphabetical list of Fellows should be drawn up, with the Council members in a separate column (Birch, History, *i, 328). Printed lists in this form appeared from that year onwards, and why this differently arranged list was made on 20 November is unclear.*

record was wholly omitted by Birch, not due to any oversight either on his part or that of the compilers of the journal book but because the meetings in question were not of the whole Society, but of committees of it which functioned between 1663 and 1665. Nevertheless, these carried out work that was comparable with that of the Society as a whole, to which it was reported, and they deserve to be more widely known as a supplement to Birch. In addition, they have one feature that the minutes of regular meetings of the Society lack (though it is shared by the minutes of the Council), namely that a full record was kept of attendance. This makes it possible to comment on members' participation in the Society's activities in a manner not normally feasible, with implications to be explored later.

This, however, is almost incidental to the main interest of this material, which illustrates a notable episode in the Society's early evolution. The bulk of the minutes date from the months following the spring of 1664, when a whole series of permanent committees were set up to deal with different aspects of the Society's work, thus representing a significant aspiration to specialization in the furtherance of the Society's programme, although in practice it proved only shortlived. Before their demise, however, the proceedings of these groups throw important light on the Society's programme for the reform of knowledge — its Baconian commitment, its encyclopaedic ambitions, and not least its hope to achieve results of immediate practical utility. The intention of this paper is therefore to accompany the text of these minutes with an account of the episode from which they originated and its implications for the Society's early development.

The use of the term 'committee' in its modern sense had become familiar in the context of parliamentary politics in the late sixteenth and early seventeenth centuries and perhaps particularly during the Civil War and Interregnum, when it has aptly been said that England experienced 'rule by committee'. This explains why, in introducing the Royal Society's use of the word committee 'in a Philosophical sence' in his *History*, Thomas Sprat added the hope that such usage might 'in some measure purge it from the ill sound, which it formerly had'.[2] The appropriation of the term to the context of natural philosophy is, however, arguably symptomatic of the process of adopting an institutional structure for the new science from precedents available elsewhere that has been surveyed in the Introduction to this volume: from the start the Royal Society made extensive use of committees — either when further investigation of a specific topic was delegated to a small group of Fellows, or when committees were appointed with a broader responsibility for some aspect of the Society's administration.

To take the latter first, a committee to deal with the Society's constitution was appointed at its first meeting, while another took responsibility for foreign correspondence: this is first heard of as a body specifically charged

[2] G. E. Aylmer, *The State's Servants* (London, 1973), p. 9; Sprat, *History*, p. 85.

with corresponding with Prince Leopold of Tuscany, but it quickly acquired a more general brief for such activity. There were also committees to consider such matters as a library for the Society, while on 19 December 1660 every Fellow was requested to 'bring in to a committee, to be appointed for that purpose, such experiments, as he should think fit for the advancement of the general design of the society', and two further committees were appointed in July 1661 to meet in the City and at Westminster to propound experiments.[3] In addition, a committee was instructed in June 1662 'to draw up a paper concerning the design of the society', which thus began the process of devising a defence of the institution which was to culminate in Sprat's *History*; another special group was that which met on Mondays at the lodgings of Dr Jonathan Goddard at Gresham College to make preparations for the King's visit to the Society.[4]

The Society's statutes and its charter were also deemed a matter appropriate to the responsibility of a committee. The first such group was appointed in May 1663, and a brief general report of 'the Committe for reviewing the Lawes' dated 19 December 1663 survives among the minutes appended to this paper. This came to the conclusion that it was best for the Society 'to have no compleate body of Lawes', though, in view of the seriousness with which, as we saw in the Introduction, the Society took questions relating to its legal standing, it is disappointing that all we otherwise know of this committee's discussion of this topic derives from brief references in the minutes.[5]

Equally common was the appointment of committees to pursue specific topics that arose during the Society's proceedings: as John Wallis explained in a letter to Johann Hevelius of 5 April 1664, 'if for lack of time or means the majority desire it so', experiments were 'referred to the care of a committee chosen from the whole body, who after examining them report back to the rest'.[6] The earliest such reference appears in the journal book on 16 January 1661, when nine Fellows were 'appointed a committee to draw up an account of the quicksilver experiment'; another was constituted 'for magnetical experiments' at the same meeting; while on the following 6 March four Fellows were 'appointed a committee to go to the glass-house at Woolwich', to enquire into the glass drops sent to the Society by the King.[7]

[3] Birch, *History*, i, 5, 7, 22–4, 26, 45. See also below, p. 247.

[4] Ibid., i, 85, 389, 391, 408, 457, 473. Cf., ibid., 507. In addition, from 1663 committees were appointed annually to to audit the Society's accounts, as was laid down in the statutes drawn up in that year. See *The Record*, p. 295, and Birch, *History*, i, 291 and passim.

[5] See ibid., i, 241, 249–50, 264, 346, 385, 388–9; ii, 131, 136. See also above, Introduction, p. 5, and below, p. 104.

[6] *Oldenburg*, ii, 170.

[7] Birch, *History*, i, 12, 18. On the drops, see Laurel Brodsley, Sir Charles Frank and J. W. Steeds, 'Prince Rupert's Drops', *Notes and Records*, 41 (1986), 1–26.

Subsequently, a range of comparable groups are heard of, especially in the Society's first few months, dealing with topics ranging from the generation of insects to experiments with a pendulum.[8]

In addition, committees were often formed to consider books or proposals that were brought to the Society. Thus we hear of a committee to evaluate John Graunt's *Natural and Political Observations . . . upon the Bills of Mortality* (1662), for instance, to inspect 'Mr Towgood's water-works', or to consider the musical theories of John Birchensha.[9] A further such committee was that appointed on 7 January 1663 in response to a proposal for the propagation of cider made in a letter from the Somerset virtuoso, John Beale. This met the following Monday at the lodgings of Charles Howard, an active member of the Society with a strong interest in agrarian projects; it subsequently reported back to the Society as a whole, which thereupon urged members to prosecute the work, requesting further material from Beale, which he provided and which was subsequently to be digested into the compendium which materialized in 1664 as John Evelyn's *Sylva.*[10]

'Committees' of this kind were clearly often quite informal, overlapping with the Society's practice of appointing 'Curators' jointly to prosecute specific experiments, and it is hardly surprising that most have left no record beyond the brief reports that sometimes appear in the Society's minutes.[11] Occasionally, however, fuller documentation survives. One instance of this concerns the queries concerning the preservation and planting of timber that the Society received from the Commissioners of the Navy at its meeting on 17 September 1662. A committee was appointed to look into this and to abstract 'a brief and methodical answer' from the material on related topics that had by this time been collected for *Sylva*. This paper survives among the Society's Classified Papers, and in it Dr Jonathan Goddard responded with general comments 'as to what is of naturall consideration' (including the recommendation that the habitat in which different trees flourished should be further investigated), while the New Englander, John Winthrop, appended detailed practical considerations on timber planting.[12] A further such report was that compiled by a committee of Fellows of the Society resident in Ireland 'concerning the structure and sayling of Sir William

8 Birch, *History*, i, 20, 23, 25, 53 (Goddard added to a pre-existing committee), 70.

9 Ibid., i, 75, 76, 81, 84, 87, 416, 418. See also ibid., 87, 296, 327, 384, 388, 424 (in some of these cases, the title 'committee' is not used). On Thomas Toogood or Towgood and his various water-raising engines, see Bennett Woodcroft, *Titles of Patents of Inventions, Chronologically Arranged* (London, 1854), nos. 132, 139, 142, 173, 179.

10 Birch, *History*, i, 172, 176 – 7, and Beale to Oldenburg, 21 Dec. 1662, *Oldenburg*, i, 483. See also Birch, *History*, i, 179, and Evelyn, *Diary*, iii, 351 and n. 4.

11 E.g. Birch, *History*, i, 31, 115, 212, 250, 322.

12 Ibid., i, 111, 117 (Winthrop's name is included on the first occasion; Wilkins's on the second); Cl.P. x(3)20 – 1 (which also includes Winthrop's proposal that ships for the navy might be built in New England: cf. Birch, *History*, i, 112 – 3).

Petty's new ship', in other words his famous 'double-bottomed' boat. The committee was appointed on 18 November 1662, and its report, which was read to the Society on 28 January 1663, was printed by Birch.[13]

Thirdly, on 18 March 1663, Henry Oldenburg reported to the Society his receipt of a letter from a Somerset gentleman, John Buckland, 'wherein was proposed a way of preventing famine, by dispersing potatoes throughout all parts of England'. A committee of thirteen was appointed to consider the matter, and of these seven met as arranged at Howard's lodgings at Arundel House on 20 March, together with the President, Lord Brouncker (who had not been named as a member). The committee reported to the Society on 25 March, summarizing the gist of their discussions and quoting their chief conclusions, and as a result the Society 'recommended this work to their members' and thanked Buckland 'for his respect to the society, and care of the public good'. In addition, however, a manuscript report survives of the discussions that took place at the meeting of the committee, and this has been reprinted here, as the earliest of the surviving sets of minutes around which this paper is focussed.[14]

The Buckland scheme is an interesting one, as very much the kind of public-spirited project that had been associated with the Hartlib Circle in the Interregnum, and which it might have been hoped that the infant Royal Society would continue to encourage: it was symptomatic that, in reporting on Buckland's proposals and soliciting support for them, Oldenburg told John Evelyn how they were 'tending to universall good'.[15] Buckland himself appears to have been a kind of projector, since we also hear of a paper of his 'concerning the way of brewing beer without malting the corn, and of turning rich cider into a kind of canary'.[16] Clearly interest in his scheme was heightened by the fact that the harvest in 1661 had been an unusually bad one, the worst since the 1590s. Its impact was exacerbated by the fact that it came at the end of a succession of five poor harvests, and its worst effects were felt in 1662, when there were severe food shortages. It is thus hardly surprising that some of the discussion that ensued at the meeting

[13] Ibid., i, 131, 183 – 92, 249. The original of the report is in E.L. P.1.38. It is also printed in the Marquess of Lansdowne, ed., *The Double Bottom or Twin-Hulled Ship of Sir William Petty* (Oxford, 1931), 49 – 53. See also ibid., p. 35 and passim.

[14] Birch, *History*, i, 207, 213. Cf. *Oldenburg*, ii, 24 – 5. See below, pp. 102 – 4. There are slight differences, in that the fifth of the committee's recommendations is omitted; the wording concerning Evelyn is slightly briefer and specifies its insertion in an appendix; while recommendations 2 and 3 differ, mentioning Howard as well as Boyle and inserting the phrase concerning 'the seminal virtue thereof'. The number of those attending has here been derived from the minutes, which reveal that the following who were nominated to the committee failed to attend: Brereton, Wilkins, Tuke, Evelyn, Merrett and Henshaw.

[15] Oldenburg to Evelyn, 9 March 1663, *Oldenburg*, ii, 30. For the Hartlibian background, see Charles Webster, *The Great Instauration* (London, 1975), passim.

[16] Birch, *History*, i, 207 (and, for other references to Buckland, ibid., i, 198, iii, 74).

of the committee concerned other means of feeding the poor; in addition, consideration was given to agricultural improvement more generally, reference being made to an engine for improved sowing devised by John Wilkins and to new methods for increasing crop yields by fertilization.[17]

The committee decided that efforts should be made to promote the use of potatoes through the example of Fellows of the Society, and for this purpose Buckland was requested to send up some potatoes for members to try. In addition, it was suggested that a note about them might be inserted in Evelyn's forthcoming *Sylva*, and that the practice might be promoted through advertizing it in the 'Diurnalls', the newspapers. That the committee was so readily convinced of the value of potatoes may have owed something to Robert Boyle, who regaled the committee with detailed knowledge evidently acquired on his estates in Ireland. Indeed, at a meeting of the Society on 8 April 1663, a letter was read from Boyle's gardener in Ireland which accompanied a box of tubers for the Fellows' use; the letter comprised a lengthy disquisition on the best methods of cultivating the plant. Subsequently, on 15 April 1663, one of the members of the committee, Jonathan Goddard, received a letter from a 'domestic' of Buckland's with further potatoes for distribution.[18]

After that, however, nothing seems to have come of this novel and imaginative scheme. No entry on potatoes was inserted in *Sylva* — where it would have been rather out of place in any case — and complete silence ensues in the Society's records. Though the indefatigable John Beale is known to have continued to be interested in the subject, his colleagues had evidently been discouraged, for reasons that are unclear.[19] Possibly they encountered the problem that was so often to confront the Society in its technological work in its early years — namely the sheer difficulty of disseminating improvements:[20] but we have no evidence that such propagation was even attempted, and the fact that it was proposed that the Society's name be suppressed in the advertisement in the diurnals could be taken to show a slight loss of nerve at the outset. All in all, it is a curious and rather disappointing episode.

[17] See W. G. Hoskins, 'Harvest Fluctuations and English Economic History, 1620 – 1759', *Agricultural History Review*, 16 (1968), 16, 18, 21, 29; Barbara Shapiro, *John Wilkins, 1614 – 72* (Berkeley and Los Angeles, 1969), p. 197. See R. N. Salaman, *The History and Social Influence of the Potatoe* (Cambridge, 1949, reprinted 1970), pp. 446 – 8, who quotes sections of these minutes on pp. 228 (with a slight omission) and 447.

[18] Birch, *History*, i, 216 – 7, 219; E.L. B.1.79; Salaman, op.cit. p. 238. On the background to Boyle's interest, see ibid., ch. 12. See also Birch, *History*, i, 243.

[19] Ibid., pp. 226 – 7, 447 – 9 (Salaman is incorrect in supposing (p. 447) that the Georgical Committee was to take over this interest). One might also note the advocacy of potato-growing by John Forster in *Englands Happiness Increased* (1664), which shows no awareness of the Royal Society's activities. See also Birch, *History*, i, 494.

[20] Cf. Hunter, *Science and Society*, ch. 4.

The Buckland committee, like most of those referred to so far, was a one-off affair, intended only to meet over a relatively brief period; only those dealing with administrative matters seem to have been expected to have a longer life. By contrast, the main group of minutes with which this paper is concerned were of committees that were intended at the same time to deal with the content of the Society's work and to be permanent. They bear witness to a feeling which had evidently emerged by 1664 that the Society was not going to be able to achieve all the tasks that it had ambitiously set itself through its regular weekly meetings: instead, the allocation of topics to smaller groups of Fellows with a special interest in these fields was a useful way of supplementing the Society's general activity. These committees thus illustrate an urge towards specialization in the pursuit of the Society's corporate goals which is significant in itself.

The germ of this idea is perhaps to be traced to the formation of a committee to deal with medical matters on 27 January 1664, when it was moved that all the physicians who were Fellows should constitute a committee 'constantly to consider, what is necessary to be prosecuted in anatomy and chirurgery'. At first, the concern seems to have been particularly with the dissections that were carried out under the Society's aegis, but at a meeting on 17 February, 'Mr Boyle mentioning, that the committee of physicians formerly appointed for anatomical administrations might be desired to consider of and bring in an account of the *Desiderata* of anatomy, it was ordered, that they should take this matter into their consideration, and give an account of it to the society'.[21]

A more direct stimulus was provided by John Beale, whose major role in the Society in its early years is often underestimated due to his residence in Somerset and his effusive eccentricity as a correspondent. At a Council meeting on 23 March 1664, 'It was ordered . . . that several committees be appointed for the consideration and improvement of several subjects of philosophy', and the incentive to this was divulged at the ordinary meeting that followed: 'Mr Beal having offered by letter to communicate several observations on agriculture, if the society should please to appoint a committee to receive and examine them; it was thought good, that the council should consider of this and other committees for several subjects; and having appointed them, and nominated fit persons for them, present them to the society for their approbation or alteration'. Somebody worked fast, for at the Council meeting a week later 'A draught of eight committees for the consideration of several subjects belonging to the cognizance of the society was read, and ordered to be offered to the society at their meeting this afternoon for their approbation'. The 'draught' apparently comprised a list of the committees and their suggested composition which was inserted in the journal book at the end of the minutes for the meeting in question, and it is accurately reproduced in Birch's *History*.[22]

[21] Birch, *History*, i, 375 – 6, 384, 386. Cf. ibid., 388, 390.

[22] Ibid., i, 397, 402 – 3, 406 – 7. On Beale's role see also Hunter, *Science and Society*, pp. 194 – 7, and, on the setting up of the committees, Oldenburg to Evelyn, 24

A number of observations may be made on the committees and their composition. First, there was some disparity in their size, ranging from the 'Mechanical Committee, to consider and improve all mechanical inventions', with sixty-nine members, to the Astronomical and Optical Committee with fifteen, two of them, Hevelius and Huygens, Fellows who resided abroad and could therefore only have participated as correspondents. The Anatomical and Chemical Committees were allocated 'all the physicians of the society', and, since these comprised twenty-five at this time, the Anatomical Committee had twenty-eight members and the Chemical thirty-two.[23] The other committees, for Agriculture, History of Trades, 'collecting all the phaenomena of nature hitherto observed, and all experiments made and recorded', and Correspondence were of comparable size, with respectively thirty-two, thirty-five, twenty-one and twenty members.

It seems most improbable that, in the week between Beale's proposal and the drawing up of the list, there had been time to consult more than a few Fellows as to which committee they would like to be on. The nominations may therefore be taken to derive partly from knowledge of members' interests, and partly from an optimistic hope that, if nominated to a group specializing in a topic in which they might plausibly be expected to be interested, those hitherto inactive might be persuaded to become involved in the Society's enterprise. The latter is clearest in the case of the immense Mechanical Committee, a disproportionate number of those nominated to which were peers, politicians and suchlike Fellows who were otherwise barely involved in the Society's proceedings: it almost seems like a wry joke on the Council's part. Other nominations, however, clearly reflected knowledge of members' relevant expertise, as was clearly the case with Robert Hooke, Christopher Wren and Sir Paul Neile and the Astronomical Committee, Boyle and the Chemical, Evelyn and Beale and the Agricultural

March 1664, *Oldenburg*, ii, 147. Although Birch reproduces the list in the journal book accurately, the following insertions in the original manuscript may be noted: in the Mechanical Committee, Sir P. and Sir C. Wyche appear at the very end out of alphabetical order, presumably having been accidentally omitted; the Anatomical Committee was initially only to comprise the physicians of the Society, and the names of Boyle, Wilkins and Hooke have been written in retrospectively in Oldenburg's hand; in the Agricultural Committee, the name of Holder has been inserted (out of order) in Oldenburg's hand, while the names of Sir Robert Moray and Edward Smith have been inserted in the hand of the amanuensis, Smith's at the very end of the list; in the History of Trades Committee, Wilkins's name has been inserted in Oldenburg's hand (and Potter is incorrectly given as 'Dr'); in the Committee for Correspondence, 'Dr Balle' has been altered from 'Mr', while Wilkins's name has been inserted in Oldenburg's hand.

In October 1666, a 'Scrowle of the Committees of the R.S.' — presumably listing the names of those on each — was 'received from Dr Goddard' by Hooke for placing in the Society's repository (City of London Guildhall Library MS 1758, last leaf).

23 See Hunter, *Royal Society*, p. 115, and above, plate 3.

and Oldenburg, Theodore Haak and Sir Robert Moray and that for Correspondence. In other cases, there are hints that less well-known figures had relevant concerns, as for instance with Thomas Henshaw and Sir Robert Paston and chemistry, since both had alchemical interests, while the presence of Dudley Palmer on the Astronomical and Optical Committee is presumably not unconnected to the presence of 'good *Telescopes*' among Palmer's curiosities, as noted by Evelyn after a visit to him in 1661.[24]

Some active Fellows, it should be noted, were on several committees: Boyle topped the list by being nominated for seven, followed by Sir Robert Moray and John Wilkins, who were each on five, while Beale and Hooke were on four. Virtually all of the 145 members of the Society at this stage were placed on at least one committee, even including three members, Gilbert Burnet, James Carkesse and Hevelius, who had been elected at the very meetings on 23 and 30 March at which the idea of committees had been suggested and the lists of members drawn up.[25]

Turning to the subject areas that the committees were appointed to deal with, various criteria seem to have come into play: the desire to cover as many of the Society's areas of interest as possible may have been balanced by a wish to give prominence to topics which aroused special enthusiasm among the Fellows or which were thought to deserve particular attention because of their relevance to contemporary needs. In addition, since the committees were intended to complement rather than to supersede the Society's main meetings, it is possible that it was felt that they should concentrate on matters which required more attention than they were receiving there.

Be that as it may, it might be felt that technology received a disproportionate amount of attention, with three out of a total of eight committees dealing with such matters — the Mechanical Committee, the Agricultural Committee and that for the History of Trades; these were also the three biggest in terms of personnel. The collecting of information was also well-represented, with two committees being devoted to it — those for collecting the phenomena of nature and for Correspondence, the latter, as we shall see, being concerned more with promoting links with far-away places than with correspondence of the kind that had been dealt with by a committee in the Society's earliest years and was by now increasingly being taken over by Henry Oldenburg. On the other hand, the astronomical and optical work which is often retrospectively perceived as lying at the core of the Society's activity was covered by the smallest committee, possibly in view of the scale of related research that the Society was already pursuing.

24 See British Library MS Sloane 2222, esp. fols. 128vf.; K. T. Hoppen, 'The Nature of the Early Royal Society', *British Journal for the History of Science*, 9 (1976), 243 – 7; Evelyn, *Diary*, iii, 293. On the Mechanical Committee, see Hunter, *Royal Society*, p. 83.

25 Ibid., p. 190. Fellows who fail to appear on any of the committees are Kincardine (catalogue entry 3), Hammond (entry 70), Brooke (entry 99), Graunt (entry 105) and More (entry 114).

In addition, two other areas of the Society's experimental work were covered by the Anatomical and Chemical Committees.

Though thus organized early in 1664, however, the committees did not immediately begin to meet, and on 1 June 'It was referred to the next meeting, to consider of the best way of putting the committees lately established upon action; in order to which the secretary was directed to have then in readiness such papers, as have been addressed to him concerning several matters of the cognizance of those committees, to recommend them to their consideration; and also to cause the amanuensis to have ready a list of the said committees, and of all the persons nominated for them, to be put up in the public meeting-place of the society'. As a result, on 8 June 'the lists of the several committees for the improvement of several philosophical subjects' were considered and an order was evidently drawn up nominating a chairman for each and ordering that a meeting of any three or more members should take place fortnightly at the chairman's lodgings at an appointed time — mostly at 2 or 3 o'clock in the afternoon, though in one case at 5 p.m. and in another at 9 a.m. on a Saturday morning. None of the meeting days for the different committees clashed with one another, doubtless in view of the fact that several Fellows were members of more than one. The chairmen named were the President, Lord Brouncker, for the Mechanical Committee; Jonathan Goddard for both the Chemical and the Astronomical and Optical; the distinguished physician, Dr (later Sir) George Ent, for the Anatomical; Charles Howard for the Agricultural; Dr Christopher Merrett for the History of Trades; the lawyer and virtuoso, John (later Sir John) Hoskins, for the group for collecting the phenomena of nature; and the government official and connoisseur, Thomas Povey, for the Committee for Correspondence.[26]

After this, there are occasional references to the committees in the minutes. On 15 February 1665, a few additions were made to their membership, with a new Fellow, Thomas Blount, elected to the Society on 8 February, joining the Mechanical and Agricultural Committees, and three other Fellows, Sir Edward Harley, Theodore Haak and John Pell, being

[26] Birch, *History*, i, 434, 439. For the order — which does not survive — see D.M. 5.62 (printed below). Birch (i, 439) quotes some details from a manuscript in the hand of John Pell which is now British Library Additional MS 4441, fol. 57, entitled '8 Committees of the Royall Society'. It is not as complete as the minutes, giving no convenor, place or time for the Chemical or Correspondence Committees and no place or time for the Anatomical, and it differs from them in the following respects, in addition to those noted by Birch: it gives precise meeting times, rather than 'about'; it describes Goddard and Ent as 'Pr.' of their committees; in connection with the 'Astronomical & Opticall' Committee, the meeting place was initially given as Goddard's 'chamber', but changed to 'lodgings', whilst the meetings were stated to be on the second and fourth, as against the first and third, Monday in the month; the Agricultural Committee was said to meet on the second and fourth Thursday of the month, not the first and third. At the end, Pell added extra information concerning the committee for collecting natural phenomena, which he was on: see below, n. 55. At the bottom of the page is a note of the title, format and price of Power's *Experimental Philosophy* (1664), which had presumably just appeared.

added to the Agricultural, presumably because they had expressed interest in the subject.[27] Then, on 10 May 1665, we hear of accounts being given by their respective chairmen of the work of the Agricultural Committee and those for 'composing of the histories of nature and art'; the first two of these may have comprised the reports preserved in the Society's archives alongside the minutes, though that of 'art', evidently presented by Christopher Merrett, the chairman of the History of Trades Committee, has not survived. After hearing the reports, the Society 'recommended to them the care of making farther progress in their respective work'.[28]

In addition, on 7 December 1664, 'It being suggested, that there were several persons of the society, whose genius was very proper and inclined to improve the English tongue, and particularly for philosophical purposes; it was voted, that there be a committee for improving the English language'. This seems to have been intended to be comparable to the other committees: twenty-one Fellows were named for it and meetings were to be held at the lodgings of Sir Peter Wyche, evidently its convenor, 'once or twice a month'. Subsequently, on 18 January, John Wilkins (who was not one of those named) was ordered to 'meet the first time (at least) with the committee for improving the English tongue' to 'intimate to them' the Council's intention that they should be concerned 'chiefly to improve the philosophy of the language', while John Evelyn refers to the committee in an entry in his diary for 24 January, presumably because it met on that day.[29] All that is known about its activities after that, however, is a letter which Evelyn wrote to Wyche on 20 June 1665, in which he apologized for not being able to attend the Tuesday afternoon meetings because they clashed with 'an employment, which being of public concernment, I can in no way dispense with'. At this time, Evelyn was serving as a Commissioner for Sick and Wounded Seamen and Prisoners in connection with the Second Dutch War, and this might suggest problems with extra weekday meetings for other Fellows who did not live wholly leisured lives.[30]

At this point we may turn to the surviving minutes and reports, which form part of a collection of administrative papers from the Society's early years now preserved in volume 5 of its Domestic Manuscripts. Apart from (1) the minutes of the Buckland Committee and (2) the report of the Legal Committee from 1663 which have already been referred to, what we have are: (3) the report on the meetings of the committee for collecting the phenomena of nature, but no minutes of its meetings; (4) minutes of eight meetings of the 'Georgical' or Agricultural Committee between 23 June

27 Birch, *History*, ii, 16. In fact, Sir Edward Harley was already on the Agricultural Committee. In the journal book, the name given here was originally 'Sir Robert Harley', which was then changed to 'Edward'.

28 Ibid., ii, 47.

29 Birch, *History*, i, 499 – 500, ii, 7; Evelyn, *Diary*, iii, 396.

30 Evelyn, *Diary and Correspondence*, iii, 159 (where the contraction has wrongly been expanded as 'with' rather than 'which': see the original in Christ Church Evelyn Collection MS 39, no. 245), and ibid., 159 – 62, passim.

1664 and 23 February 1665, together with the report on its activities; (5) minutes of six meetings of the Mechanical Committee between 16 July 1664 and 12 May 1665; and (6) minutes of two meetings of the Committee for Correspondence, of 19 August and 23 September 1664. All this material is printed below.

No papers relating to the other committees survive, in at least some cases because the committees probably never met, though in others a record may once have existed which is now lost. If an account of the work of the Committee for the History of Trades *was* given to the Society on 10 May 1665, then there must have been meetings of which we have no documentation. Certainly the History of Trades was a project which attracted a good deal of interest in the Society's earliest years, and it could be significant that it was during the months when the other committees were meeting that the chairman of this committee, Christopher Merrett, brought to meetings of the Society as a whole catalogues of trades from which members were asked to choose which they would like to write about.[31]

As for the Anatomical Committee, it was ordered on 5 October 1664 that this should meet on the following Monday afternoon at the house of its chairman, George Ent, 'to consider of the particulars to be observed in the next anatomical administration'. At the next meeting of the Society Walter Charleton recorded the discussion that had taken place, naming three specific areas of inquiry for 'the next dissection at Gresham College', which evidently took place on 22 October and which, since it was not a regular meeting of the Society, was possibly a meeting of this committee.[32] Whether further such gatherings occurred is unclear; that they did *not* is rather suggested by the fact that on 15 February 1665 'It was ordered, that a body be procured the next sessions, and that Dr Charlton endeavour to get a meeting of some physicians of the society, in order to consider of experiments and inquiries', an almost direct echo of the earlier plea for method in the Society's medical work.[33] Again, however, no more is thereafter heard.

Turning to the committees whose records survive, the most active was the Agricultural Committee, perhaps not entirely surprisingly, in view of the fact that it was Beale's suggestion to set up this that stimulated the inauguration of the committees as a whole. It is also the only one to have attracted serious attention in modern times: Reginald Lennard used its minutes for his pioneering article on the Royal Society's early agricultural

[31] Birch, *History*, i, 476, 502. Cf. Hunter, *Science and Society*, ch. 4; K. H. Ochs, 'The Royal Society of London's History of Trades Programme: an Early Episode in Applied Science', *Notes and Records*, 39 (1985), 129 – 58.

[32] Birch, *History*, i, 474 – 5; Oldenburg to Boyle, 22, 27 October 1664, *Oldenburg*, ii, 269, 273. Cf. also Oldenburg to Boyle, 6 Oct. 1664, ibid., ii, 248. Note, however, that the meeting was not on the day appointed for meetings of this committee, which was Friday.

[33] Birch, *History*, ii, 16.

work, published in the *Economic History Review* in 1932.[34]

The eight meetings of this committee took place at intervals of four, five or six weeks over a period of eight months. It began its work by considering its rationale and requesting members to peruse books on agriculture: on this basis it was intended that inquiries should be drawn up and sent all over Britain as a means of identifying and disseminating improved practices. One work was referred to on this occasion, Samuel Hartlib's *Legacy* (1651) — symbolic, perhaps, of the continuity from Interregnum to Restoration preoccupations which was also exemplified by the central role in getting the committees going played by Hartlib's former colleague, John Beale. The composition and dissemination of inquiries was to dominate the committee's work, but in addition a wider range of topics was brought up at the first, as at later, meetings — from drying fruit and boring for earth samples (about a device for which the Halifax natural philosopher, Henry Power, was to be consulted) to the improvement of waste ground. A setting for the latter idea may be provided by a report at a meeting of the Society on 3 June 1663 of a petition lately presented to the King 'for a patent to practise a secret of making all grain grow plentifully in any barren ground, without laying on any dung or compost' and for encouraging vines, orange trees and other plants 'to a sudden growth': possibly the committee was at this point alluding to this otherwise unrecorded project.[35] Lastly, responsibility for investigating agricultural practice in different counties of England and in Scotland and Ireland was allocated to the various members of the committee and other Fellows of the Society.

At the next meeting, lists of authors were brought in, one from Beale, who, at the behest of Sir Robert Moray, had evidently sent the committee voluminous papers on their chosen topic which 'hinted at' a possible agenda for its work.[36] But as yet no inquiries were drawn up, something which took place at the subsequent meeting, on 25 August. On the basis of a paper by Moray, a set of queries was drawn up on that occasion covering arable land and meadows. Oldenburg sent a copy of this in a letter to Robert Boyle at Oxford on 1 September, requesting him to solicit responses through his acquaintance there, which, as he explained, 'will much tend to the compiling of a Good History of Agriculture' and would encourage the committee 'to proceed to the other parts of that Employment'. Though the minutes do not

[34] R. V. Lennard, 'English Agriculture under Charles II: the Evidence of the Royal Society's 'Enquiries'', *Economic History Review*, 4 (1932 – 4), 23 – 45.

[35] Birch, *History*, i, 252 (an occasion on which related topics were also discussed). Cf. Hooke to Boyle, 3 June 1663, Boyle, *Works*, vi, 483; Oldenburg to Boyle, 10 June 1663, *Oldenburg*, ii, 66 (which states that the discussion took place on that date). For a patent that dealt, among other things, with improving land, but which dates from 1661, see PRO SP 29/39, no. 103, and British Library MS Sloane 856, fol. 40 (I am indebted to Christine MacLeod for this reference).

[36] E.L. B.1.41f., dated 20, 27 June and 1, 9 July 1664. See also Beale to Evelyn, 25 June 1664, preserved in a folder of loose letters in a box of 'Evelyn Period' papers in the Evelyn Collection at Christ Church, Oxford.

record it, there must have been further discussion at the meeting on 22 September, since in a letter of the 29th of that month Oldenburg sent Boyle some 'additionall Inquiries'.[37] Later, these extra queries were incorporated into the main series in a revised version which was published in the *Philosophical Transactions* in July 1665. In the meantime, however, a number of answers came in and were brought to subsequent meetings of the committee, and these and later replies have provided modern scholars with valuable information on agricultural practice at the time, despite the fact that the projected history of agriculture never materialized.[38]

At the meeting on 22 September, two new topics arose. Christopher Merrett produced a list of 'what is cultivated in England', thus stimulating a drive to collect data on the culture of such plants as woad, flax, hemp or buckwheat which is referred to at subsequent meetings. There was also discussion of where each of these grew best, and it may have been this that led Merrett to suggest that to the existing inquiries should be added the question 'What plants severall grounds produce naturally?' Though it is not noted in the minutes, Oldenburg reported this as having arisen 'at our late Committee at Mr Howards' in letters to Boyle of 22 and 27 October, adding: 'This he thought might not a litle conduce to discover the nature of the soyle'. Moreover this further query does indeed appear in a manuscript copy of the inquiries by Charles Howard that survives, though it is not in the published version.[39]

Also on 22 September, Charles Howard was desired to collect information on plants for kitchen gardens and on winter greens, and this topic recurred at various later meetings, as he brought in drafts and was requested to improve and extend them. He also produced inquiries which, like the agricultural ones, were distributed to members, and copies of both his papers and his inquiries survive among the manuscripts of Thomas Birch now in the British Library.[40]

These themes were accompanied at subsequent meetings — as earlier — by more miscellaneous discussion, as on 1 December 1664, when Colonel (later Sir James) Long gave a real gardening enthusiast's disquisition on the

[37] *Oldenburg*, ii, 224 – 6 (in this version there are 14 + 7 queries, since no. 3 combines two points that appeared separately in other copies), 242. For a copy in the hand of Charles Howard of the original 15 + 7 inquiries (the existence of which Lennard postulated on the basis of the returns, but could cite no copy of: 'English Agriculture' (n. 34), pp. 26 – 7n.), see British Library Additional MS 4458, fols. 137v, 140. To this, Howard added the additional queries that Oldenburg sent to Boyle, together with a sixth, on which see n. 39.

[38] *Phil. Trans.*, 1 (1665), 92 – 4. For the replies, see Lennard, 'English Agriculture' (n. 34), p. 29 and passim, and *Oldenburg*, ii, 256 – 61, iv, 11 – 16. See also Hunter, *Science and Society*, pp. 92 – 3, 102.

[39] British Library Additional MS 4458, fol. 140; *Oldenburg*, ii, 270, 273. In a letter of 20 October, ibid., ii, 266, he refers to his attendance at the meeting on that day.

[40] British Library Additional MS 4458, fols. 134 – 7. Beale sent information on winter greens for Howard in a letter to Oldenburg of 18 Jan. 1665: *Oldenburg*, ii, 351 – 2.

available varieties of cherry tree, or on 12 January, when the curative or culinary properties of a range of herbs were canvassed by different members. On 23 February, Sir Edward Harley, a Fellow of high status attending for the first time, monopolized the conversation in a manner reminiscent of the correlation between social eminence and loquaciousness observed by Robert G. Frank in proceedings at the Society's ordinary meetings.[41] (Oldenburg had possibly ceased to concentrate as the worthy knight droned on about the property of stringed bumble-bees in driving away mice, since he notes: 'This I took imperfectly'.) Indeed, it is in general notable how, although the improvement of agricultural practice was the stated objective of the committee, this did not preclude a wider curiosity about all sorts of overlapping topics, as seen, for instance, in the conversation that took place at the meeting on 12 January about mistletoe and its mode of propagation.

But practicality was clearly the main thrust of the proceedings, and an ambition for large-scale improvement of the kind seen in 1663 in the Buckland proposal — and perhaps also in the putative patent for improving waste ground — recurred at the meeting on 12 January 1665. On that occasion Robert Boyle produced proposals for encouraging the planting of fruit and timber trees which the Oxford gardener, projector and former protégé of Hartlib, Ralph Austen, wished to present to parliament. Two copies of these proposals survive among the Society's Classified Papers, one of them including fairly draconian measures for supervision which are omitted from the other.[42] They had been noted at an ordinary meeting of the Royal Society on 14 December 1664, when they were referred to this committee; but its consideration of the matter was pre-empted by further discussion which evidently occurred at the Society's meeting on 21 December, when it was ordered that the proposals should 'be recommended to such persons of the Society, as were members of the parliament'. The matter finally came to the committee the following month, when the lawyers on it were asked for their advice as to how the propositions should be 'best ordered' to be effectual: this, together with the other main heads of the committee's work, was noted in the report presented to the Society which accompanies its minutes.[43]

Although it has hitherto attracted far less attention than the Agricultural Committee, and though its minutes are brief by comparison, the Mechanical Committee was almost as active, with six meetings over a slightly longer

[41] See R. G. Frank, jr., 'Institutional Structure and Scientific Activity in the Early Royal Society', *Proceedings of the 14th Congress of the History of Science, 1974* (Tokyo, 1975), iv, 96.

[42] Cl.P. x(3)7 and 8. See also Royal Society Boyle Papers, 37, fols. 134 – 5. For Austen's Hartlibian associations, see Webster, *Great Instauration* (n. 15), esp. pp. 477 – 8.

[43] Birch, *History*, i, 504, 509. See also Oldenburg to Boyle, 10 Dec. 1664, *Oldenburg*, ii, 332.

period: there were meetings each month between July and November 1664, with a final meeting occurring on 12 May 1665, some months after the Agricultural Committee had ceased to convene.[44]

The break in its activities reflects the fact that early in 1665 the Mechanical Committee seems to have been displaced by committees dealing with more specialized topics falling within its general area of competence. One was 'a committee, to consider of the improvement of artillery', set up on 15 March 1665, presumably to offer technical advice in the context of the Second Dutch War, which had broken out earlier that month. This mainly comprised active members of the Mechanical Committee (Brouncker, Boyle, Moray, Petty and Hooke), together with the Earl of Northampton (a hitherto inactive member of the committee) and the virtuoso Thomas Henshaw, who belonged to the Chemical, but not the Mechanical Committee.[45] It is possibly significant that gunnery was the first topic discussed by the Mechanical Committee at the meeting held in May 1665 after its six-month break; it was also central to the patent obtained in March 1665 by Abraham Hill, evidently on behalf of the Society as a whole, about which more will be said shortly.

In addition, on 12 April 1665, a group of Fellows were requested to bring to the Mechanical Committee 'experiments for improving chariots', another topic, incidentally, mentioned in the patent application. These comprised the five most active members of the committee — Brouncker, Moray, Petty, Wilkins and Hooke — together with Colonel Thomas Blount, who had been added to it on 15 February after being elected to the Society a week earlier. Blount had, in fact, already been known to the Society for his work on 'chariots' before his election, since early in 1664 Moray and John Evelyn had been asked to visit him 'and confer with him concerning the improvement of land-carriage'. Shortly after his election to the Society, and partly in response to comments on the subject from the Dutch scientist, Christiaan Huygens, Blount was requested on 1 March 1665 to bring in 'a model of his conceptions' about such improvements, and a visit was arranged by Moray, Wilkins and Evelyn to a place suitable for trials with carriages near Blount's house in Deptford.[46]

The referral to the Mechanical Committee followed further such discussions, but in fact the meeting on 21 April to which Blount's proposals were supposed to be submitted did not materialize. Instead, the smaller committee took on a life of its own, meeting at Blount's house to carry out experiments, the results of which were reported and discussed at a meeting of the Society as a whole on 3 May. Thereafter, further work on carriage design continued after the Mechanical Committee ceased to meet, into the

44 For virtually the only reference hitherto, see Hunter, *Science and Society*, p. 94.

45 Birch, *History*, ii, 24. For an earlier reference to gunnery, see ibid., i, 396.

46 Ibid., i, 385, ii, 19 – 20, Cf. ibid., ii, 22, 27 – 8, 30, 32. On Blount, see also ibid., i, 105, Evelyn, *Diary*, iii, 170, 196, Pepys, *Diary*, vi, 94, 213, vii, 20 and n.

period when the Society went into recess during the Great Plague of 1665 and beyond.[47]

The early months of 1665 thus represented a high point in the Society's attempt to use the skills of its members to produce inventions which might be of immediate utility: it is no coincidence that, as already noted, in March that year a patent was obtained for a whole series of inventions, of which the Royal Society as a corporate body was to be the beneficiary, by Abraham Hill, Treasurer of the Society from 1663 to 1665 and an active Council member throughout this period. The patent was to run for fourteen years (though there is no evidence that the Society derived any benefit from it during this time) and it itemized a number of new carriage designs; various novel types of guns giving 'speedier and more effectual discharge'; a spring-opening powder box; an engine for breaking flax and hemp and washing linen; and a new type of pendulum clock for use at sea.[48] The latter may have been Hooke's 'proposition of giving the discovery of the longitude', which is referred to in the Society's minutes at this time and the possibility of seeking a patent for which had been considered in January.[49] The machine for breaking flax and hemp was conceivably linked to the Agricultural Committee's interest in the cultivation of these and other useful plants: the man behind it might have been Charles Howard, who was to obtain a patent for a new method of preparing hemp and flax in 1678.[50]

The background to this takes us back to the previous February, when Hooke had been asked to try to help the Society's image and its finances by devising 'some experiments, that might be acceptable and useful to the public'. He had obliged by suggesting that consideration should be given to 'the experiment of land-carriage, and of a speedy conveying of intelligence', and it was evidently this that stimulated the Society's interest in Blount's work on carriages later that month, when Hooke was also requested to produce his proposals for 'speedy intelligence'. Hooke also continued to work on carriages, being told on 23 March to expedite a design for a one-wheeled vehicle, while at a Council meeting on 31 August 'It was resolved,

47 Birch, *History*, ii, 30, 41, 45, 53, 56, 59, 66, 89, 95, 123 – 4, 139 – 40, 142, 150, 154, 157, 163; Oldenburg to Boyle, 16, 27 Jan. 1666, *Oldenburg*, iii, 24, 32 and n. 7. For related papers by Blount, see Cl.P. iii(1)20 – 1 (the former, dated 8 April 1665, comprises the experiments concerning load-bearing referred to in Birch, *History*, ii, 30; the latter is dated 26 March 1667).

48 See Woodcroft, *Titles of Patents* (n. 9) no. 143, and British Library Additional MS 4441, fols. 101 – 2. This document and its background would merit further study.

49 Birch, *History*, ii, 7, 24. See also A. R. Hall, 'Horology and Criticism: Robert Hooke', *Studia Copernicana*, 16 (1978), 262 – 3; J. A. Bennett, 'Robert Hooke as Mechanic and Natural Philosopher', *Notes and Records*, 35 (1980), 38 – 9; and Michael Wright, 'Robert Hooke's Longitude Timekeeper', in Michael Hunter and Simon Schaffer, eds., *Robert Hooke: New Studies* (Woodbridge, forthcoming), where relevant source-material is printed.

50 See Woodcroft, *Titles of Patents* (n. 9), no. 201; see also Cl.P. xxv.109, British Library MS Sloane 4021, fols. 132 – 3, and Bodleian Library MS Rawlinson A 191, fol. 147.

that the patent for the several new-fashioned chariots be drawn up for Mr Hooke and his assistants'.[51] As a result, Hooke had taken out such a patent in September, though this was to be subsumed into the omnibus patent obtained by Hill the following spring.[52]

The Mechanical Committee clearly fits into this context, for a significant part of its discussions were underlain by a similar ambition to produce useful inventions, often relying on the services of Hooke. At its first meeting, he proposed making 'an instrument, by which may be found a ship's motion through the water', which it was topically suggested might be tried out on Sir William Petty's 'double bottom', itself a further example of the utilitarian work characteristic of the Society at this stage. Topics subsequently deemed worthy of investigation included the raising of water — with Hooke being asked to demonstrate that towers were unnecessary for the purpose — while there may have been similarly practical reasons for Hooke being 'desired, To consider the way of producing wind by the fall of water' at the meeting on 14 November.

This concern to be of public benefit was perhaps seen most strikingly when the committee took up Sir William Petty's (otherwise unrecorded and therefore hitherto unknown) proposal for 'propelling with more quicknes the water of Islington river'. By this was meant the New River constructed by Sir Hugh Myddleton in the early seventeenth century to supply water from Hertfordshire to New River Head in Clerkenwell, whence it was distributed to the metropolis. As London grew, the supply provided by the New River proved increasingly inadequate: during this period there were various schemes to supplement it, either by raising water from the Thames or bringing it from elsewhere on the periphery of London, while in the early eighteenth century an attempt was made to improve the supply from the New River itself by using pumps to raise the elevation of the water at New River Head.[53] Petty's scheme appears to have been a precursor of the latter, but, though some papers concerning raising water survive among Petty's manuscripts, these probably date from the 1680s, and they refer only in passing to the New River.[54]

More to the point is the confidence, typical of such schemes, that was displayed in the suggestion that Petty's idea might be proposed 'to those, that are interessed in the worke', in the hope that in return 'a share in the

51 Birch, *History*, i, 379, 385, 397, 463. Cf. also ibid., i, 370, and, for earlier work by Hooke on carriage design, ibid., i, 330, 333.

52 *Calendar of State Papers Domestic, 1664 – 5*, p. 9, and PRO SP 44/14, fol. 35v (which overlaps verbally with the Hill patent). On Hooke's later interest in such matters see Hooke, *Philosophical Experiments and Observations* (London, 1726), pp. 142 – 67.

53 See H. G. Dickinson, *Water Supply of Greater London* (Leamington Spa, 1954), chs. 3 – 4.

54 The Marquess of Lansdowne, ed., *The Petty Papers* (London, 1928), ii, 135 – 7. No other relevant material appears to survive among the Petty MSS: see M. D. Slatter, *Calendar of Literary and Personal Papers of Sir William Petty (1623 – 87) at Bowood House, Calne, Wilts.* (London, 1980).

profit, or a consideration' might be forthcoming. One has to record, however, that this optimistic technological venture came to nothing: it may be significant that in the minutes the 'share in the profit' or 'consideration' was toned down to the hope that the interested parties would 'consider' the projectors 'for it'. It is nevertheless appropriate that this scheme formed the climax to the main sequence of this committee's meetings.

As with other committees, this ambition to be of immediate utility was combined with an attempt to scour the extant published literature on the subject to discover what had been done in the field and what needed to be improved. Quite a comprehensive list of ancient and modern 'Authors of Mechanicks' was drawn up, which is itself worthy of note for the evidence it gives of the Society's catholic terms of reference. These works were then allocated to different members, while a strategy for approaching them was outlined, involving an interesting demarcation between what was recorded in books and practiced, and what was done in practice 'and not found in Books'.

Turning to the 'committee for collecting the phaenomena of nature and all philosophicall experiments hitherto observed, made and recorded', we have only the report made by its chairman, John Hoskins, rather than the minutes of its meetings. A memorandum by a member of the committee, John Pell, shows that it met at Hoskins' chambers at Essex Court in the Middle Temple on 24 June and 8 July 1664, but since the report refers to 'severall' meetings and was evidently not submitted till the following May, these may only have been the first two. If so, however, nothing is known of the rest, and neither do we know which members attended, apart from the chairman and Christopher Merrett, who is mentioned in the report.[55]

This, however, is a fascinating document, very characteristic of the Society's encyclopaedic ambitions for the reform of knowledge in its early years, and it was appropriate that Thomas Sprat should have referred to this committee in his *History of the Royal Society* in connection with the Society's 'purpose of *collecting* divers patterns of all *Natural*, and *Artificial* things'. Sprat explained with typical grandiloquence how the Society had '(amongst others) appointed a *Committee*, whose chief employment shall be to read over whatever *Books* have been written on such subjects', continuing: 'By this means they hope speedily to observe, and digest into *Manuscript volumes*, all that has been hitherto try'd, or propounded in such studies'. It is interesting to juxtapose with this the earlier ambition of the Oxford group of experimental philosophers in the 1650s, who had seen it as their 'first businesse' 'to gather together such things as are already discovered and to make a booke with a generall index of them, then to have a collection of

55 British Library Additional MS 4441, fol. 57 (cf. above, n. 26); Pell also refers here to the instructions dated Wednesday, 8 June 1664 that the committee should 'make a Report of their progresse therein to the Society, once a quarter', while he states that the meetings were on the second and fourth Friday, not Thursday, of the month, and that they were at 3, not 2 p.m. For the submission of a report the following May, see Birch, *History*, ii, 47.

those which are still inquirenda and according to our opportunityes to make inquisitive experiments'.[56]

In view of the Royal Society's professed Baconian rationale, it is perhaps not surprising to find that it was to Bacon's *Parasceve* or *Preparative towards a Natural and Experimental History*, with its attached *Catalogue of Particular Histories*, that the committee turned for their inspiration.[57] But it is worth pointing out that the committee's report shows a more literal adherence to Baconianism even than is normal in the Society's early years: it comprises little more than a paraphrase of Bacon's seminal work and an attempt to demarcate the responsibilities of the different committees in terms of it. Its parting shot tried to impose the same framework on the other committees 'that so they may all agree', and this implies a rigidity which is at odds, for instance, with the rather different arrangement for natural history suggested by Boyle (who was not on this committee) in a proposal put to Oldenburg in the following year.[58] Here again we see evidence of contrasting styles of Baconianism in the Society in its early years, as documented in chapter 6 of this volume.

Almost everything that the committee said can be exactly paralleled in Bacon, including their basic classification and their view that, though essentially distinct, 'generations' and 'praetergenerations' might be considered together; their instruction that philological and superstitious matters should be omitted and superfluity of description avoided; and their priorities in the study of arts. Virtually the only departure from Bacon is their curiosity about what might almost be called the sociology of knowledge in their wish to know 'the reason why those things which are false were ever believed to bee true': Bacon's interest in such falsehoods was only in their explicit refutation 'that the sciences may be no more troubled with them'.[59]

Having summarized Bacon, the committee went on to suggest a recent bibliography of relevant books, namely the compendium *De Scriptis Medicis* (1637) by the Dutch author, J. A. Van der Linden, which had reached a third edition in 1662. This certainly provided a vast bank of data about published works on medicine and ancillary topics, including natural history, chemistry and agriculture, though its medical focus limited its usefulness in relation to natural philosophy as a whole: Descartes, for instance, appears in it only for his *Passions de l'Ame*.[60]

[56] Sprat, *History*, p. 252; H. W. Robinson, 'An Unpublished Letter of Dr Seth Ward Relating to the Early Meetings of the Oxford Philosophical Society', *Notes and Records*, 7 (1949), 69.

[57] See Bacon, *Works*, iv, 249 – 71.

[58] Boyle to Oldenburg, 13 June 1666, *Oldenburg*, iii, 164 – 5.

[59] Bacon, *Works*, iv, 260. Cf. ibid., 253, 255 – 6, 261, 262. The report takes for granted Bacon's general rationale for such collecting (sect. 2), and ignores some of his further caveats (sects. 6, 9, 10).

[60] J. A. Van der Linden, *De Scriptis Medicis* (3rd ed., Amsterdam, 1662), p. 552 and passim.

However, a start had to be made somewhere, and, though one might consider the committee's preoccupation with published works a little bookish for a Society with the motto 'Nullius in Verba', this objection was foreseen by Sprat, who explained how 'This is the only help that an *Experimenter* can receive from *Books*: which he may still use, as his *Guides*, though not as his *Masters*'. As we have seen, Sprat also explained how they intended to digest their findings into '*Manuscript volumes*': but, though his *History* was not published till 1667, this statement evidently reflects hopes of 1664 – 5 rather than actualities, and, apart from the report printed below, nothing further is known of the committee's work, probably because nothing materialized.[61]

Lastly, we have the Committee for Correspondence. This had apparently not convened by 10 August 1664, when its chairman, Thomas Povey, had to name a day for it to meet, and at the same time either fixed or confirmed the time and place of meeting. Its agenda, however, had been set at the previous meeting of the Society, when, following the reading of the Treasurer, Abraham Hill's, 'queries for Guinea', 'It was ordered, that the committee for correspondence consider of drawing up both general and particular heads of inquiries for all the parts of the world'.[62]

As this reveals, clearly those who conceived the committee saw 'correspondence' as being less to do with letter-writing like Oldenburg's than with making contact with out-of-the-way places where data might be collected. Indeed, the name is slightly surprising, since this committee had a precursor in the form of a 'committee for foreign inquiries' which had been established in February 1661 and which was separate from the 'committee of correspondency' set up at much the same time. There is no evidence, however, that this group achieved anything, and a further committee 'to view the propositions for inquiries in foreign parts' was established in October that year, who may have been responsible for the 'inquiries for foreign parts' which existed by January 1662.[63] By contrast, the 1664 committee was quite active for a while, and, more than any other committee, this one was preoccupied with scrutinizing books, in this case travel-books published in different parts of Europe at various times during the previous century, of which the committee's list comprises a veritable compendium. From these it was intended that inquiries for distant lands should be compiled, and it may well be that some papers in Oldenburg's hand comprising notes from books by such authors as Samuel Purchas, J. H. van Linschoten and Jacob Bontius, together with 'Inquiries' stemming from these concerning various strange natural phenomena that they reported, are to be linked to the work of this committee.[64]

Indeed, this concern on the part of the Correspondence Committee epitomizes an interest in such writings to be found among members of the Society more generally, and not least men like Boyle and Hooke, who

61 Sprat, *History*, p. 252.
62 Birch, *History*, i, 456, 458. For Hill's 'Enquiries for Guinny', see Cl.P. xix.56.
63 Birch, *History*, i, 15, 16, 17, 24, 47, 68. See also ibid., i, 79.
64 Cl.P. xix.83 – 6.

clearly hoped that, from such books, information might be derived from which knowledge of the workings of nature in different circumstances might be extended and hypotheses thus tested.[65] In the case of the committee, various such works were allocated to different members with a view to drawing up queries, while in addition consideration was given to some 'Generall inquiries'. Beyond this, the only other piece of business recorded concerned the acquisition of a collection of rarities for the Society, which seems to have been seen as a related enterprise, evidently being associated in the minds of the Fellows with contacts with distant lands.[66]

This committee was apparently distinguished by the lavishness of the hospitality that it received from its host, Povey, whose wine cellar was renowned. In a report on the first meeting in a letter of 25 August 1664 to Robert Boyle, whose absence he regretted, Henry Oldenburg continued: 'This was our Entertainment above ground: I leave you to guesse, what our correspondence and entertainment was under ground, in the Grotto, and neer the well, that is the Conservatory of so many dozen of wine bottels of all kinds'.[67] Possibly the conviviality that followed the subsequent meeting went so far that people were discouraged from further attendance: the minutes for 23 September end, 'The Committee never mett since Mr Povey's Feasting of them', and no further meeting of the committee is known, despite the fact that on 9 November its members were requested to meet with representatives of the East India Company who were prepared to assist with the Society's inquiries.[68]

The early demise of the Correspondence Committee was a portent of what was in store for the committee system as a whole. In a letter to Boyle written the day after the last known meeting of this group, 24 September 1664, Oldenburg complained 'that our meetings are very thin; and that our committees fall to the ground, because tis not possible, to bring people together; tho I sollicite, to the making myselfe troublesome to others, not to say much of the trouble, which I create to myself, good store'.[69] Oldenburg's gloom may have been premature at this point, but in the long term it was borne out, for even the most successful of these groups failed to last as much as a year, and most of them less, even allowing for the hints that we have of the activity of the committees for which no minutes or reports survive. It would be charitable to be able to follow R. L. Lennard in attributing at least part of the blame for the demise of the committees on the disruption caused by the Great Plague in 1665, which clearly had a very harmful effect on the continuity of the Society's activities: in fact, however,

65 For Hooke, see e.g. Cl.P. xx.79 – 80; City of London Guildhall Library MS 1757, items 3, 4, 7, 10. For Boyle, see Royal Society Boyle Papers 39, passim, and, e.g., ibid., 8, fols. 63 – 91; 25, pp. 228 – 40, 269 – 70.

66 See below, 'Between Cabinet of Curiosities and Research Collection', p. 126. See also the juxtaposition of instructions to Povey on both subjects in Birch, *History*, i, 458.

67 Oldenburg, ii, 209. On Povey's cellar, see Pepys, *Diary*, iv, 18, 298; v, 161.

68 Birch, *History*, i, 487. Cf. ibid., i, 478.

69 Oldenburg to Boyle, 22 Sept. 1664, *Oldenburg*, ii, 235.

the committee system appears already to have been defunct before the plague struck in the summer of that year.[70]

There are a couple of allusions to the earlier committees after the plague. Sprat's reference to the committee for collecting the phenomena of nature in his *History of the Royal Society* has already been noted, while the Anatomical Committee was mentioned in April 1667, when as many of its members as possible were requested to forgather at the old venue of Sir George Ent's house. But 'very few' met on that occasion and 'little was done', and, though the group was ordered to reconvene later that month and apparently did so, carrying out experiments at Ent's house on lobsters and skates, it is soon lost from sight.[71] None of the other committees of 1664 – 5 appear to have reconvened at all, and, instead, what recurs in the later 1660s was the summoning of committees on a one-off basis to consider specific projects, as in the Society's earliest years. Already in April 1665, we find a committee being set up to deal with a specific topic which might have come within the cognizance of the Agricultural Committee — to which it had, in fact, been referred after that committee had ceased to meet — and this was the consideration of experiments about vegetables and particularly Dr Daniel Coxe's inquiries on the subject.[72] Similar groups were appointed to supervise Hooke's telescopic observations in 1666, for instance; to orchestrate the Society's experiments with blood transfusion in the same year; or to 'consider of a way of determining the measure of a degree of the earth' in 1669.[73] Other committees dealt with administrative matters: the search for a new meeting place when Gresham College was commandeered by the City in 1666, the bequest of the Arundelian library by Henry Howard of Norfolk, and the Society's rights to Chelsea College.[74] Somewhere between the two fell committees appointed in 1669 'for directing of experiments' and to consider the future direction of the Society's business.[75]

Some of these later, one-off committees have left documentation which survives among the Society's papers. Thus the volume of Domestic Manuscripts which contains the main series of committee minutes also includes the report of the committee deputed to deal with the Arundel Library, and the order for setting up the committee to consider how to measure a degree, both of which were also included in the minutes of the meetings of the Society as a whole and printed by Birch.[76]

70 Lennard, 'English Agriculture' (n. 34), pp. 27 – 8. For a reference to those Royal Society members who met at Oxford during the plague as a 'committee', see Birch, *History*, ii, 68.

71 Birch, *History*, ii, 166, 168, 173, 176, 181.

72 Ibid., ii, 29, 32. Cf. ibid., ii, 54 – 5, 57, for a committee 'formerly appointed' to observe magnetic variation.

73 Birch, *History*, ii, 98, 112, 117, 388 – 9, 398, 400. Cf. ibid., ii, 189.

74 Ibid., ii, 73, 113 – 4, 345 – 6, 351, 371, 377 – 8, 480. Cf. also the committee to supervise Sprat: ibid., ii, 6, 47.

75 Birch, *History*, ii, 344, 346, 378.

76 D.M. 5.69 (in the hand of Hoskins), 70; printed respectively in Birch, *History*, ii, 351-2 (though 'council' is a misprint for 'committee'), 398.

Elsewhere, a record survives of the deliberations of a committee appointed to consider some papers on the use of coloured dyes dating from the 1630s, which were produced at a meeting of the Society on 23 May 1666.[77] These survive among the Royal Society's Classified Papers, and with them are the minutes of the committee meeting at Sir George Ent's house on 28 May at which they were discussed. These almost entirely comprise a list of the papers with names noted against a number of items, presumably of which committee member was to deal with them. In addition, the names of those in attendance are given: Ent, Sir Theodore de Vaux, Hooke, Oldenburg, Daniel Coxe, Dr William Quatremaine, and Daniel Colwall, the latter in spite of the fact that his name had not been included among those ordered to constitute the committee in the Society's journal book.[78]

Lastly, a report survives of a meeting on Thursday, 3 February 1676, of a committee responsible for the management of the Society's repository, orders to which also survive in manuscript among the Society's papers. This report again records who was present — Sir Jonas Moore, Daniel Colwall, William Croone, Abraham Hill and Robert Hooke — and of these, Colwall, Croone and Hill signed a record of the decisions taken, which are dealt with elsewhere in this volume.[79] As will be seen there, this was a purely administrative matter, and it illustrates by way of contrast the abnormal interest of the minutes for the committees of 1664-5 printed here, with their broad concern with major issues in the Society's scientific and technological programme, and their attempt to deal with them on a sustained and systematic basis.

Returning, therefore, to the committees of 1664 – 5 and looking at them as a whole, what was their significance, and why did they not survive? Most obviously, they represent an impulse towards specialization, an attempt to subdivide the Society's business and to delegate aspects of it to groups of Fellows: this clearly reflects a growing awareness on the part of the Society's organizers of the sheer scale of the task they had set themselves in their grandiose ambitions for the reform of knowledge. What is equally important about the committees of 1664 – 5 — in contrast to those established both earlier and later to deal with specific topics over a limited period — was that they were to be permanent. The result was that the three most active, those for Agriculture, Mechanics and collecting the phenomena of nature, devoted a considerable amount of energy at the outset to defining their 'method', while the latter seems to have been particularly worried

77 Ibid., ii, 93, 97. See also Oldenburg to Boyle, 8 June 1666, *Oldenburg*, iii, 154.

78 Cl.P. iii(1)27 – 41. Further related material is to be found in Cl.P. xxiv.80 – 2. See also Birch, *History*, ii, 199.

79 Royal Society Miscellaneous Manuscripts 16.39. The report appears in Birch, *History*, iii, 310. See also ibid., iii, 310 – 11, for the appointment of a further committee; an MS which overlaps with this is D.M. 5.71: see below, 'Between Cabinet of Curiosities and Research Collection', pp. 140, 142.

about the demarcation between its own activity and that of other committees.

It was in this spirit of permanence that their inauguration was noted in the *Philosophical Transactions* in 1665: 'the *Royal Society*, in prosecuting the *Improvements of Natural knowledge*, have it in design to collect *Histories of Nature and Arts*, and for that purpose have already, according to the several Inclinations and Studies of their members, divided themselves into divers *Commitees* to execute the said design'.[80] Sprat's sentiments about the 'committee for collecting the phaenomena of nature and all philosophicall experiments hitherto observed, made and recorded' were similar, while the very title of that group well illustrates the Society's encyclopaedic ambitions in its early years, its hope to give definitive surveys of whole fields of knowledge. Hence not only this but also other committees set out to record all that had already been written on their chosen subject as a means of laying the foundations for new developments, a significant part of the programme of the new science in this period, which is also to be observed in the Oxford group and which, for all its apparent impracticality, is not to be retrospectively underestimated. In addition, as has already been noted, the preoccupations of the two most active committees, those for Agriculture and Mechanics, illustrate how keen to be of real and immediate utility the Society was in its early years.

It should also be pointed out how central the idea of committees remained for all those who considered how the Society's business might be more effectively managed during the late seventeenth and early eighteenth centuries. The notion of reinstating committees recurred repeatedly in the proposals for restructuring the Society dealt with elsewhere in this volume, and these often echoed the terminology of the committees established in 1664, as with William Neile's idea in the late 1660s of a committee responsible for making 'a Catalogue of all former experiments made by whomsoever, and also of all the Phaenomena of nature', or Wren's plan in January 1681 for a series of committees including a 'georgical committee' and one intended 'to register all things, that should be remarkable'.[81] Indeed, it is possible that the Agricultural Committee reconvened at this point, to judge by a laconic entry in Robert Hooke's diary, though nothing is known of what transpired.[82]

If in theory the committee system seemed highly attractive, in practice it was to work far less well than had been hoped. Indeed, many of the difficulties involved are apparent from the minutes printed below, and it is worth elaborating on them here, because in many respects the failure of the

80 *Phil. Trans.*, 1 (1665), 91, and see above, p. 91.
81 See below, 'Towards Solomon's House', pp. 210 – 11, 225; Birch, *History*, iv, 65.
82 City of London Guildhall Library MS 1758, fol. 74v (entry for 26 May 1681). (The categorical statement to the contrary made on p. 198, below, should be modified accordingly.) The revival of the Agricultural Committee is referred to by John Houghton in his *Collection of Letters For the Improvement of Husbandry & Trade*, no. 1 (8 Sept. 1681), 2; he also reprinted the agricultural enquiries from vol. 1 of *Phil. Trans.* in ibid., pp. 6 – 9.

committees reflects in miniature the general problems that the Society encountered in its attempts to organize scientific activity at this time.

To start with, none of the committees met more than once a month, as against twice, as anticipated, and neither did they report to the Society as a whole quarterly, as initially requested.[83] In addition, often either individual members or whole committees failed to do what they promised. The minutes of the second meeting of the Agricultural Committee on 28 July 1664 display a clear sense of frustration that William Croone had neither attended nor provided what he had promised, while on 6 August, when the whole Mechanical Committee failed to do any homework, the meeting was abandoned. It is similarly symptomatic that the minutes of the last meeting of the Correspondence Committee consist almost entirely of orders to people to do things.

Considering their known interest in related topics, some Fellows were surprisingly inactive: John Evelyn, for instance, though retrospectively famous as the author of *Sylva* and an expert on the topics in which the Agricultural Committee specialized, attended only two out of its eight meetings and failed to respond to requests for assistance. In Evelyn's case, this was partly due to his official duties during the Anglo-Dutch War, but such lack of participation was clearly discouraging to those who were active. That this was the case is suggested by the parting shot of the report of the Agricultural Committee: this noted, in connection with the improvement of waste ground, how Charles Howard had 'promised, to contribute his part therein, but desired withall, that others might give their assistance to it; which he is in expectation off'. Presumably he was disappointed, and no notes on this subject either by himself or anyone else survive in the Royal Society's records. A similar air of frustration is arguably in evidence at the last meeting of the Mechanical Committee, with its return to first principles in the form of a reconsideration of how the committee's work should be prosecuted.

An equally revealing facet of the minutes printed here is the complete record of attendance that they include: as already noted, this is without parallel as far as the ordinary meetings of the Society at which its scientific business was transacted are concerned, though similar data is available concerning Council meetings.[84] In the case of the Agricultural Committee, this shows that a total of twenty-seven attended at least one meeting. Two (Howard and Oldenburg) attended all eight meetings; one (Hill) attended five; six (Moray, Lord Brereton, John Austen, Merrett, Wilkins and Daniel Colwall) attended four; two (Boyle and Croone) attended three; six (Goddard, William Balle, Evelyn, Dr George Smith, Long and Edward Waterhouse) attended two; and the remainder only one, perhaps hardly surprisingly including all those who were nominated to the committee only in 1665. As for the Mechanical Committee, only Oldenburg and Hooke attended every meeting; but three others attended five out of the six

83 See above, p. 82, and n. 55.

84 Cf. Hunter, *Royal Society*, pp. 84 – 5.

(Brouncker, Wilkins and Petty), one (Moray) attended four, one (Hill) three, and one (Palmer) two. A further six (Howard, Boyle, Balle, William Erskine, Colwall, Sir Paul Neile) attended one meeting. Lastly, with the Correspondence Committee, four attended both meetings (Povey, Moray, Oldenburg, Erskine), while four more attended one (Wilkins, Colwall, Haak, Henry Slingsby).

Various interesting points emerge from this record of attendance. One is that the hope that the committees would attract the participation of rank and file members who were otherwise not very active in the Society was almost wholly unrealized. Though virtually all the Society's Fellows were nominated to one committee or another, most ignored this: if the names of those who attended are collated with the record of activity at the Society's ordinary meetings published by Birch, hardly anyone attended even a single committee meeting who was not regularly in evidence at those. The only exceptions to this are two members of the Agricultural Committee, Sir Edward Harley and John Austen, but the former attended only once, and only the latter became a regular participant.[85]

In general, activity was dominated to a quite disproportionate extent by a handful of men who were the most active in the Society in any case — figures like Brouncker, Howard, Moray, Oldenburg, Hill, Wilkins, Colwall, Petty and Hooke, all of whom attended one committee or another six or more times and all of whose names recur frequently in the records of the Society's ordinary meetings.[86] This therefore illustrates in microcosm a problem which afflicted the Royal Society throughout its early years, namely that too much was done by too few: the result was that if their enthusiasm flagged or if they were prevented by other duties from sustaining their commitment to the Society, the enterprise was threatened. In this connection it is perhaps worth noting that, by comparison with its earlier meetings, the membership of the Agricultural Committee diversified at its meetings from December 1664 onwards, which may suggest that this problem had been noted even at that stage and a conscious effort made to bring more people in before the project was abandoned.

There is also evidence that the attempt at specialization which the committees represented did not commend itself to the Society's active members, who preferred to retain an involvement in a broad range of subjects. Indeed, Fellows seem to have felt quite at liberty to attend meetings of committees to which they did not belong, in spite of the fact that the trouble was taken to record changes in membership early in 1665

85 Ibid., catalogue entries 25, 144. It should be noted, however, that the lack of a record of attendance at ordinary meetings means that Austen might have been silently present at them. See ibid., pp. 84 – 5. It is perhaps also worth noting that Slingsby (catalogue entry 14), who attended one meeting of the Correspondence Committee, was otherwise barely active after 1663.

86 Boyle only attended five; W.Balle, Merrett, Austen and Erskine attended four. This includes attendance at the Buckland committee.

and that some active Fellows had been nominated to several committees in the first place. Instances of this are the attendance at the Mechanical Committee of Charles Howard, William Balle, Daniel Colwall and Abraham Hill, the latter three times; and of William Erskine, Petty, Hooke and Croone at the Agricultural Committee, the latter again attending three times although not a member.[87] Indeed, apart from those who only attended one or two meetings, the norm was for Fellows to be involved in committees on more than one subject, the only exceptions to this being Lord Brouncker, who limited himself to the Mechanical Committee, and John Austen and Lord Brereton, who stuck to the Agricultural (in the latter case, despite being on the Mechanical Committee as well).

Hence it could be argued that what the committees achieved was not so much a division of labour between active Fellows as an increase in the number of times per month that such men met. The committees thus represented an intensification of the high pitch of activity which characterized the Society in its first few years, the problem being how long this could be sustained. For whereas in the years up to and including that in which the committees were set up the Society had some fifty-two meetings per year, by 1667 this had fallen to forty, and by 1670 to thirty-two. Up to a point, it could be argued that the demise of the committees represented the first symptom of the Society's difficulty in sustaining the burst of enthusiasm which had followed its foundation. As its organizers were forced to acknowledge, it was hard enough to keep the Society's main activities going without encouraging members to dissipate their energies in attending subcommittees.[88]

Underlying this, however, was a more significant issue, of whether corporate activity was necessarily the best means of achieving the Society's goals. The committees were a logical part of the Society's initial presumption that all of its activities should be carried out jointly, a presumption also reflected at this stage in the form of corporate projects like Evelyn's *Sylva* — jointly composed, though published under Evelyn's name — or more strikingly still in the corporate patent application made by Abraham Hill.[89] For patents were quintessentially sought by individuals, and, though the Society may have felt that by covering its work jointly, it might simplify the procedure of application, minimizing the expense involved and the likelihood of disagreement among Fellows, it seems

87 In addition, Slingsby attended the Correspondence Committee although not on it; while Palmer was not officially a member of the Mechanical Committee. Oldenburg was also not named as a member of the Mechanical Committee, but presumably attended in a secretarial capacity.

88 Hunter, *Royal Society*, Table 3 (p. 113), and ch. 3, passim.

89 Hunter, *Science and Society*, p. 93; on the patent application, see above, pp. 88 – 90. For background, see Christine MacLeod, 'The 1690s Patent Boom: Invention or Stock-jobbing?', *Economic History Review*, 2nd series 39 (1986), 549 – 71.

equally likely that this represents a wish to subsume individual profit into a corporate enterprise.

Later, of course, patent applications reverted to an individual level, while *Sylva* lacked successors as a joint project and was increasingly identified exclusively with Evelyn, and this was symptomatic of a realization that corporate enterprise might not be as universally effective as had originally been hoped. Hence one finds Fellows doing more of their work on their own and simply reporting it to the Society, and the demise of the committees is arguably a reflection of this process as well as of a simple dwindling of activity as a whole.

Lastly, it is worth recalling how much of the committees' time was taken up by technological matters, since here too the discontinuance of these groups apparently reflects a broader shift to be discerned in the Society's early history. For, as technical change proved more difficult to implement than had initially been optimistically hoped, the Society instead increasingly stressed improvements of pure knowledge of the kind which — so far as we can tell from the sources that have survived — were more tangential to the deliberations of the committees than of the Society's ordinary meetings.[90]

Hence from the breakdown of the committee system one can arguably learn much of significance about the direction in which the Society evolved in its formative years. In addition, as has been indicated, the data now first made available has considerable intrinsic interest in its own right. It enlarges our knowledge of the Royal Society's proceedings; it provides an insight into the kind of experimentation in organizing knowledge in which the Society indulged; and it throws light on various facets of the institution's corporate aspirations in the first few years of its existence.

[90] Hunter, *Science and Society*, chs. 2, 4, passim.

DOCUMENTS

Minutes and Reports of Committees of the Royal Society, 1663 – 5 Royal Society Domestic Manuscripts 5.60 – 68[1]

(1) Buckland Committee [D.M. 5.60]

March 20 1662[/3].

The Committee appointed by the Society March 18 to consider of Mr Bucklands proposition to plant potatoes through all the parts of England, did meet at Mr Howards lodgings in Arundell house: where were present My Lord Brounker, Mr Howard, Mr Boyle, Sir Rob. Moray, Dr Goddard, Dr Whistler, Mr Ball, Mr Oldenburg.[2]

Mr Bucklands letter, containing the said proposition, together with the whole way of planting, and ordering potato's, and the benefit thereoff in time of scarcity of food, was read; and the substance thereoff considered and approuved, with this result. 1. That all those members of the Society, as have land, should be desired, to begin the planting of this root, and to persuade their friends to doe the same. 2. That in order thereunto Mr Buckland should be desired to send up,[3] what quantity he could of the smaller Potato's, to furnish those, that have conveniency to plant them: in order to which Mr Boyle offred to provide as many of them, as he could: as also, to communicate such notes of his, concerning this root and their diffusivenes, as he could recover. 3. that Mr Evelyn should be desired to insert this proposition and the approbation thereoff, together with the management of planting and spreading them, into the Treatise, which he is

[1] D.M. 5.60 and 63 – 8 are in the hand of Oldenburg; item 62 is in the hand of Hoskins, but the endorsement on the conjugate sheet is by Oldenburg; item 61 is in a scribal hand; items 63 – 4 and 66 – 8 are folded sheets each of which therefore comprises two folios. It is clear from the MS that Oldenburg changed his mind about wording as he wrote; these alterations have been recorded in footnotes, as have all deletions, but insertions have been silently ignored. Where Oldenburg changed his mind more than once, so that within his deletion one word replaces another, this has been indicated by the insertion of slashes.

[2] For the personnel of this and the other committees, see Hunter, *Royal Society*. 'Mr Ball' is clearly William Balle, catalogue entry 9; Mr Henshaw is Thomas Henshaw, entry 15; Mr Aerskins is William Erskine, entry 79; Dr Smith is George Smith, entry 120.

[3] 'Send up' replaces 'furnish', deleted; 'what quantity he could' replaces 'a Boxfull', deleted; 'if he could' has been deleted between 'up' and 'what'. After the colon, 'in order to which' replaces '3.', deleted, while '3.' in the next sentence has been altered from '4.' In the second line of this paragraph, 'the' has been deleted before 'potato's'.

now publishing by order of the Society concerning the planting of Trees[4] 5. that provision being made of these roots, the way and usefulnes of planting them should be further publisht and recommended to the Nation, in the Diurnalls,[5] without naming the Society in it, and that therein direction should be given to certain places, where they may be had for those, that have a mind to plant them.

Occasion being given to discourse of the way of setting corn, it was thought good, that Dr Wilkins should be called upon for his Engine formerly made for that purpose, vid. to set corn at due distance and depth, and withall to cover it, when tis thus sett. [In margin: [i]t saveth 3/4 of the corn.][6]

Mr Boyle mention'd a certain Author, that hath made a collection of the severall things, that may be a supply, in times off scarcity of usuall food.[7] And offred to communicate the same, when he should find it, in order to be reprinted, it being out of print.

Mr Boyle mentioned a certain neighbor of his at Chelsey, who had a way of ordering corn so by an imbibition, that it produced severall eares of one grain; sometimes 15. sometimes 20.

He mentioned likewise, that there was a person in Holland, of whom he was very credibly informed, that he had, by an imbibition of corn, made it grow in a meer gravelly sand. And that he intended to make a further inquiry thereof.

It being suggested, that salt, mingled with earth, or dissolved in water, would serve for dung, with great advantage; Dr Whistler mentioned, that whereas /verso/ [8]one acre required 40 loades of dung, at 4 or 5 sh. the load, there would be required but 2 bushels of salt for that proportion of land.

Mr Boyle[9] related, that he knew, that in a time of famine in Ireland there were kept from starving, thousands off poor people, by potatoes: And that this root would make good bread, mixed with wheaten-meale: that it will yield good drink too, but of no long duration: that it will prosper after the plow, as well as by delving: that it feeds poultry, and other animals well: that any refuse will keep them from frost: that the very stalkes of them,

[4] I.e., *Sylva*, published in 1664. At this point, a mark appears on the MS, indicating the end of the passage transcribed into the Society's minutes: cf. Birch, *History*, i, 213.

[5] Before 'Diurnalls', 'weekly' has been deleted, as earlier in the sentence have the words 'it mi[ght?]' after 'these roots,'; later in the sentence, 'may' replaces 'might', deleted.

[6] See Birch, *History*, i, 213, 220. In this sentence in the MS, 'for' has been deleted before 'that', and 'th..' before 'tis thus sett'; later 'may' replaces 'might', deleted.

[7] Probably Sir Hugh Plat, *Sundrie New and Artificiall Remedies against Famine* (1596). After 'collection of', 'all' has been deleted; later in the sentence, 'of food' has been deleted after 'a supply,' and 'in' replaces 'at', deleted.

[8] Before 'one acre', 'now' has been deleted; in the next line, 'if' has been deleted before 'there would be'.

[9] After 'Boyle', 'by way of discourse' has been deleted. Later in the paragraph, 'will' has been deleted before 'thrown into the ground'. On the Irish Famine see R. N. Salaman, *History . . . of the Potatoe* (Cambridge, 1949), pp. 225 – 6.

thrown into the ground, will produce good roots: that the planting of them doth not hinder poor people from other imployment.

> NB. the Society to be advertised, that Sir H.Bennet[10] hath got the King's hand to the Patent, and that My Lord Brounk[er] may be desired to give him the thanks of the Society for it.

Endorsed: 'Orders made concerning the planting of Potatoes all over England'. Adjacent to this are some scribbles apparently by a child.

(2) Legal Committee [D.M. 5.61][11]

The nineteenth of December 1663

At the Committe for reviewing the Lawes:

This Committe is of Opinion. That it is the Interest of the Society, to have no compleate body of Lawes. But that such Orders bee from time to time made as the State of Affaires shall necessarily require, And that the Lawes already made bee kept private.

John Hoskyns

(3) Committee for Collecting Natural Phenomena [D.M. 5.62]

According to an order made at a meeting of the Royall Society the eighth of June 1664 appoynting the persons therein named or any three or more of them to bee a committee for collecting the phaenomena of nature and all philosophicall experiments hitherto observed made and recorded; A competent number of them have severall times met and Resolving it best to divide the worke among the severall members of the Society they have thought fit to recommend to each person that shall take a share the directions given for that end by my Lord Bacon in a treatise written by him called *Parasceve ad historiam naturalem*[12] and that accordingly they will divide the observations of the phaenomena of nature into observations of generations, pretergenerations and arts but the last is to bee left to the care of the Committee for Trades. The history of praetergenerations is to bee

[10] I.e., Sir Henry Bennet, created Earl of Arlington in 1663, who had been Secretary of State since 1662. This sentence evidently comprises a memorandum from Oldenburg to himself about an unrelated matter, the granting of the Royal Society's second charter, news of which was reported to the Society at the same meeting as that to which the report of this committee was submitted, on 25 March 1663 (Birch, *History*, i, 214). At the end of this sentence, 'his' has been deleted before 'it'.

[11] This report is printed in Birch, *History*, i, 389, but has been included here for the sake of completeness.

[12] Cf. above, p. 92. Earlier in this sentence, 'into'[?] has been deleted before 'among the severall members'; 'have' in the previous line appears to have been altered from 'hath'.

ioyned in every distinct subject with that of Generations which is fivefold. First of coelestiall bodyes the recording of whose observations This committee conceives belongs properly to that of Astronomy and Opticks[.] Secondly of the ayre and metoeors a part of which that considers comets belongs to the same last mentioned committee[.] 3^{ly} of the earth and sea so far forth as they are integrall parts of the Universe[.] Fourthly of the Elements as they are great masses that is of their substance[.] Fifthly of particular bodyes or lesser species and these are to bee divided into Animalls vegetables and fossills.

This Committee likewise thought it most convenient that the collectors of these observations would let passe all matters[13] of philology, tedious and overexact descriptions, and parts of magick and superstition and that among arts they chiefly enquire into those that alter their subjectmatter and in the history of all arts take notice of such things as may make for the increase of naturall history though of no great moment to that which is the end of the art[.] That they bee as exact as may bee in recording proportion and that they set downe things certaine succinctly and give notice of the doubtfulnes of others but not the arguments pro & con unlesse it bee a matter of consequence[,] and if they can the reason why those things which are false were ever believed to bee true. And further this Committee were of opinion that the best choyce of Authors out of whose writings these observations were to bee gathered was in *Vander Linden de scriptis medicis*[14] whom together with a catalogue of such other good authors of the same subjects as are since published Dr Merret was pleas'd to promise hee would bring in when desired

This Committee likewise make it their request to the society that other Committees bee desired upon occasion to consider the same method and rules of my Lord Bacon that so they may all agree.

Verso blank; attached is a conjugate part-leaf, endorsed: 'Mr Hoskins Committee/ concerning the Collecting/ the Phaenomena of Nature/ hitherto observed etc.'

(4) Agricultural Committee (a) Minutes [D.M. 5.63-4]

D.M. 5.63, fol. 1

June 23 1664.

The Committee for Agriculture met according to the order of the Society; and there were present

[13] Replacing 'pieces', deleted. At the end of the next line of the MS, the word 'parts' was begun to be written at the end of a line, and has been deleted.

[14] I.e. J. A. Van der Linden, *De Scriptis Medicis Libri Duo* (1637, and subsequent editions). In the next paragraph, 'This Committee' has been altered from 'The Committee', and 'mak[e]' has been deleted before 'consider'.

Dr Croon was desired to send to Dr Power for an account about the Boring instrument, for fetching up of Earth from great depths, to discover the severall sorts of Earths, Minerals, etc.[19]

It was intimated, that Sir R. Moray also was very well acquainted with this kind of Borers, and that he should be desired to communicate what he knew thereof.

It being thought necessary, that for the composing of a good History of Agriculture[20] as many of the members of this Committee, and of the Society in generall, should be desired to chuse and nominate such Shires or Counties, as they could procure an account of their Husbandry-practise from; the Committee began with themselves, and severall of them undertook to get such accounts; videl.

Ld. Brereton, for Cheshire and Staffordshire.
Mr Howard, for Surrey and Barkshire.
Mr Ball, for Devonshire.
Mr Evelyn and Mr Hill for Kent.
Mr Austen, for Lincolnshire.
Dr Merret for the Vale of Evesham in Gloucestershire.
Dr Smith and Mr Smith for Gloucestershire.
Dr Croon and Sir Thom. Not for Surrey.

And further considering, who there might be amongst the absent members of this committee, as also among the other Fellows of the Society to doe the like, they thought upon the persons following;[21]

Mr Boyle, for Dorsetshire and Ireland.
Lord Craford[22] and Sir R. Moray, for Scotland.
Sir Wm. Petty and Mr Southwell, for Ireland.
Mr Beale, for Somersetshire and Herefordshire.
Dr Cotton, for Devonshire and Cornwall.
Sir Edw. Harley and Mr Hoskins, for Herefordshire.
Col. Long, Mr Awbrey and Mr Newbury, for Wiltshire and Dorsetshire
Mr Henshaw, for Midlesex.
Dr Wilkins and Dr Power for Yorkshire.
Mr Ford, for Devonshire /fol. 2
Mr Waller for Buckinghamshire.

19 Power's instrument has not been identified, though it is possibly relevant that he had done experiments in mine shafts in 1662: see Birch, *History*, i, 133f.; Power, *Experimental Philosophy* (London, 1664), pp. 171f. In this sentence, 'co. . .' has been deleted after 'account', 'the' before 'for fetching', and 'and' before 'Minerals'; the 'etc.' has been repeated and the first deleted.

20 After 'Agriculture,' 'that' has been deleted.

21 'they thought upon' replaces 'they soe[?] guessed, that', deleted, while after 'following' has been deleted: 'might employ themselves for this purpose in the/these places [replacing 'shires', deleted] videl.'. Earlier in the sentence, 'of' has been deleted after 'who', and 'of the' before 'amongst'.

22 I.e., the Earl of Crawford and Lindsay, F.R.S. For his answers to queries concerning trees and his comments on the agricultural inquiries, see Cl.P. xix.11.

Mr Waterhouse for Essex.
Mr Matthew Wren and Dr Holder for the Ile of Ely.
Mr Wylde for Shropshire, Worcestershire and Bedfordshire.

Adjourned till July 28.

July 28 1664.

Present at the Committee for Agriculture

Mr Howard.	Dr Merret
Mr Boyle.	Mr Hill.
	Mr Oldenburg.

According to the order of the former meeting, some lists of Georgicall Authors were brought in, one from Mr Beale, the other from Dr Croon by Mr Hill: both were read. But no Inquiries being yet drawn up by any of this Committee, and Dr Croon not appeared [sic], who had undertaken[23] in the precedent meeting to extract inquiries out of Mr Hartlibs Legacy, Dr Merret was desired, to reduce all what belongs to Husbandry to certain comprehensive heads, such as were hinted at by Mr Beale in a paper, that was read.

Mr Howard acquainted the Company with what he had begun to doe concerning the practise of Husbandry in Surrey and Barkshire; and he was desired to perfect his undertaking, if he could conveniently, against the next meeting.[24]

Mr Boyle did suggest these two Authors for Agriculture, L'ufficio del Cittadino in Villa, and Carlo Stephano.[25]

Dr Merret named Le 20. giornate dell'Agricoltora del Agostino Gallo.[26]
Adjourned till Aug. 25. /fol. 2v

Aug. 25 1664.

The committee met. Present

Lord Brereton.	Mr Colwall.
Mr Howard	Mr Hill.
Sir R.Moray	Mr Oldenburg.
Dr Wilkins	
Dr Goddard.	

Mr Howard offered a Liste of what was to be considered in composing a History of Agriculture: it was read, and the Author receaved thanks.

23 Replacing 'promised', deleted.

24 See Howard's 'A briefe discription of the Soyle and Husbandry of Surrey and Berkshire', Cl.P. xix.21. But see R. V. Lennard, 'English Agriculture under Charles II', *Economic History Review*, 4 (1932 – 4), 29 – 30n, and Birch, *History*, ii, 122.

25 Charles Estienne (d. 1564) was the author of various works on agricultural and horticultural topics; the other work has not been identified.

26 I.e. Agostino Gallo, *Le Dieci Giornate della Vera Agricolturae Piaceri della Villa* (1566).

There were also read some inquiries concerning Arable and Meddows, drawne up by Sir R. Moray: and these two papers furnisht the Company with matter, to drawe up inquiries for the newly mentioned two parts of Husbandry; as they are to be found in another paper;[27] And these inquiries were ordered to be distributed among the members of the Society, and by them to be recommended to their Friends, that are Experienced Husbandmen, all over England, Scotland and Ireland, for a full account upon them.

Adjourned till Sept. 22

Sept. 22 1664.

The Committee met. Present

Mr Howard.	Dr Smith.
Sir R. Moray.	Mr Colwall.
Dr Wilkins.	Mr Oldenburg.
Dr Merret.	

Dr Merret produced a List of what is cultivated in England, for which he receaved thanks: And the Company thought fit, that such as had opportunity, should endeavòur to procure the History of severall of the vegetables, mentiond in Dr Merrets Liste, as that of Woad, Hopps, Flax, Hemp, Madder, Buckwheat, Tobacco, Licorish etc.

It was mentioned, that woad did grow in Oxfordshire. Madder in Lincolnshire. Licorish the best in Yorkshire. and sauce[?][28] in Northamptonshire.

Mr Howard was desired to collect out of Parkinsons Kitchengarden[29] such heads, as may be a ground to make olitory inquiries upon; as also, to give in a List of all the wintergreens here in England.

Dr Smith brought in an Account upon the Inquiries out of some parts of Glocestershire.[30]

Adjourned till Oct. 20. /D.M. 5.64, fol. 1

Oct. 20 1664

Present at the Committee for Agriculture,

Mr Howard.	Mr Colwall.
Mr Aerskins.	Dr Croon.

[27] This no longer survives in the archive of the Royal Society, presumably because the queries were published in *Phil. Trans.*, I (1665), 92 – 4. But see also above, p. 86, n. 37. In the previous paragraph, 'it' replaces 'which', deleted.

[28] Perhaps Sauce-alone, the hedge-weed *Sisymbrium Alliaria*: see *OED*. I am grateful to Christine MacLeod for this suggestion.

[29] I.e. John Parkinson, *Paradisi in Sole* (London, 1629), pp. 461 – 533: 'The Kitchen Garden'. After 'Parkinson', 'and' has been deleted; at the start of this paragraph, 'Ordered, that' has been deleted before 'Mr Howard'.

[30] Replacing 'the vale of Evesham', deleted. At the start of the paragraph, 'Dr' replaces 'Mr', deleted.

Sir R. Moray.	Mr Hill.
Dr Wilkins.	Mr Oldenburg.

Mr Howard brought in a liste of the materials for Kitchengarden. He was desired, to consider further, whether it might not be made more compleat; as also to adde to it[31] an Account concerning Orchards, and to raise inquiries upon both these parts: Moreover, to bring in a Liste of all such wintergreens, as are knowne to him or described in books, whether in England already, or such, as he judgeth, may grow well therein.

The Secretary was ordered to speake with M. Evelyn, and to inquire of him, what Olitary Vegetables he hath taken notice off.

The same was ordered, to desire Mr Beale, that he would please to procure for the Society the History's of the Culture of Flaxe, Hemp, Hopps, Madder, Woad, Buckwheat, Rapeseed.[32]

Dec. 1 1664.

Present at the committee.

Mr Howard.	Mr Waterhouse.
Lord Brereton.	Mr Awbrey.
Sir Wm Petty.	Mr Colwall.
Col. Long.	Mr Hill.
Mr Austen.	Mr Haak.
Dr Goddard.	Mr Oldenburg.

Mr Howard brought in a fuller Liste of the Kitchengarden-materials, together with a Catalogue of winter-greens, and inquiries raised upon them: it was read, and the Author receaved the thanks of the Committee.

Occasion being given to discourse of Cherries, Col. Long affirmed that he was acquainted with the following sorts of Cherrytree.[33] 1. The Neapolitan, round, dark, and big, a good bearer. 2. The Carvon-Cherry. 3. Chrystall or Amber-Cherry. 4. Pigs-heart. 5. White Gascon-cherry. 6. Great Carnation-cherry. 7. Duke Cherry. 8. White-heart, very sweet. 9. the ordinary black-heart. 10. Flanders-cherry, to which he said did belong the Kentish-cherry.

Mr Waterhouse mentiond, that at Lee in Essex there is a great abundance of Rape-seed; and promised, to procure the history of making rape-oyle. /fol. 1v

Januar. 12 1664/5.

Present at the Committee,

Mr Howard.	Mr Waterhouse.

31 After 'add to it', 'the' has been deleted, as has 'whe[n]' after 'grow' in the last line of this paragraph.

32 In fact, Beale does not seem to have obliged.

33 Replacing 'them', deleted.

Mr Boyle.
Sir R. Moray.
Lord Brereton.
Col. Long.
Dr Merret
Mr Ball.
Mr Austen.
Mr Packer.[34]
Mr Hoskins.
Dr Ball.
Mr Oldenburg.

Resolved, that Copies should be made of Mr Howards Catalogue of materials for Kitchengarden and wintergreens, and the Inquiries thereupon; and that they should be communicated to such members of the Society, as could recommend them to skilfull Gardiners, for their Answers to them.

The Secretary was desired, to procure from Italy the Culture of Tortafoli, a kind of underground fungus.[35]

Oakslungs were commended as excellent for a Consumption. Red sage, in Rye-dough, baked in a bread oven, and putt into bier, was commended by Mr Waterhouse as very good for a Dropsy.[36]

The juyce of Quinces was affirmed to yield an excellent Cider. Dogbryer, having a white Flower, was said to yield tops, which putt into the ordinary drink, make it very effectuall against the scurvy.

Lavor, a seaheard, found in the Severn-Sea, growing upon small stones, was affirmed to be eaten in the Contry for Colewort, when boyled. Mr Hoskins.

Sir R. Moray mentiond a Scottish seaherb, called Dills, good to eat, being boyled.

It was suggested, that it should be inquired, how Misletoe was propagated, whether by seed, or otherwise.

Mr Boyle intimated, that there was great difference in Misletoe, as to shape and taste.

Molosses was affirmed to be good to preserve bier, and to make it strong.

Wormwood and Ginger, were commended to be a good succedaneum to Hopps.[37]

Mr Boyle producing a paper of Mr Austens, gardner, wherein are contained proposals for the improving and planting of Timber and fruit-trees;[38] the paper was read; whereupon the Committee ordered, that Mr Waterhouse, Mr Packer and Mr Hoskins should be desired, to consider, what the Law had already pro/fol. 2/vided in this particular;[39] and withall to

34 Altered from 'Dr'. At the end of the next paragraph, there is a tiny deletion between 'could' and 'recommend'.

35 'Teretifolium' literally means 'round leaf'.

36 'as' replaces 'for very savory, and', deleted. After 'commended' 'f..' has been deleted, and the full-stop after 'Dropsy' replaces a deleted comma. In the next paragraph, 'are/go' has been deleted before 'make', and in the following one, 'be[s?]' before 'was affirmed'.

37 'succedaneum' is Latin for 'substitute'.

38 For these proposals, see Cl.P. x(3)7 – 8. Before 'wherein', there is an illegible deletion: 'at Ox'? 'Improving' has been altered from something else.

39 Replacing 'matter', deleted; earlier, 'and' has been deleted before 'whereupon'.

bring in their advise, how the propositions concerning this matter, may be best ordered, to make them effectuall.

Febr. 23 1664/5.

Present at the Committee,

Mr Howard.	Dr Pell.
Mr Boyle.	Dr Wilkins.
Sir Edw. Harlow [sic]	Dr Croon.
Mr Evelyn.	Mr Hook.
Mr Austin.	M. Oldenburg.

Mr Howard desired a Copy of the Agriculture inquiries for the Duke of Albemarle.[40]

Sir Edw. Harlow affirmd, that corne did not thrive by Stratton in Shropshire, where Barbary-trees abounding, and wind blowing from them upon the corne, was conceived to be the cause of blasting it.

It was mentiond, that one Mr Gosnall at Beaconsfield had a Gras, of such a sort, that he knew not, what seed it came from: that it had a yellow flower; was neither trefoile, nor cinqfoile; that he intending to buy Clovergras, by a mistake lighted upon another sort, which was this unknowne one; which yet proved good, and yielded a good crop.[41]

The Secretary was desired, to inquire of Mr Waller, who hath a house at Beaconsfield, concerning this particular.

Sir Edw. Harlow mentiond, that humble-bees being stringed would drive away mice. This I took imperfectly.

The same mentiond that Buzzards will clear the ground of mice.

It was ordered again, that the Kitchingarden-inquiries should be recommended to experienced Gardiners.

Whereupon the Secretary gave the company this account, that he had already delivered severall copies of them to such persons, as had promised the recommending of them to good gardners: vid. to Lord Brereton, Mr Beale, Mr Read.[42]

fol. 2v blank, except for endorsement: 'Transactions of the/ Georgicall Committee'.

[40] I.e. George Monck, who had been made an honorary F.R.S. in January 1665.

[41] In this paragraph, 'but k. . .' has been deleted before 'of such'; 'ki. . .' before 'sort'; and an illegible word ['noe'?] before 'intending'. In the next paragraph, there is a small illegible deletion before 'inquire'. For a Buckinghamshire landowner of the mid seventeenth century called George Gosnold or Gosnald, see *Victoria County History of Buckinghamshire*, iii (1925), 25; Waller was the poet, Edmund Waller, F.R.S.

[42] I.e. Richard Reed, a Herefordshire gentleman and cousin of John Beale. Cf. *Oldenburg*, iv, 115, ix, 385. Before 'Mr Beale' 'Mr[?] Mr Beale' has been deleted, as has 'se. . .' after 'that he had'. At the beginning of the previous paragraph 'ordere. . .' was written and deleted after 'It was'.

(b) Report [D.M. 5.65]

An Account given to the Society
of
The Performances of the Georgicall Committee,

This Committee having met the first time June 23 1664 at Mr Charles Howards lodgings in Arundell house, they began to consider[43] the End of their Meeting, which being the Composing of a good History of Agriculture and Gardening, in order to improuve the practise thereoff; And resolved, that, for this end, Lists of Georgicall Authors should be brought in, out of which, inquiries should be collected, and digested; and then sent by the Fellows of the Society into all the Counties and shires of England, Scotland and Ireland, where they have[44] a speciall interest, to procure by this means from their Acquaintance, experienced in Husbandry, good answers to such Inquiries; whereby it may be knowne, what is practised already, and every place be enrich'd with the aides, that are found in any place.

This Liste of inquiries was accordingly drawne up, as to Arable[45] and Meadows, and many Copies of them were made, and distributed among the Fellows of the Society at severall meetings thereoff. Some returns have been since made, by Mr Howard, Dr Smith, Mr Beale, both the Secretaries: and 'tis hoped, that others will awaken their Friends in the Contry, to send in theirs also.

Moreover, Dr Merret having brought in a Liste of what is cultivated in England, and the Committee thereupon thinking[46] fit, that such, as had opportunity, should procure the History of the Culture of severall of the vegetables, mentiond in the said Liste, as that of Woad, Hopps, Flaxe, Hemp, Madder, Buckwheat, Licorish, Rapeseed: it was ordered, that these particulars should be recommended to Mr Beale by H. Oldenburg.

Further, Mr Howard brought in a Liste of materials for Kitchingarden, and of wintergreens, together with inquiries concerning their Culture and use: of both which it was ordered that severall Copies should be made, and those recommended to such Fellows of the Society, as had acquaintance with skilfull Gardiners, to procure from them answers thereunto. [In the margin adjacent to this sentence are listed: Dr Wilkins, Mr Henshaw, Dr Holder.] Which is begun to be done accordingly.

Mr Boyle having recommended to this Committee some proposals of Mr Austin, touching the planting both of Timber and Fruit-trees; it was ordered, that Mr Packer, Mr Waterhouse and Mr Hoskins should /verso/ be desired, to consider first, what the Law had already provided in this matter,

43 Altered from 'considering'; 'to' replaces 'with', deleted.

44 Before 'have', 'may' has been deleted, as has 'desiring' before 'to procure' and 'a. . .'[?] before 'good answers'.

45 Replacing 'Agriculture', deleted. Later in the sentence, 'the' has been deleted before 'severall'. For answers to the inquiries, see above, p. 86.

46 Replacing 'thought', deleted. After 'Rapeseed:' 'and that' has been deleted. The 'H.' in 'H. Oldenburg' appears to have been altered from 'Mr' or 'mee'. In the next paragraph, 'Liste and inquiries' has been deleted after 'of both which'.

and also to bring in their advice, how these proposals may be best ordered, to make them effectuall.

Mr Howard moving, that it might be considered by this Committee, how waste Lands, Heathy Grounds and Boggs might be well imployed and improved, he was desired, to bring in, what himselfe knows of the wayes of such improuvements; who promised, to contribute his part therein, but desired withall, that others might give their assistance to it; which he is in expectation off.[47]

(5) Mechanical Committee [D.M. 5.66 – 7]

D.M. 5.66, fol. 1

July 16 1664,

The Mechanicall Committee met the 1st time at the Lord Brounkers house: present Lord Brounker

Sir R. Moray. Mr Old.
Sir Wm Petty.
Mr Hook.

Proposed, that a method should be thought of, for this Committee to proceed in.

Proposed by Mr Hook, to make an instrument, by which may be found a ships motion through the water.[48]

Ordered that such an instrument be made, and experimented, as soon it shall be ready, upon Sir Wm Petty's skiffe.

Aug. 6 1664.

Present at the Committee.
Lord Brounker
Mr Boyle. Mr Old.
Sir R. Moray.
Dr Wilkins.
Mr Hook.

Inquiry was made, whether any of the committee had thought upon the method, formerly mentiond, and it being found, that it had not yet been

[47] The last section is written in a different ink, and is possibly an addition. The word 'be' has been deleted after 'others might', and 'of' before 'which he'.

[48] Evidently a form of 'waywiser' or mechanical log: for such a device by Hooke, see J. A. Bennett, 'Robert Hooke as Mechanic and Natural Philosopher', *Notes and Records*, 35 (1980), 38 and n. 56; Birch, *History*, iv, 230 – 1 (where Hooke records showing it to the Society c. 1663); and Hooke, *Posthumous Works*, pp. 561, 562. A 'Ship waywiser' is also mentioned in a list of inventions and experiments among Hooke's papers in Cl.P. xx.54. Petty's 'skiffe' is his 'double-bottom': see above, pp. 30, 76 – 7. In this sentence, 'a ships' has been deleted before and 'shown' after 'may be'.

considered by any, the Company was desired, to take it into consideration against next meeting. /fol. 1v

Sept. 3 1664,

Present at the Committee,
Sir R. Moray.
Sir Wm Petty. Mr Old.
Dr Wilkins.
Mr Hill.
Mr Hook

An Enumeration was made of some Authors of Mechanicks, and the Company was desired, to think upon a good Liste against next Meeting. /fol. 2

Oct. 17 1664.

Present at the Mechanicall Committee
Lord Brounker. Dr Wilkins.
Sir Wm Petty. Mr Hook.
Mr Palmer Mr Old.
Mr Hill.

Authors of Mechanicks were considered off; as
Schottus.[49] Cardanus.[53]
Ramelli.[50] R. de Fluctibus.[54] De Arquer[?].[57]
Ubaldus.[51] Agricola[55] Brahe.[58]
Mersennus.[52] Furnier.[56]

[49] I.e. Gaspar Schott (1608 – 66), author of numerous works on technical subjects.
[50] I.e. Agostino Ramelli, *Le Diverse et Artificiose Machine* (1588).
[51] I.e. Guidobaldo del Monte (1545 – 1607), author of *Liber Mechanicorum* (1577) and other works.
[52] I.e. Marin Mersenne (1588 – 1648): it is not clear which of his works are referred to, but it is perhaps likeliest to be the 'Mechanicorum Libri' which forms part of his *Synopsis Mathematica* (1626).
[53] I.e. the sixteenth-century Italian author, Girolamo Cardano (1501 – 76), whose main work on mechanics was his *Opus Novum de Proportionibus* (1570).
[54] This appears to be the name Robert Fludd in Latinized form, but, if so, if is not clear what work is referred to.
[55] I.e. Georgius Agricola: cf. his *De Re Metallica Libri XII* (1556).
[56] I.e. Georges Fournier (1595 – 1652), a French Jesuit who wrote works on navigation and fortification published in the 1640s.
[57] Perhaps *L'Avis de Monsieur [Pierre d'] Arquier . . . sur le Dessin du Canal Royal pour la Communication du Deux Mers* (1667, but also circulated in manuscript).
[58] I.e. the astronomer, Tycho Brahe (1546 – 1601); probably his *Astronomiae Instauratae Mechanica* (1598) is referred to.

Kircher.[59]
Bettinus.[60]
Bessonius.[61]
De Molendinis[62]
Bate's Nature and Art.[63]

The Method for this Committee to proceed in, was thought to be, to collect what is delivered in Books, and practised; to take notice of what is practised and not found in Books; and then to consider what may be further done for the improuvement of mechanicks.

It was resolved on, that Authors being distributed among the members of this Committee, and other members of the Society, the busines of raising of water should be first of all taken into consideration; vid. what hath been done in it hitherto, and whether and how it may be improuved?

Here Mr Hook was desired, to shew by an Experiment, that there is no necessity of Towers for the raising of water. /fol. 2v blank

D.M. 5.67, fol. 1

Nov. 14 1664.

Present at the Committee for mechanicks

Lord Brounker.
Sir P. Neile.
Sir W. Petty.
Mr Howard.
Mr Palmer.
Mr Ball.
Dr Wilkins.
Mr Colwall.
Mr Hill.
Mr Hook.
Mr Oldenburg.

The Mechanicall Authors, formerly brought in, were recommended to the perusall of the members of the Society.

Sir P. Neile did chose Schottus for Mr Neile, his son.[64]

Ld Brounker was pleased to take Mersennus, Tycho Brahe, and Mr Bates Nature and Art.

Sir Wm Petty, Fournier.

Sir R. Moray should be desired to take the Author de Molendinis.

Mr Ball, Kircher de consonis et dissonis, and de Magnetica.

Dr Wilkins, Sir Hugh Plat and Mathematicall Magick.[65]

59 I.e. Athanasius Kircher (1602 – 80): see below for references to his *Magnes, sive de Arte Magnetica* (1641), *Ars Magna Lucis et Umbrae in Mundo* (1646) and *Musurgia Universalis sive Ars Magna Consoni et Dissoni* (1650).

60 Presumably Marius Bettinus, *Apiaria Universae Philosophicae Mathematicae, in Quibus Paradoxa, et Nova Pleraque Machinamenta ad Usus Eximios Traducta* (1642).

61 I.e. Jacques Besson, a late sixteenth-century French author on scientific instruments and related topics.

62 'Molendinum' is Latin for 'mill-house', but the work in question has not been identified. See also below, minutes for 14 Nov.

63 I.e. John Bate, *The Mysteryes of Nature, and Art* (1634). In the next paragraph, 'take notice of' replaces 'consider', deleted.

64 I.e. William Neile, F.R.S. In the previous sentence, 'severall' has been deleted before 'members'.

65 I.e. Wilkins' own book, published in 1648, and Sir Hugh Plat's *Jewell House of Art and Nature* (1594). For the other works mentioned here, see above, nn. 49 – 63. In the previous sentence, 's' has been deleted at the end of 'Kircher'.

Mr Hill, Ars magna Lucis et Umbra of Kircher.

Mr Hook, Vitruvius, Vegetius, Galilaei, Besson, Riccioli, Schouterus Conick Sections.[66]

Oldenburg, Agricola.

Mr Palmer promised to look what Authors of this kind he hath, and then to chuse a taske.

Sir P. Neile was desired to lend Hero de Spiritalibus[?][67] to Lord Brounker, to peruse.

Every one was sollicited to reduce what he meets with in the Author he hath undertaken, to certain Mechanicall heads./fol. 1v

Sir Wm Petty, Dr Wilkins and Mr Hook were desired to consider of the way, proposed by Sir W^m, of propelling with more quicknes the water of Islington river, that supplieth London.[68]

Lord Brounker mentioned the raising of a dam, to make a declivity.

Dr Wilkins was desired to propose to those, that are interessed in the worke of bringing the water of Islington river to London, whether in case any persons can quicken the course of the water, they will consider them for it.[69]

NB. Mr Hook was desired, To consider the way of producing wind by the fall of water. /fol. 2

May 12 1665.

Present at this Committee,
Lord Brounker.
Mr Aerskins.
Sir R. Moray.
Sir Wm Petty.
Dr Wilkins.
Mr Hook.
Mr Oldenburg.

The Company discoursed a good while about Gunnery, and in particular of

[66] I.e. the ancient authors Vitruvius and Vegetius; Galileo Galilei; Jacques Besson (see n. 61); Giambattista Riccioli (1598 – 1671), the Italian Jesuit astronomer and geographer. 'Scouterus' is perhaps Schott (see above, n. 49), but, if so, it is not clear what work is referred to. For the other works mentioned here, see above.

[67] Hero's *Spiritalium Liber* had appeared in more than one edition in the late sixteenth century. In the next sentence 'the' has been altered from 'these'.

[68] Before 'London', 'a part of' has been deleted, and 'with more' after it. In the previous line, 'Islington' replaces 'the', deleted. For the background, see above, p. 90.

[69] 'for it' replaces 'a share in the profit, or a consideration', deleted. Earlier in the sentence, 'the' has apparently been deleted before 'Islington'; 'any persons' replaces 'we', deleted; 'promote and' has been deleted before 'quicken'; and 'consider' replaces 'give', deleted.

the knowledge concerning the randome of Guns.

And then they renew'd their former order, of taking into serious considerations, the prosecution of the work[70] of this Committee; and ordered, that Mr Hook should draw up a scheme of mechanicall matters, unto which, what hath been done hitherto in Mechanicks, may be referred.

fol. 2v blank except for the endorsement: 'Transactions of the/ Mechanicall Committee'.

(6) Correspondence Committee [D.M. 5.68]

fol. 1

Aug. 19 1664.

The Committee for Correspondence met at Mr Poveys.
Present, Mr Povey. P.
Sir R. Moray.
Mr Aerskins.
Mr Slingsby.
Dr Wilkins.
Mr Haak.
Mr Oldenburg.

The Committee named some Books of Voyages, to be perused, for inquiries: E.g.

Purchas.[71]	Barros, of the Eastindies, in
Acosta.[72]	4 volums: continued by
Olearius.[73]	Conto, in 8 volums more: the
Bontius.[74]	2 first translated out of Portugual
Piso.[75]	into Italian by Ulhoa; which are
Prito.[76]	said to be meerly politicall and of commerce;

[70] Before 'work', 'desse[ign]' has been deleted; earlier, 'did' has been deleted before 'renew'd', while 'the prosecution of' replaces 'the Method, for the improving of', deleted. Subsequently, 'shoul[d]' has been deleted after 'Mr Hook' and 'ref. . .' [?] before 'referred'. In the previous paragraph, 'the' has been deleted after 'and'.

[71] I.e. Samuel Purchas's *Pilgrimage: or Relations of the World and the Religions observed in all Ages* (1613)

[72] I.e. José de Acosta, *The Naturall and Morall Historie of the East and West Indies* (Eng. trans., 1604).

[73] I.e. Adam Olearius, *The Voyages and Travels of the Ambassadors sent by Frederick Duke of Holstein, to the Great Duke of Muscovy, and the King of Persia* (Eng. trans., 1662).

[74] I.e. Jacobus Bontius, *De Medicina Indorum* (1642).

[75] I.e. Gulielmus Piso (1611 – 1678), author of various works on Brazil.

[76] Not identified.

Martinius.[77]
Vincent le Blank.[78]
De Laet.[79]
Linschoten.[80]
Ligon.[81]
History of the Antiles.[82]
Rhamnusius.[83]
Eusebius Nierenbergensis.[84]
Relation de la Groenlande.[85]

the rest, philosophicall.[86]
Trigaltius[87]
History of the yncas of Peru.[88]
History of the Kingdom of Chili, by Alphonso de Ovallo, in Roma, 1646. Hispan.[89]
Antonii de Herrera descriptio indiae occidentalis. Portug.[90]
Antonij de Herrera descriptio of the Acores: Portug.[91]
Voyage de Levant par le Sieur de Hayes.[92]
Viaggi di Pietro de la Valle.[93]

77 I.e. Martinus Martini (1614 – 1661), Italian Jesuit missionary and author of two books on China published in the 1650s.

78 I.e. *The World Surveyed, or the Famous Voyages & Travailes of Vincent de Blanc* (Eng. trans., 1660)

79 I.e. Johann de Laet, author of *De Imperio Magni Mongolis* (1631), *Persia* (1633), *Novus Orbis* (1633), etc.

80 I.e. J. H. van Linschoten, *Discours of Voyages into the Easte & West Indies* (Eng. trans., 1598). For notes on his work and that of Bontius and others, see above, p. 93.

81 I.e. Richard Ligon, *A True & Exact History of the Island of Barbados* (1657).

82 I.e. [J. B. Du Tertre], *Histoire Naturelle et Morale des 'Isles Antilles de l'Amerique* (1658). For its allocation to Oldenburg, see below and Oldenburg to Boyle, 6 Oct. 1664, *Oldenburg*, ii, 248.

83 I.e. Giovanni Battista Ramusio, *Raccolta di Navigazioni e Viaggi* (1550 – 9).

84 I.e. Juan Eusebio Nieremberg, *Historia Naturae Maxima Peregrina Libri XVI* (1635).

85 I.e. Isaac de la Peyrère, *Relation du Groenland* (1647)

86 I.e. *Asia de Joam de Barros, dos Fectos que os Portugueses Fizeram no Descobrimento e Conquista dos Mares e Terres do Oriente* (1552 – 63 and 1615). In the description of this work, '4' replaces 'severall', deleted.

87 I.e. Nicolas Trigault (1577 – 1628), a Jesuit missionary who published various books on China and Japan.

88 Probably Agustin de Zárate, *The Strange and Delectable History of the Discoverie and Conquest of . . . Peru* (Eng. trans., 1581).

89 I.e. Alonso de Ovallo, *Historica Relacion del Reyano de Chile* (1646)

90 Presumably Antonio de Herrera Tordesillas, *Historia General de Los Leches de Los Castelanos en Las Islas* (1601 – 15).

91 I.e. id., *Cinco Libros de la Historia de Portugal y Conquista de las Islas de los Açores en los Años de 1582 y 1583* (1591).

92 I.e. Louis de la Haye, Baron de Courmenin, *Voyages du Levant, Fait par le Commandement du Roi en 1621* (1624).

93 I.e. *Viaggi di Pietro della Valle. . . cioè la Turchia, la Persia, e l'India* (1650 – 3).

Leonis Afri totius Africae de-
scriptio Antverp. 1556.[94] /fol. 1v

Some of the Company undertook to peruse some of these Authors;
Sir R. Moray, Acosta,
Mr Aerskins, Flacour.[95]
Dr Wilkins, Purchas.[96]
Mr Haak, Olearius.
Mr Oldenb. The History of the Antiles.

Some Generall inquiries, to be sent into all the parts of the world, being red, it was ordered, that the Secretary[97] take care to have them transcribed, and reviewed by some of this Committee, as Sir R. Moray, Dr Wilkins.

Ordered likewise, that a Copy be made of the particular inquiries for the East-indies, and the Directions for Seamen, both contained in the Society's Register-book, and that by Dr Wilkins they be recommended to the Lord Berkley.[98]

It was recommended to Mr Povey, to treat with Mr Hubbard about his Collection of Curiosities.[99] /fol. 2

Sept. 23 1664.

Committee met,
Sir R. Moray.
Mr Povey
Mr Aerskins.
Mr Colwall.
Mr Oldenb.

94 I have not been able to trace this work, though Boyle in his *Experimental Discourse of Some Unheeded Causes of the Insalubrity and Salubrity of the Air* cites 'Leo Africanus' (*Works*, v, 57). On the next line, 'Rela . . .' has been deleted.

95 I.e. Etienne de Flacourt, *Histoire de la Grand Isle Madagascar* (1658).

96 After 'Purchas.' 'and' has been deleted.

97 This has been altered from 'Secretaries'. For some 'General Heads of Inquiries for all Countries' in Oldenburg's hand, see Cl.P. xix.43: these appear to date from 1669, but may be based on an earlier exemplar. Reporting on this meeting in a letter to Robert Boyle of 23 August 1664, Oldenburg says that General Inquiries were 'drawn up' at it: *Oldenburg*, ii, 209.

98 The Earl of Berkeley, F.R.S., had well-known colonial interests. For the 'Inquiries for the East Indies' and the 'Directions for Sea-men, bound for far Voyages', see Royal Society Copy Register Book, i, 153 – 6, 216 – 21 (from Cl.P. xix.2 – 5), Sprat, *History*, pp. 158f., and *Phil.Trans.*, 1 (1665), 140 – 3. In this sentence, 'both' replaces 'be already', deleted, and 'be,' has been deleted before 'and that'.

99 On this see below, 'Between Cabinet of Curiosities and Research Collection', pp. 126 – 7, 129f. Before 'Collection', 'rarit[ies]' has been deleted.

To let the Amanuensis make another copy of the inquiries for Sir R. Moray.

Generall inquiries for Mr Povey.

To mind the persons charged with drawing up of inquiries, of their taske.[100]

Mr Colwall, to make inquiries out of V. le Blank.

The Committee never mett since Mr Povey's Feasting of them.

Fol. 2v blank except for the endorsement: 'Transactions of the/ Committee for Correspon/dence'.

[100] 'drawing up of' replaces 'making', deleted.

Daniel Colwal Armiger Musæi Regalis
Societatis Fundator

MUSÆUM REGALIS SOCIETATIS.

OR A

Catalogue & Description

Of the Natural and Artificial

RARITIES

Belonging to the

ROYAL SOCIETY

And preserved at

Gresham Colledge.

MADE

By *Nehemjah Grew* M. D. Fellow of the *Royal Society*, and of the *Colledge* of *Physitians*.

Whereunto is Subjoyned the

Comparative Anatomy

OF

Stomachs and Guts.

By the same AUTHOR.

LONDON,

Printed by *W. Rawlins*, for the Author, 1681.

4

BETWEEN CABINET OF CURIOSITIES AND RESEARCH COLLECTION: THE HISTORY OF THE ROYAL SOCIETY'S 'REPOSITORY'*

The Royal Society's museum or 'repository' was one of the most celebrated and 'visible' aspects of the Society in its early years. Whereas only through personal introduction was it possible actually to attend a meeting of the Society, the repository became one of London's tourist attractions in the late seventeenth and early eighteenth centuries: many evidently visited it and formed an impression of the Society not least on this basis.[1] In addition, one of the most lavish publications that emanated from the Society in its early years was the catalogue of the collection compiled by Nehemiah Grew and published in 1681 as *Musaeum Regalis Societatis. Or a Catalogue & Description of the Natural and Artificial Rarities Belonging to the Royal Society and preserved at Gresham Colledge*, a grandiose folio published by subscription and replete with engraved illustrations of objects in the collection. This book became part of the literature of museums at the time — joining works like Ole Worm's *Museum Wormianum* (1655) or Ulisse Aldrovandi's catalogues of his collection — and, as was the case with these books and the objects which they chronicled, it reified the collection in book-form so that its fame spread even more widely than it did from travellers' reports. The German scholar,

* On the relationship between this chapter and my earlier essay, 'The Cabinet Institutionalized: the Royal Society's "Repository" and its Background', in *The Origins of Museums: the Cabinet of Curiosities in Sixteenth- and Seventeenth-century Europe*, ed. Oliver Impey and Arthur MacGregor (Oxford, 1985), pp. 159 – 68, see above, Preface.

[1] See, for instance, Lorenzo Magalotti, *Travels of Cosmo III, Grand Duke of Tuscany, through England (1669)* (London, 1821), pp. 188 – 9; Evelyn, *Diary*, iv, 34, 133, 195; Joseph Hunter, ed., *The Diary of Ralph Thoresby* (London, 1830), i, 298, 340; Ned Ward, *The London-Spy Compleat*, ed. Ralph Strauss (London, 1924), pp. 60 – 1; Edward Hatton, *A New View of London* (London, 1708), ii, 666-86; C. Erndtel, *De Itinere suo Anglicano, 1706* (Amsterdam, 1711), p. 59; *British Curiosities in Nature and Art* (London, 1713), pp. 43 – 6; John Macky, *A Journey through England* (2nd ed., London, 1722), i, 260 – 2.

Plate 4. The title-page of Nehemiah Grew's Musaeum Regalis Societatis *(1681), with a facing frontispiece celebrating Daniel Colwall as 'Founder' of the collection. The copy reproduced here is John Evelyn's, now British Library shelf-mark Eve.b.46; at the foot of the title-page is Evelyn's shelf-mark, 'C.71'. Reproduced by permission of the British Library.*

Zacharias Conrad von Uffenbach, for instance, visiting England in 1710, came with expectations formed by reading Grew's book in addition to the *Philosophical Transactions*.[2] The repository therefore contributed significantly to the Society's early image, and this is one good reason for investigating its history.

But it also worthy of study because it illustrates some important themes in the Society's early development. It gives an interesting insight into the aspirations and performance of an early scientific institution, and into a transitional phase in the evolution of the concept of institutional — as against private — collections. Equally important is what one can learn from the history of the collection about what might be described as a conflict between 'virtuoso' and 'scientific' values in the scientific community of the day. The virtuosi formed the clientele on which a voluntary institution like the Royal Society depended, and, because of this, the Society was under pressure to mould itself in their image, even if this clashed with the scientific programme of some of its leading organizers. Moreover this was a matter of special significance in connection with the matter under study here, for a cabinet of curiosities was the quintessential appendage of a successful virtuoso, and these cabinets exemplified in a pronounced form the characteristics of the virtuoso movement classically outlined by W. E. Houghton in his study of 1942.[3] As he pointed out, virtuosi were typified by their love of the rare, the exotic, the marvellous and the inexplicable; they were also characteristically snobbish and elitist, and, as has more than once been pointed out, such traits were fundamentally antipathetic to many of the aims espoused by Baconian scientists of the late seventeenth century. The history of the Royal Society's 'repository' illustrates some of the tensions that could result.

Thirdly, and perhaps most important, the museum's history illuminates what seems to me a significant and neglected theme, the way in which ideas about the function of a scientific institution and the facilities which it could usefully provide were modified through experience, which revealed some approaches as more feasible than others in a way that could not have been predicted without actually trying them out. In the case of the repository, this is illustrated with particular clarity on the question of whether it was realistic to aspire to a 'complete' collection of natural things — as did a significant core of the Society in its early years — or whether this was beyond the capacity of a voluntary body with other demands on its limited resources: as it was, the assemblage remained relatively selective, and it became apparent to those who actually used it for scholarly purposes how

[2] W. H. Quarrell and M. Mare, trans. and ed., *London in 1710: from the Travels of Zacharias Conrad von Uffenbach* (London, 1934), p. 98. On earlier catalogues, see Oliver Impey and Arthur McGregor, eds., *The Origins of Museums* (Oxford, 1985), and Sir David Murray, *Museums, Their History and their Use* (Glasgow, 1904).

[3] W. E. Houghton, 'The English Virtuoso in the Seventeenth Century', *Journal of the History of Ideas*, 3 (1942), 51 – 73, 190 – 219. See also Lawrence Stone, *The Crisis of the Aristocracy, 1558 – 1641* (Oxford, 1964), pp. 715 – 7; Hunter, *Science and Society*, pp. 66 – 8.

the rarities that it contained could be supplemented by common objects from other sources to build up a comprehensive classification of natural things.

The history — or prehistory — of the Society's museum begins in its earliest years. From the first, the Society accumulated equipment relevant to its research interests — for instance, Robert Boyle presented an air pump in May 1661 — while it also early began to acquire items of the kind common in the private cabinets of virtuosi. In June 1662, for instance, Lord Berkeley presented a bird of paradise, and John Evelyn records the gift of a piece of elephant skin in September of the same year.[4] In this diary entry, Evelyn uses the word 'Repository' to describe the Society's collection, although this word first appears in the minutes only in October 1663. I should perhaps note that from that point 'repository' was the Society's standard term for referring to its collection, with 'Museum' being used as the Latin equivalent of this.[5]

The occasion on which the word was first used in the minutes was a Council meeting on 19 October 1663, when Robert Hooke, the Society's Curator of Experiments, was ordered to 'have the keeping of the repository' and the west gallery of Gresham College, where the Society met, was 'appointed' for it. At a meeting of the Society two days later, Sir Robert Moray, one of the leading lights of the institution, promised a piece of copper ore from Sweden and all Fellows were 'desired to bring in ores of several kinds, to be put in their repository', while a week later another prominent member, John Wilkins, Dean of Ripon, swelled the collection by a donation of seventeen objects: these are listed in full in the minutes and they ranged from burning glasses and a wind-gun to an ostrich egg and 'a strange bone, with a rib in the middle'.[6] Indeed, it may well be that it is at this point in late October to early November 1663 that the repository should be seen as first being established.

Further developments occurred early in 1664, when the Society seems to have given serious thought to how best to build up a collection of specimens appropriate to its enterprise. At a meeting on 9 March that year it was ordered that two medical Fellows of the Society, Christopher Merrett and Walter Charleton, should 'consider and make a catalogue of what is most desirable of all sorts of animals for the repository of the Society, both exotic and domestic; and withal, to give directions, how to prepare them as to their skins, when dead'. In addition, four other Fellows, Thomas Povey, Daniel Colwall, Peter Balle and Abraham Hill, were requested 'to take care of sending into foreign parts for such animals, as they should be directed, from time to time, by Dr Merret and Dr Charleton' — that Povey was a civil servant and Colwall and Hill both London citizens may suggest that they

4 Birch, *History*, i, 23, 85; Evelyn, *Diary*, iii, 334.

5 Birch, *History*, i, 316. For 'Museum', see, e.g., ibid., iv, 171.

6 Ibid., i, 316, 321, 324. Cf. also ibid., i, 322.

were thought to have suitable contacts for this — while Charleton went on to suggest that the Society should obtain the carcasses of any of the exotic birds in the King's aviary which might die. In addition, Merrett moved 'that in the first place notice should be taken, and a collection made, of all the rare productions of England, as to beasts, birds, fishes, vegetables, minerals, &c.', and he followed this up on 30 March by proposing someone to act not only as an assistant to the Society's 'operator' but also as 'a collector of the curiosities of England, as to plants, birds, and fish'.[7] No appointment was to occur at this stage, but these references are significant, as they suggest that what at least some Fellows had in mind at this stage was a fairly comprehensive collection, of domestic as well as exotic phenomena, and arguably something rather different from the somewhat haphazard collection of curiosities that the Society had acquired so far.

Then, on 10 August 1664, Thomas Povey, who had been nominated Chairman of the Committee for Correspondence that the Society had set up earlier that year, 'was requested to procure for the society a collection of all sorts of curious woods, minerals, and petrified substances; which he promised to do'; he also promised to bring in 'shells found in mines with minerals in them' from his brother, who had recently returned from Jamaica.[8] It was evidently because the Correspondence Committee was mainly concerned with the compilation of enquiries concerning exotic places that this topic seemed to fall within its remit, and one might feel that there was some potential conflict between this and the concern for domestic items evidenced earlier in the year. The minutes of the first meeting of this committee, on the 19th of that month, record that 'It was recommended to Mr Povey, to treat with Mr Hubbard about his Collection of Curiosities'. Mr Hubbard's collection was evidently a large and celebrated cabinet of rarities, that of Robert Hubert, alias Forges, which was then on display in London, and about which more will be said shortly, since the Society was indeed to acquire it in 1666. Moreover it may well be that Povey was asked to make this approach because he himself was among Hubert's benefactors, who had gratefully acknowledged in print the generosity of 'that Honourable Mr. *Povey*, one of the Royal society of Philosophers, and Treasurer to his Royal Highnesse the Duke of *York*'.[9]

Little is then heard of the repository for the rest of 1664 and the whole of 1665, in the latter year not least because of the intermission of the Society's activities imposed by the Great Plague in the latter part of 1665 and the beginning of 1666, as members prudently retired to the country. Early in

[7] Ibid., i, 392, 393, 403. Cf. also 396. The person whose services Merrett suggested might have been Thomas Willisel, who instead helped Merrett prepare his *Pinax Rerum Naturalium Britannicarum* (1666): see C. E. Raven, *John Ray, Naturalist* (2nd ed., Cambridge, 1950), p. 77.

[8] Birch, *History*, i, 458. For the committee and its minutes see above, 'An Experiment in Corporate Enterprise'.

[9] *A Catalogue of part of those Rarities Collected in thirty years time with a great deal of Pains and Industry, by one of his Majesties sworn Servants, R.H. alias Forges, Gentleman* (London, n.d.), p. 13. Cf. ibid., p. 11.

1666, however, when the Society reconvened, the repository was high on the agenda. As the Secretary, Henry Oldenburg, reported to Robert Boyle in a letter of 27 January, 'those of the Society, that are now in London, doe endeavour to gett a good Collection of Naturall and Artificiall Curiosities for the Societies repository; and they hope, to make shortly an acquest of a very good stock of that kind, which will looke as something towards a foundation, and will invite generous men to increase it from time to time'. On 3 February Robert Hooke comparably reported to Boyle how 'I am now making a collection of natural rarities, and hope, within a short time, to get as good as any have been yet made in the world, through the bounty of some of the noble-minded persons of the Royal Society'.[10]

Sure enough, this was the first topic to be considered by the Council when its meetings reconvened on 21 February. On that occasion, it was decided that the sum of £100 which has been given to the Society by its Treasurer, Daniel Colwall, partly at this stage and partly in December 1663, should be used 'to pay for the collection of rarities formerly belonging to Mr Hubbard'. Hence, through this acquisition, the Society's collection suddenly received an artificial boost which was retrospectively regarded practically as a foundation: Colwall was celebrated as the 'Founder' of the museum in Grew's catalogue, to which his engraved portrait is prefixed.[11] By 21 March, the collection was evidently installed: Hooke told Boyle how 'Our collection of rarities at *Gresham* college is now very well worth your perusal, and I hope to increase it every day', while at a meeting of the Society on that date, a committee of ten Fellows was 'appointed to take care of the well ordering, preserving, and increasing the stock of the said repository'.[12]

That this should have occurred when it did was not coincidental, for this 'foundation' was a deliberate response to the problems which resulted from the eight-month break in activities imposed by the Plague, in the course of which the Society's initial momentum had been lost. Clearly the need was felt for a gesture which would assist in visibly galvanizing the Society and showing its vitality, and we learn from a further letter from Oldenburg to Boyle that various facilities were planned at this time, 'as the Collecting a Repository, the setting up a Chymicall Laboratory, a Mechanicall operatory, an Astronomicall Observatory, and an Optick Chamber', though he complained how 'the paucity of the Undertakers is such, that it must needs stick, unlesse more come in, and putt their shoulders to the work'.[13]

10 Oldenburg to Boyle, 27 Jan. 1666, *Oldenburg*, iii, 32; Hooke to Boyle, 3 Feb. 1666, in Boyle, *Works*, vi, 505.

11 Birch, *History*, ii, 64. For the earlier gift, see ibid., i, 337, and, for Colwall's portrait, Grew, *Musaeum*, frontispiece. It is perhaps worth pointing out here that it is apparently due to pure misunderstanding that James Granger in his *Biographical History of England* (2nd ed., London, 1775), iii, 402 – 3, claims that Colwall was himself the collector of the museum that Grew was to catalogue.

12 Birch, *History*, ii, 73; Hooke to Boyle, 21 March 1666, Boyle, *Works*, vi, 506.

13 Oldenburg to Boyle, 24 Feb. 1666, *Oldenburg*, iii, 45. On the 'operatory', see also Hooke to Boyle, 13 Feb. 1666, Boyle, *Works*, vi, 505.

Of these possibilities it was the repository which — thanks to Colwall's generosity — came to fruition, and the significance of the establishment of the greatly enlarged museum for the fortunes of the Royal Society itself is made clear by various reports of the purchase in the correspondence of the Society's members. John Evelyn, for instance, told his sister-in-law: 'We are meeting afresh at Gresham College: and have purchased for us, since these days of separation, the fullest, and certainly noblest collection of natural raritys of all kinds that is this day in Europe to be seen: Tell Mr Bohun [the family tutor], The Royall Society is not at an end, florit floreat'. Oldenburg comparably referred to the purchase of this 'very handsome Collection of Naturall things' as something 'which may be some part of an Establishment', thus alluding to the need for an institutional base of facilities for the Society which much preoccupied its organizers at this time.[14]

Various reasons may be suggested as to why it was the repository which materialized, rather than the various other facilities which Oldenburg told Boyle were being considered in February 1666. It may have had something to do with the availability of Hubert's collection at the time, as we shall see, while the Society may rightly have sensed that they were getting a bargain. £100 seems extraordinarily cheap for the thousands of items in Hubert's collection, considering that John Evelyn was to value the not dissimilar collection of William Courten at £8,000 only a few years later (it may be, however, that, when it came to a sale, complete collections were less valuable than their component parts: it is interesting that the University of Oxford could have had the rarities of the Bristol collector, William Cole, 'scandalously cheap' for about £60 at the end of the century).[15] In addition, the Society's organizers may have seen a collection of rarities as a higher priority than the other facilities whose installation was also considered, in the context of the appetite for Baconian collecting and recording which bulked so large in the Society's programme at this stage.[16]

They may also have seen the collection as a likely source of public acclaim for the Society, and in this they were justified by the high opinions of it immediately expressed. Evelyn's and Hooke's views have already been quoted, while perhaps most striking of all is the estimate of Walter Charleton, the medical Fellow whose role in the early planning of the Society's collection has already been noted. For in his *Onomasticon Zoicon* of

[14] Evelyn to Mrs G. Evelyn, 29 Jan. 1666, quoted from a nineteenth-century bookseller's catalogue which I have not been able to locate in L. G. Sharp, 'Sir William Petty and Some Aspects of Seventeenth-century Natural Philosophy' (Oxford D.Phil. thesis, 1977), p. 256, n. 4; Oldenburg to Boyle, 24 Feb. 1666, *Oldenburg*, iii, 45.

[15] Evelyn, *Diary*, iv, 532; Anthony Turner, 'A Forgotten Naturalist of the Seventeenth Century: William Cole of Bristol and his Collections', *Archives of Natural History*, 11 (1982), 35.

[16] See above, pp. 36 – 7, and 'An Experiment in Corporate Enterprise'. See also P. B. Wood, 'Methodology and Apologetics: Thomas Sprat's *History of the Royal Society*', *British Journal for the History of Science*, 13 (1980), 1 – 26.

1668, Charleton not only referred to a number of specimens in the collection but also gave it the following glowing description in his preface : '*Hic* enim tantus ostenditur Rerum Naturalium cumulus, ut vix dies sufficere possit ad immensam earum varietatem vel perfunctoriè visendum; ad singulatim diagnoscendum, vix annus'.[17]

Since the Society's purchase in spring 1666 undoubtedly swamped the scattering of objects that it already possessed, we should turn at this point to consider what is known of the ready-made collection that it thus acquired. The cabinet of Robert Hubert, alias Forges, had been on display 'at the place formerly called the *Musick House*, near the West end of St Paul's Church': in fact, this music house which Hubert had run in London House Yard, adjacent to St Paul's, was the first such enterprise in London, and the curiosities may initially have been a side attraction. By the early years of the Restoration, however, we are told that the collection was attended by a 'great Concourse of people', and passing references in letters — such as one to Oldenburg from the virtuoso, William Balle — suggest that it was quite well-known.[18]

Indeed, after the famous cabinet of the Tradescants which was to form the basis of the Ashmolean Museum, Hubert's was perhaps the most celebrated collection of curiosities in England at this time, and it was the only one apart from the Tradescants' to be dignified by a printed catalogue: of this various recensions were issued, reaching a climax with a 76-page version published in 1665.[19] Hubert is an elusive but intriguing figure, possibly of immigrant stock.[20] From the printed catalogue of his collection it is apparent that, like

17 Walter Charleton, *Onomasticon Zoicon* (London, 1668), sig. a3v, and ibid, pp. 84, 96, 112, 113, 114, 115, 116, 186, 246 – 7, 290.

18 See *A Catalogue of Many Natural Rarities, with Great Industry, Cost, and thirty Years Travel in Foraign Countries, Collected by Robert Hubert alias Forges, Gent.* (London, 1665), title-page; Sir John Hawkins, *A General History of the Science and Practice of Music* (London, 1776), iv, 378 – 9; J. R. Magrath, ed., *The Flemings at Oxford, Vol. 1, 1650 – 1680*, (Oxford Historical Society, 44, 1904), 62, n. 3; Hubert, n.d. (n. 9), p. 27; Balle to Oldenburg, 14 April 1666, *Oldenburg*, iii, 90.

19 Careful collation of the different versions suggests that the earliest recension is the undated catalogue of part of the collection referred to in n. 9, of which a copy survives in the Bodleian Library, Oxford (MS Ashmole 967): an undated facsimile of this was issued in 1922 with a prefatory note signed 'D.M.' [David Murray?] which explains that it was reprinted by Professor John Ferguson for distribution to his friends. In addition, there are two dated versions of the catalogue: that dated 1665 has been cited in n. 18; that dated 1664 has an almost identical title-page except that, in describing where the collection was on display, it omits 'formerly' (probably accidentally: cf. p. 27 of the undated catalogue), and adds 'at the *MITER*' after '*Musick House*'. Of this 1664 version there are two issues, of which the later has an addendum describing new acquisitions (a copy of this will be found in the British Library, shelfmark 957.e.13); the content of the addendum is then incorporated in the text of the final, 1665, recension, to which is appended a catalogue of the rarities on show at the university garden in Leiden.

20 R. E. G. and E. F. Kirk, *Returns of Aliens in London* (Huguenot Society, 1900 – 8), index s.v. 'Hubbard'.

Plate 5 (above). A Grosbeak or Hawfinch, Coccothraustes vulgaris, *from Walter Charleton's* Onomasticon Zoicon *(London, 1668), p. 85, drawn from the Royal Society's specimen. For the gift of this specimen to the Society, see Birch,* History, *ii, 139. See also Grew,* Musaeum, *p. 61, and Charleton, op. cit., p. 84. Reproduced by permission of the British Library.*

Plate 6 (opposite). 'Guaperva cauda forcipata, pinnis maculosis e M[usei] R[egalis] S[ocietatis]', a plate drawn from a specimen in the Royal Society's repository for Francis Willughby's De Historia Piscium Libri Quattuor *(Oxford, 1686), plate 122. This was one of fifty plates paid for by Samuel Pepys, then President of the Society (see Birch,* History, *iv, 428). Reproduced from a copy in the British Library, shelf-mark 457.e.10, formerly owned by Sir Joseph Banks, whose faint pencil notes appear at the top of the page. Reproduced by permission of the British Library.*

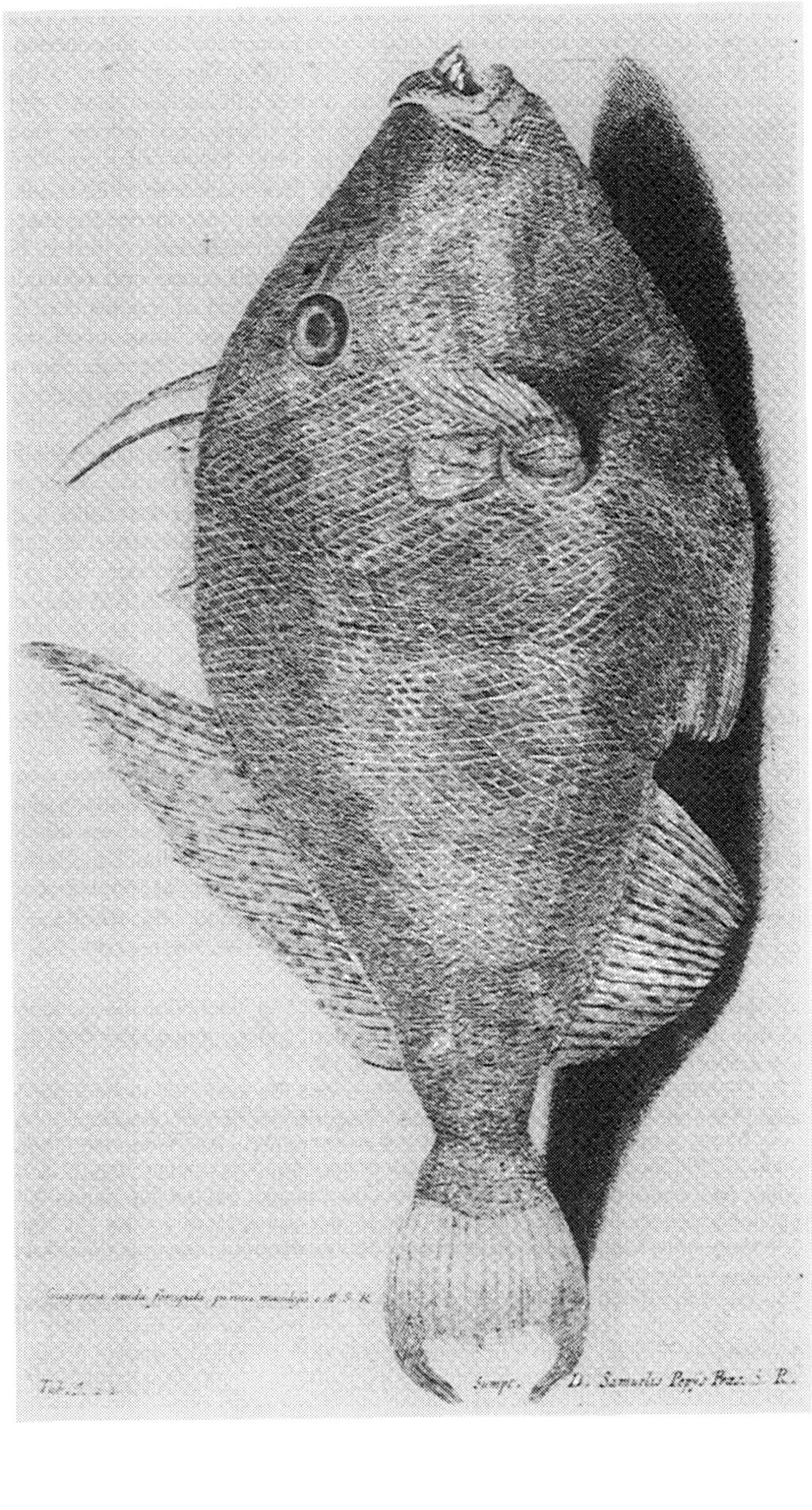
Sumpt. D. Samuelis Pepys Praes. R. S.

such other collectors as the Tradescants, he had travelled widely — he claimed for thirty years — to the West Indies, Constantinople and elsewhere.[21] He was also sufficiently well-connected to have been given at least one rarity by King Charles I, and it may have been because of these courtly connections — in the 1660s he was to call himself 'Gent. and sworn Servant to His Majesty' — that when the Republican regime reached its zenith he took his rarities abroad and in 1651 exhibited them at Leipzig, Hamburg and probably elsewhere, where they made a marked impression.[22] How long he stayed abroad is unknown, but, in contrast to the English flavour of the Tradescants' list of benefactors, those who appear in Hubert's catalogue include a disproportionate number of rulers and officials in the German-speaking countries — such as the Electors of Saxony and Cologne or the Emperor Frederick III — perhaps suggesting a prolonged stay there. Other provenances are also cosmopolitan, spread through Europe and evidently bearing out his claim that his rarities were 'collected in many foreign courts during his exile'.[23]

At the Restoration, Hubert returned to England. At first, he brought over only part of his collection, which was visited by Charles II, who apparently promised Hubert the post of Gentleman Usher on this occasion, 'but he lost it by his absence, when sent into divers parts of Germany to collect the remainder of his rarities'. In 1662, on his return, Hubert is to be found petitioning the King for the place of 'Page of the back stairs or Groom of the Privy Chamber to the Queen'. His allusion to the Queen as 'one of the choicest rarities of her sex in this our age', however, evidently backfired, and his application proved unsuccessful.[24] Instead, he probably devoted himself entirely to the display of his rarities, as the details given in the published catalogues suggest.

What is evidently the earliest published catalogue explained how, apart from the basic tour of the collection, 'those that are more curious, and will be at some more Charge' could come for additional sessions: on Monday and Thurday he showed sea plants, corals and shells; on Tuesday and Friday, 'things of the Land' (metals, minerals, crystals, 'things of strange nature and operation'); and on Wednesday and Saturday, 'things of the Sea, Land and Ayre', 'strange Bones', teeth, eggs and small creatures. He also

21 Hubert, 1665 (n. 18), title-page, pp. 13, 25, 37, 63. On the travels of other collectors, see Arthur MacGregor, ed., *Tradescant's Rarities* (Oxford, 1983); Impey and MacGregor, *Origins of Museums* (n. 2), ch. 18.

22 Hubert, 1664 (n. 19), p. 41 (this item does not reappear in the 1665 version); Hubert, 1665 (n. 18), title-page; P. J. Sachs, *Gammarologia* (Wratislaw, 1665), p. 53; J. D. Major, *Dissertatio Epistolica de Cancris et Serpentibus Petrefactis* (Jena, 1664), pp. 63, 85; *See-Farth nach der Neuen Welt ohne Schiff- und Segel* (Hamburg, 1683), p. 109; Murray, *Museums* (n. 2), i, 127. In his *Dissertatio* (p. 63), Major claims that Forges published a catalogue in German, but I have not located a copy; he also says of Forges' collection 'quod ex Britanniae Regis (durantibus intestinis cladibus) thesauris depromptum ferebatur'.

23 Hubert, 1665 (n. 18), pp. 69 – 71. Cf. John Tradescant, *Musaeum Tradescantianum* (London, 1656), pp. 179f.; *Calendar of State Papers Domestic 1661 – 2*, p. 390.

24 Loc.cit.

promised that 'if any Nobleman or person of quality be desirous to see alone with their Families and Friends, all, or most of them [the rarities]: The owner of them will endeavour to give satisfaction for half a day together' by keeping away the crowds, if given a day's notice. He also offered tours in three or four languages for foreign ambassadors.[25]

At least by 1664 Hubert seems to have been looking for a purchaser for his collection, as is shown by a passage in that year's recension of his catalogue in which he wrote of his miscellaneous, undescribed curiosities: 'if the owner of this collection of Rarities does sell them to any Noble minded party, he then, God willing, will write at large a more ample declaration to the expressing of each thing in particular, to honour that vertuous person that shall buy them'.[26] It may well have been the hope of a sale that stimulated him to publish an inventory of his collection, though the addendum to the 1664 catalogue and the extended version published in 1665 show that he was still adding to it at this point: indeed, some of his most spectacular acquisitions — such as 'an extraordinary great *Cameleon*, about 27 or 28 inches long' — reached the cabinet now. He may also have been selling individual items, like other collectors later who also acted as dealers, such as William Courten, since a few items which appeared in earlier recensions of the catalogue do not recur in the last.[27] Be that as it may, however, the fact that he issued these catalogues means that we not only know more about Hubert's collection than most at the time, but also the values that underlay it.

As his published remarks display, Hubert almost epitomizes those characteristics of the virtuoso movement which contemporary commentators noted and criticized, its cult of 'Rarity *for Rareness-sake*' and its deliberate pursuit of 'all things opposite to the vulgar sort'.[28] Hubert's was a classic virtuoso cabinet, trading on the rare and exotic nature of its content and on the social cachet of his august benefactors and patrons. As his earliest catalogue put it, 'You may see every afternoon, that which hath been seen by those that are Admirours of Gods Works in Nature, with other things that hath been seen by Emperours, Empresses, Kings and Queens, and many other Soveraign Princes'. Later there is reference to 'Several sorts of Rare shells of great Princes: as of the Emperor, the Queen of *Sweeds*, and of other Kings and Soveraigns', while Hubert elsewhere assured his clients how they were inspecting 'a Rarity that was esteemed very much by divers persons beyond Seas'.[29]

Mainly, however, the stress was on the marvellous quality of the items. 'First of Fishes Heads, the least of these Heads are as big as a mans Head', his earliest list opened, while a section of 'Other Rarities' included 'A Gyants

25 Hubert, n.d. (n. 9), pp. 25 – 7.

26 Hubert, 1664 (n. 19), p. 59. Cf. Hubert, 1665 (n. 18), pp. 55, 68.

27 Ibid., pp. 36 – 7. For items that disappear, see, e.g., that said to have been presented by Charles I, above, n. 22; on Courten, see British Library MSS Sloane 3961 – 2.

28 See Houghton, 'The English Virtuoso' (n. 3), pp. 204, 213 and passim.

29 Hubert, n.d. (n. 9), pp. 1, 24; Hubert, 1664 (n. 19), p. 46.

Thighbone 4 foot long, it is an Extraordinary Rarity' or 'A Tamanduaquaca, or the Aunt-bear, it is as big as a Spaniel, and participates of the shape of five or six Creatures; it is a very strange wild Beast in *Brasil*, and hard to be procured'. So often is the strange and wondrous quality of items protested that it becomes almost wearisome. Again and again we return to 'a very strange Rarity', such as 'A *Poyson-fish* of East *India*, it is so venemous that thirteen men in a ship dyed by eating one of them', or 'A great *Crab*, having its shell covered with Oysters that are grown to it, a Rarity, worth considering'. Only occasionally is this varied by biographical detail, as with 'a sprig or large bush of black Feathers', for which the Emperor Mathias was said to have given 2000 Reichs dollars — almost £500 sterling — 'in Rarity and beauty exceeding that of the great Turks or Sultans, which the Master of these Rarities saw at *Constantinople*'.[30]

Moreover, though Hubert's collection included some native items, it is clear that what he valued most highly were exotic objects from far-off places: armadillos (one of them formerly owned by King James I), chameleons, crocodiles, humming birds, saw-fish, turtles, lizards or tropical shells. In this he was again typical, for these were the characteristic desiderata of virtuoso cabinets, prestige accruing from ownership of such far-off things in a way that would not have been the case with others more commonplace. Also typical of such collections was the interest in specimens in which art and nature seemed miraculously to come together. Hubert had a whole section concerning slabs of polished stone which appeared to depict landscapes — 'Natural Landskips in Stone' — as well as items like 'A cross of wood grown by Nature in *Bohemia*, given by Doctor *Moretus*, professor in the Mathematicks in the University of *Prague*', or 'A Cross of wood, growing in the form of St *Andrews* Cross, given by Doctor *Pinker*, one of the privy Council to the Prince Elector of *Sax*[ony]', or even stones shaped like 'the secret parts of a woman' or a sweetmeat.[31]

Within the sequence of catalogues that Hubert put out a slight shift of emphasis may be discerned: the naive stress on oddity and on the social cachet of the objects is at its most overt in the earliest, whereas a degree of sophistication enters into the subsequent recensions, including more elaborate descriptions of the characteristics and habitat of the specimens included. In his last recension of 1665, Hubert even had the idea of stressing the scholarly potential of his cabinet by including a catalogue of the rarities that were on show in the university garden at Leiden, '*to shew the difference of both the Collections*'.[32] Possibly Hubert did this with the intention of angling at the Royal Society, who by this time had expressed interest in the collection, while an even clearer bait was the fact — stressed on the title-pages of his 1664 and 1665 catalogues — that this was a collection exclusively of 'Natural Rarities'. It thus contrasted with the Tradescant collection, for instance, with its heavy showing of ethnographical curiosities, coins and other human artefacts. Quite apart from the prestige accruing

[30] Hubert, n.d. (n. 9), pp. 1, 8 – 9, 12; Hubert, 1665 (n. 18), pp. 13, 16, 23, 60.
[31] Hubert, n.d. (n. 9), pp. 17, 18; Hubert, 1665 (n. 18), pp. 35, 58 – 9, 66 and passim.
[32] Hubert 1665 (n. 18), p. 72.

from its public acclaim and its large size — which offered a ready-made museum overnight — Hubert's collection thus had a particular appeal to a Society devoted to research into natural history.[33]

For all this, however, Hubert's catalogue remained dominated even in its later recensions by the values of the rare and the marvellous: even the descriptive material that was added to these was often of a somewhat sensationalist nature. Hence to us it may seem surprising that it should have seemed desirable for the accumulation that the catalogue chronicled to be transferred to the possession of a body which prided itself on its serious mission to reform knowledge, and still more so that it should have been described with the kind of superlatives already quoted from the correspondence of the Society's activists at the time of its purchase. It is therefore worth pausing here to consider the rationale of this acquisition. To start with, we should beware of underestimating the extent to which those active in the Royal Society shared the values that Hubert exemplified and to which he appealed: Evelyn, for one, certainly falls into this category.[34] More important, it is clear that the same exotic or abnormal objects could be approached from different points of view, as 'wonders' to be rather superficially and mindlessly admired, or as specimens worthy of serious scholarly scrutiny. There was thus much to interest the serious researcher in Hubert's, as in other, private collections of the time, and it is not surprising that such cabinets were extensively consulted by naturalists of the day like John Ray.[35]

On the other hand, it seems to have been felt that the transference of the collection from private to institutional ownership would somehow alter its character and make it more worthwhile: this exemplifies that aspiration to permanence through institutionalization which, as we have seen in the Introduction, was so central to the Royal Society in its early years. When reference was made to the repository and Colwall's benefaction to it in the *Philosophical Transactions* in October 1666, potential donors to the collection were urged on with the assurance that their gifts would there be 'preserved for After-ages, (probably much better and safer, than in their own private Cabinets)', and a background to this may be found in the attitudes of virtuosi like Evelyn and John Aubrey. We have already noted Aubrey's concern to vest his writings and other possessions in a 'publick Repository', while Evelyn, writing to Samuel Pepys in 1689 about collections of books and other rarities, bewails at more than one point the tendency of worthwhile collections to be neglected or dispersed on the death of their founder.[36] Undoubtedly the sense that institutions had a collective life beyond that of individuals seemed to offer a guarantee of security for

33 Ibid., title-page. For the Tradescant rarities, see *Musaeum Tradescantianum* (n. 23), esp. pp. 36f.

34 See Hunter, *Science and Society*, p. 67.

35 Raven, *John Ray* (n. 7), p. 229; Impey and McGregor, *Origins of Museums* (n. 2), pp. 151, 157.

36 *Phil. Trans.*, 1 (1666), 321; see above, Introduction, pp. 6 – 7; Evelyn to Pepys, 12 Aug. 1689, in Evelyn, *Diary and Correspondence*, iii, 294 – 311.

valuable accumulations which might otherwise be dispersed, and it is interesting to note the comparable proclivity of benefactors earlier in the century to vest interesting objects in public institutions like Oxford and Cambridge colleges.[37]

As far as the content of the collection is concerned, there is no reason to doubt that virtuosi like Evelyn were perfectly happy with the general physiognomy of the Hubert collection, merely wanting it to be swelled by acquisitions to become an even bigger and better cabinet of rare and exotic items. For others, however, this was not the case, for some seem to have hoped that the collection which the Royal Society would build on the basis of Hubert's might be qualitatively different from his — in other words, that institutionalization might actually produce a different kind of collection from existing private ones, which would serve a new and important role in the reform of knowledge.

This was implicit in the hope expressed in the *Philosophical Transactions* that the repository 'in progress of Time will be employed for considerable Philosophical and Usefull purposes', while the implication of this was spelt out in other statements which make clear an intention to substitute a more egalitarian and complete view of the world for the élitist and partial one of the virtuosi.[38] Though the basis of the Society's museum was a private cabinet, there is no mistaking the disdainful attitude towards such collections displayed by Thomas Sprat in a passage in his *History of the Royal Society* dealing with transplantations, where he wrote: 'the chief Progress that has hitherto bin made, hath bin rather for the *collection* of *Curiosities* to adorn *Cabinets* and *Gardens*, than for the solidity of *Philosophical Discoveries*'. Robert Hooke comparably complained how 'the use of such a Collection is not for Divertisement, and Wonder, and Gazing, as 'tis for the most part thought and esteemed, and like Pictures for Children to admire and be pleased with, but for the most serious and diligent study of the most able Proficient in Natural Philosophy'.[39]

Now there was an opportunity for a serious collection to play a major role in the promotion of the scholarly ends for which the Society had been inaugurated, and Sprat described 'a General Collection of all the Effects of *Arts*, and the Common, or Monstrous *Works* of *Nature*' as 'one of the Principal Intentions' of the Society 'as soon as they were reduc'd into a *Fix'd Assembly*'. Indeed, this falls into place among the Society's grandiose ambitions in its early years for reformulating knowledge about the natural world, and it is no coincidence that it was in the context of a major effort to promote the collaborative study of natural history at home and abroad that Oldenurg reported to Boyle on the purchase of the Hubert rarities.[40]

The link of the foundation of the repository with these high hopes on the Society's part is made clearest by the juxtaposition of aspirations for the

37 See Michael Hunter, 'The Cabinet Institutionalized', in Impey and MacGregor, *Origins of Museums* (n. 2), pp. 160 – 1.
38 *Phil. Trans.*, 1 (1666), 321.
39 Sprat, *History*, p. 386; Hooke, *Posthumous Works*, p. 338.
40 Sprat, *History*, p. 251; Oldenburg to Boyle, 27 Jan. 1666, *Oldenburg*, iii, 32.

collection with one of the most ambitious projects associated with the Society at this time — the attempt to construct a new, rational language which would accurately and clearly reflect reality. The classic text in this connection, John Wilkins' *An Essay Towards a Real Character, and a Philosophical Language* (1668), was the outcome of work closely associated with the Royal Society, who encouraged Wilkins to persevere with his task in spite of such setbacks as the destruction of the entire first impression of a substantial part of the book in the Fire of London. Moreover, in attempting to improve his work, Wilkins enrolled the help of the naturalists John Ray and Francis Willughby, requesting their 'best Assistance for the regular Enumeration and defining of the Families of Plants and Animals' for the tables that he included, in which concepts and natural phenomena were classified as a basis for the language.[41]

These tables played a central role in Wilkins' project, because he hoped that through them language could at the same time describe and define natural phenomena, thereby serving an important taxonomic purpose as well as a linguistic one. What he aspired to was a universal system of classification which would accurately mirror the natural order, and this broad ambition was as important to him and other language-planners as narrower linguistic ones: Wilkins saw these tables as 'the great foundation of the thing here designed'. Moreover, Wilkins considered this an ongoing task rather than a completed one, as he made clear in dedicating his *Essay* to the Royal Society. For, apologizing for the imperfections of his work, he expressed the hope that the Royal Society might be able to improve on it through its collaborative efforts. Moreover in this connection he singled out 'those Tables that concern the species of *Natural bodies*; which if they were (so far as they are yet known and discovered) distinctly reduced and described, This would very much promote and facilitate the knowledg of Nature, which is one great End of your Institution'.[42]

It is thus significant that when Thomas Sprat referred to the repository in his *History of the Royal Society*, he used the opportunity to bring in a reference to this aspiration to create a new, universal language. Alluding to Robert Hooke and the collection, he wrote: 'This *Repository* he has begun to reduce under its several heads, according to the exact Method and Ranks of all the *Species* of *Nature*, which has been compos'd by Doctor *Wilkins*, and will

[41] Wilkins to Willugby, 20 Oct. 1666, in John Ray, *Philosophical Letters*, ed. William Derham (London, 1718), p. 366. See also Raven, *John Ray* (n. 7), pp. 181 – 3. On Wilkins' scheme and its background see Vivian Salmon, *The Works of Francis Lodwick* (London, 1972), esp. ch. 2; J. R. Knowlson, *Universal Language Schemes in England and France, 1600 – 1800* (Toronto, 1975), ch. 3.; Vivian Salmon, 'John Wilkins's *Essay*: Critics and Continuators', *Historiographia Linguistica*, 1 (1974), 147 – 63; and M. M. Slaughter, *Universal Languages and Scientific Taxonomy in the Seventeenth Century* (Cambridge, 1982).

[42] Wilkins, *An Essay* (London, 1668), sig. a1v and p.1. See also the works cited in n. 41, especially Slaughter.

shortly be publish'd in his *Universal Language* '. Wilkins likewise referred to the repository in the Epistle Dedicatory to his *Essay*, going on from the remarks just quoted about the desirability of a proper taxonomy to add: 'And besides, the ranging of these things into such an order as the Society shall approve, would afford a very good method for your *Repository*, both for the disposal of what you have already, and the supplying of what you want'.[43]

To this end, the idea was that casual donations and even Hubert's cabinet should merely form the basis of a larger, more complete collection of natural things. Hence, though some may have been happy with the repository much as it was, others clearly saw it as needing to be converted into a tool to help achieve this universal, accurate taxonomy of nature. The notion of 'completeness' recurs in remarks about the repository. Hooke wanted 'as full and compleat a Collection of all varieties of Natural Bodies as could be obtain'd', while Sprat proclaimed, with characteristic optimism, that the Society had 'already drawn together into one Room, the greatest part of all the several kinds of things, that are scatter'd throughout the *Universe*'. Wilkins in his Epistle Dedicatory similarly referred to 'the compleating of that Collection, so generously begun of late, by the bounty of *Mr Daniel Collwal*, a worthy Member of this *Society*. And by this means, I should not doubt, but that in a very short space, you would have the most usefull *Repository* in the World'.[44]

Such ideas thus echoed the discussions of the proper content of the Society's repository that had occurred in 1664, when it had clearly been felt that 'all sorts of animals' should be included in it, and comparable attitudes were quite widely shared. Thus the view of the repository of Sir John Hoskins, an active and influential Fellow, was that 'Nothing should bee wanting, not clay, peble &c so to find likenesse and unlikenesse of things upon a suddaine'; he was enthusiastic about the preparation of an inventory so that 'when once tis knowne (to the members at least) what wee have every body will readily give what wee have not and so it will swell vastly'. Later, in 1683, the journalist and tradesman John Houghton 'brought in a bundle of seeds of different sorts, which he collected towards furnishing a particular thesaurus of seeds in the repository', thus displaying a comparable ambition for a comprehensive assemblage of specimens.[45]

Inevitably, this entailed the realization that piecemeal gifts were unlikely to suffice as the basis for such a collection, and already in 1664 the Society had considered appointing someone to add to the collection systematically, though nothing came of this. In 1669, however, the Society did employ the

43 Sprat, *History*, p. 251; Wilkins, *Essay* (n. 42), sig. a1v.

44 Sprat, *History*, p. 251; Hooke, *Posthumous Works*, p. 338; Wilkins, *Essay* (n. 42), a1v.

45 See above, pp. 125 – 6; Hoskins to Aubrey, 25 March 1674, Bodleian Library MS Aubrey 12, fol. 214; Birch, *History*, iv, 200.

botanizer, Thomas Willisel — possibly the man who had been suggested for the job in 1664 — for this purpose. His task was acquire botanical, zoological and mineralogical specimens around the British Isles, and a committee which included Hooke was asked to inform Willisel 'of such natural things, as may be had in England, and were yet wanting in the society's repository'.[46] In addition, an elaborate testimonial for Willisel was drafted, which stressed how 'tis none of the smallest importance, to search after and collect whatever Nature may have stored every respective Contry [sic] with, be they Minerals, Vegetables or Animals', and requested all to whom he showed it to give him free passage and to assist him in his work, 'which so much tends to the discovery of Nature, and the Application of the works thereof to the Benefit of Mankind'.[47]

The aspirations for the repository therefore form part of the Royal Society's broader aim to build up a complete and accurate description of the natural world through the accumulation of particulars, an enterprise which seemed all the more important when it was believed (naively, as it was to turn out) that language could be reformed to reflect this. It was for this reason that the collection could seem as important for its potential for serious savants as it did in its actuality for casual virtuosi, acquiring a potentially crucial role in the Society's corporate plans for the reform of knowledge.

What actually happened? In fact, things did not go according to plan from the start, due in part to a degree of impracticality in the aspirations themselves; in part to the Society's debilitating lack of resources, which stunted so many of its early plans; and in part to the conflicting values of those associated with the collection. It is symptomatic that the employment of Willisel proved shortlived, and instead the Society was dependent on gifts to add to the collection: as we shall see, it was these — together with the inherited physiognomy of the Hubert collection — that gave the repository the character which is apparent from Grew's catalogue of it, which will be analysed shortly and which was at odds with the high hopes of the 1660s.

Even the administration of the collection proved problematic. Contrary to the implication of Charleton's grandiose account of 1668, which has already been quoted, in that year the rarities were actually in Hooke's rooms at Gresham College 'as in a storeroom', as we learn from the Italian visitor

46 See above, p. 126; Birch, *History*, ii, 358, 371, 378 – 9, 395, 396, 398, 425 – 6, 431, 433. The Royal Society's accounts for 1669 – 70 record the payment to Willisel of an allowance totalling £30 for a journey to Scotland and 'to furnish his Journey for collecting Naturall Rarities in England &c.' On Willisel, see Raven, *John Ray* (n. 7), p. 151, and John Aubrey, *The Natural History of Wiltshire*, ed. John Britton (London, 1847), p. 48.

47 D.M. 5. 41, a different and longer version of the certificate than that given in Birch, *History*, ii, 378 – 9, with which it overlaps verbally and for which it is possibly a draft: it is in Oldenburg's hand.

Lorenzo Magalotti.[48] This was due to the Society's lack of resources and vulnerability to circumstances beyond its control. In June 1666 steps had been taken for the newly extended collection to be displayed in the west gallery of Gresham College, where the Society had met since 1660. But the situation changed in the aftermath of the Great Fire, when the college was commandeered by the City and the Society moved to Arundel House in the Strand.[49] Only after the Society's return to Gresham late in 1673 were fresh steps taken to display the rarities. Moreover it seems likely that, in the meantime, some objects disintegrated or were lost: certain items which had appeared in Hubert's catalogues failed to recur in Grew's, and, since these included fragile objects like stuffed humming-birds, this may well have been due to neglect.[50]

From 1675, however, decisive steps were taken to put things right. First, in February that year, Hooke was ordered to remove the Society's repository and library to the north (or short) gallery of the college. Hooke noted in his diary, however, that the Council 'Seemd to Quarrell' about this, and by June it had been decided that the west gallery would be more suitable, the only snag being that the room had first to be vacated by its previous users, the East India Company, which took some months to achieve.[51] Finally, on 3 February 1676, a committee who had been appointed to deal with the matter signed an order that 'the said Curiosities, now in the custody of Mr Robert Hook, be on Monday next removed into the said Galery', and by 6 March they could report to the Council that they had had the rarities moved to the gallery 'and there ranged them in order', so that all that was required was to have them catalogued and to appoint a keeper for them.[52]

Cataloguing too, however, had proved problematic from the outset. A committee had been appointed to deal with 'ordering' the rarities in spring 1666, but the task of compiling a catalogue was given to Robert Hooke, in addition to all the work that the Society expected him to do in preparing

[48] W. E. K. Middleton, ed., *Lorenzo Magalotti at the Court of Charles II: his 'Relazione d'Inghilterra' of 1668* (Waterloo, Ontario, 1980), p. 140. It was presumably in connection with this that in 1668 Hooke was paid 14. 10s 'for fitting the place in Gresham-college for the society's repository': Birch, *History*, ii, 300; cf. ibid., iii, 310 for a reference to 'the rooms where they had hitherto been'. D.M. 5. 20 has an undated note concerning Hooke's expenditure of £9 for 'making a window to the Room where the Repository was formerly kept'; the Royal Society's accounts for 1669 record the payment of £6 for four chests of drawers for the repository (and £6.3.0 to Hooke 'for work done at Gresham College').

[49] Birch, *History*, ii, 96, 113 – 4.

[50] Compare Grew, *Musaeum*, pp. 61-2, with Hubert, 1665 (n. 18), pp. 9 – 10: whereas Hubert had 'several' humming birds, Grew mentions only two, one of which lacked its head.

[51] Birch, *History*, iii, 191, 224, 227, 228, 242; Hooke, *Diary*, p. 149.

[52] Royal Society Miscellaneous Manuscripts 16.39 (minutes of the meeting on 3 Feb. 1676: this refers to an earlier Council order setting the committee up which apparently does not survive); Birch, *History*, iii, 310. See also Royal Society accounts, 1676 – 7, for bills for work done in the repository.

experiments for its weekly meetings, and his responsibilities outside the Society. Sprat's reference to Hooke's arrangement of the repository according to Wilkins' classification of nature has already been noted. In addition, in 1668 Hooke was urged to finish 'the printed list of the collection bestowed by Mr Colwall on the society' with a view to its being inserted in a new edition of Sprat's *History* which was projected at that time: nothing, however, seems to have materialized.[53] Even at a more mundane level, the documentation of the collection got off to a bad start. In 1663, Hooke had been told that, as keeper of the repository, he 'should always affix some note to the things in it, by which it might be known what they are, and by whom they were presented', while in the same year it was suggested that the names of benefactors to the Society should be kept in a register, a promise which was reiterated in connection with the repository in 1666 and again in 1674.[54] But no such record appears to have been kept: Grew was asked to make a list of benefactors in 1682, while the fact that in the early eighteenth century such a register had to be compiled retrospectively from the Society's minutes suggests that till then no such record ever existed.[55]

Efforts to produce a catalogue redoubled in the 1670s, partly in conjunction with the moves to place the objects on display at that time, but partly also because the publication of a systematic, learned catalogue remained crucial to the museum's function in the context of the ambitions of Wilkins and others. We learn from letters from John Aubrey to his friends in 1674 – 5 that Aubrey was then in the process of writing a catalogue 'according to that incomparable Method of Dr Wilkins Philos[ophical] Grammar', which he had presumably completed by July 1675, when he apologized to Sir William Petty for not having had time to write it out fairly.[56] Meanwhile, Hooke and his assistant, Henry Hunt, seem also to have been at work. On 25 February 1675 Hooke had been ordered 'to perfect the catalogue' of both the repository and the Society's library, and on 9 March that year Hooke refers in his diary to the fact that he 'set Harry about Repository', noting that Hunt had the *Museum Wormianum* and other books.[57] Nothing has survived of Aubrey's, Hooke's or Hunt's work, but it is perhaps worth noting that Grew was to produce a lengthy catalogue with

53 See above, pp. 127 and 137; Birch, *History*, ii, 266. For a reference to stones and minerals presented by Moray being 'reduced into order' by Hooke, see ibid., ii, 108.
54 Ibid., i, 322, 344; iii, 158; *Phil. Trans.*, 1 (1666), 321.
55 Birch, *History*, iv, 171; D.M. 5.85f., 101f.; Royal Society MS 416.
56 Aubrey to Anthony Wood, 31 March 1674, Bodleian Library MS Wood F 39, fol. 261v; Aubrey to Petty, 17 July 1675, Bowood House Petty MSS, Letters, vol. 6, 2nd series, no. 21. The implication is that Aubrey was answerable to Petty concerning the catalogue, which may suggest a link with the reform effort in the Society at this time, with which Petty was closely connected: Hunter, *Royal Society*, pp. 37 – 8.
57 Birch, *History*, iii, 191; Hooke, *Diary*, p. 152 (9 March 1675). Hooke also has a puzzling reference to talk of a 'salary for repository' from the London merchant, Sir John Lawrence (on whom see Hunter, *Royal Society*, catalogue entry 301).

surprising speed, as we shall see, and it is conceivable that he utilized notes on the collection that these men had already made.

Early in 1676, the committee which had just arranged for the museum to be installed in the west gallery at Gresham College turned its mind to the cataloguing and care of the collection. From overlapping texts whose status and mutual relationship is not entirely clear, it seems that they first ordered that Grew and another fellow, Abraham Hill, 'be desired to make a Catalogue of all the said Rareties' and that Richard Shortgrave, the Society's operator, should be their keeper: then, however, both tasks seem to have been given to Grew, the doctor and botanist who had been employed by the Royal Society as a researcher in the early 1670s (Shortgrave was, in fact, to die in 1676).[58] At a Council meeting on 18 July 1678 this was confirmed — possibly partly as a means of dispelling tension between Hooke and Grew in the aftermath of the death of Oldenburg — and it was 'Ordered, That Dr *Grew* be desired, at his leasure, to Make a Catalogue and Description of the Rarities belonging to this Society'.[59] Grew must have worked on this over the next few months, reading parts of it at meetings of the Society in April and May 1679, when it was stated that they were 'fitted for the press'. Then, on 5 July that year, an imprimatur was given to Grew's book at a further Council meeting which is otherwise unrecorded. Subsequently, a proposal for subscriptions to the volume was brought to a meeting in February 1680, and in 1681 it was published by subscription.[60] Finally and belatedly, therefore, a catalogue had been achieved, and it is to this book, and the collection as described in it, that I now wish to turn.

In the first place, it is significant how far Grew's aspirations for the repository in his preface echoed the broader taxonomic aims of authors like

[58] D.M. 5.71 (n.d.), an order to the same people as those listed in Birch, *History*, iii, 310 – 11, except that the MS version lacks the name of Milles, and it gives the quorum, which Birch leaves blank, as two. However, the instructions given differ. The committee was ordered to take into its custody the Society's books, rarities and other possessions; Grew and Hill were to catalogue the rarities, and the committee to make an inventory of the Society's other possessions; and Shortgrave was to be the keeper. It also adds that this committee was to take responsibility for supervising all bills concerning experiments, etc. In the minutes of the committee meeting on 3 Feb. 1676 in Miscellaneous Manuscripts 16.39, Grew had already been 'ordered to have the keeping of the same'.

[59] See Grew, *Musaeum*, unsigned last page of prelims: there is no reference to a Council meeting on that date in Birch (but cf. Hooke, *Diary*, p. 367, where Hooke records how Grew requested 'An Establishment' on that date). See also below, 'Early Problems in Professionalizing Scientific Research', pp. 275 – 6.

[60] Birch, *History*, iii, 480, 481, 486, iv, 16. The imprimatur to the book is dated 5 July 1679 (there is again no corresponding entry in Birch). For a copy of the proposal for the book, see British Library Harleian MS 5946, fols. 162-3. See also below, 'Early Problems', n. 72.

Plate 7. Tab. I from Grew's Musaeum Regalis Societatis *(1681), illustrating various specimens in the Society's collection. Reproduced by permission of the British Library.*

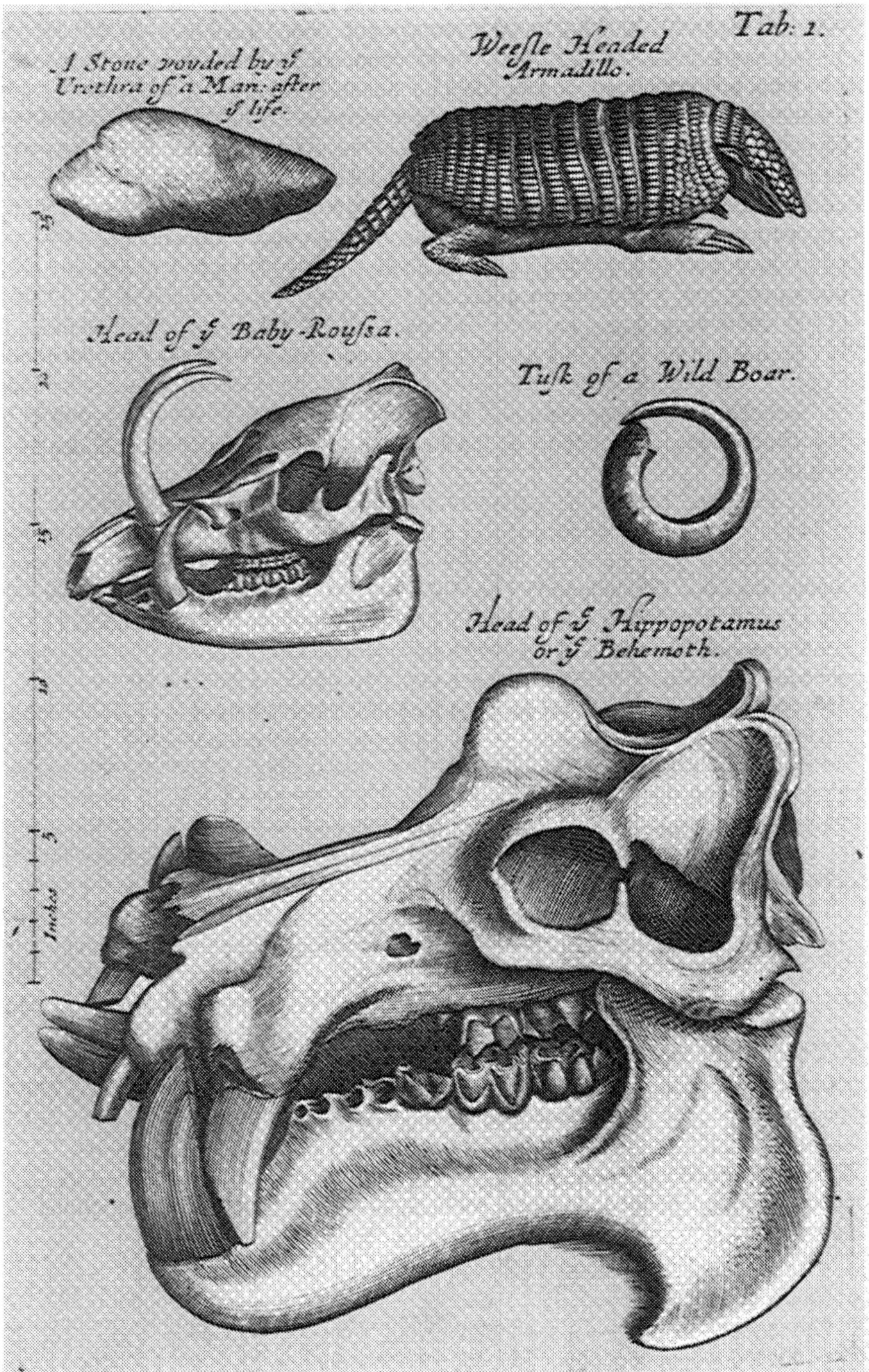
Tab: 1.
A Stone voyded by ye Urethra of a Man: after ye life.
Weesle Headed Armadillo.
Head of ye Baby-Roussa.
Tusk of a Wild Boar.
Head of ye Hippopotamus or ye Behemoth.
Inches

Wilkins (whose protégé Grew, in fact, was). Grew was optimistic that if sufficient data about natural things and their functions and characteristics was collected, 'a better History of them might be written in five years, than hath hitherto been done in two Thousand', and this was directly linked to a concern for taxonomy. 'It were certainly a Thing both in it self Desirable, and of much Consequence;' he wrote, 'To have such an Inventory of Nature, wherein, as on the one hand, nothing should be Wanting; so nothing Repeated or Confounded, on the other. For which, there is no way without a cleer and full Description of Things'. Significantly, he stressed that 'it were also very proper, That not only Things strange and rare, but the most known and common amongst us, were thus describ'd'. In addition, he echoed language-planners like Wilkins in his belief 'that the Names of Things should be always taken from something more observedly declarative of their Form, or Nature. The doing of which, would much facilitate and Improve the Knowledge of them many ways. For so, every Name were a short Definition. Where as if Words are confus'd, little else can be distinctly learn'd'.[61]

Perhaps the most obvious characteristic of Grew's catalogue is its erudition: its long descriptions of objects contrast with the brief entries in Hubert's catalogue or the *Musaeum Tradescantianum*, as he brought wide reading in the relevant scientific literature to bear on the objects in his custody. The Royal Society actually bought a number of standard works for Grew's use in his task: the *Museum Wormianum*, Ludovico Moscardo's *Note, ovvere Memorie del suo Museo* (1656), Lorenzo Legati's *Museo Cospiano* (1677), Giorgio de Sepi's 1678 catalogue of the collection bequeathed by Athanasius Kircher to the Jesuit College at Rome, and a volume of John Jonston's *Historiae Naturalis . . . Libri* (1650 – 3). Grew used these, together with older authors like Aldrovandi, Moffett and the Bauhins, and the most up-to-date contributions on particular topics — Lower, Kerckring and Swammerdam on anatomy, Willughby on birds, Lister on shells, and articles in *Philosophical Transactions* throughout.[62]

Equally significant is the detail into which Grew went in his lengthy and painstaking descriptions of specimens, which often involved measurements, and he justified the care he took to potential critics on the grounds that 'perhaps they have not so well considered the necessity hereof, for the cleer and evident distinction of the several Kinds and Species, in so great a variety of Things known in the World'. As a result, he was frequently able to criticize such predecessors as Jonston, Aldrovandi and Worm for inaccurate descriptions and misidentifications, thereby making a contribution to the accumulation of reliable information which — like such contemporaries as Willughby and Ray — he saw as playing a crucial role in the advancement of natural history. Indeed, in some ways this was the most valuable feature of his book.[63]

[61] Grew, *Musaeum*, preface.
[62] Ibid., passim; Royal Society accounts, 1679 – 81. Cf. Birch, *History*, iii, 450. On these museum catalogues, see Murray, *Museums* (n. 2), i, 84, 89, 104 – 7.
[63] Grew, *Musaeum*, preface and passim.

It was in certain of his lengthy descriptive passages that Grew's task arguably came closest to his existing research interests. Grew had made his name with the studies of plant physiology that were published separately in the 1670s and then collected together as *The Anatomy of Plants* in 1682. At the time when he undertook his catalogue of the Royal Society rarities, however, he was transferring his attention to animals. As an appendix to his *Musaeum*, he published a pioneering 'Comparative Anatomy of Stomachs and Guts Begun', a series of lectures given to the Royal Society in 1677, while in various displays at the Royal Society in the late 1670s he showed various parts of animals — skulls, tongues, gullets and the like — and commented on their functions. Moreover his sentiments on these occasions overlapped with a passage in the preface to the *Musaeum*, in which he justified description partly on the grounds that it stimulated new questions about the workings of animal organs and the like, thereby illustrating 'the Providence of Nature'.[64]

In his preface he also expressed his wish not to meddle with 'Mystick, Mythologick, or Hieroglyphick matters . . . as some have done: I thought it much more proper, To remarque some of the Uses and Reasons of Things'. Not surprisingly, he was hostile to the stress on the strange and inexplicable typical of collectors like Hubert and so well exemplified by the earlier catalogue. Quite apart from recognizing Hubert's giant's thigh-bone as the bone of an elephant, he was sarcastically dismissive of phenomena that had preoccupied virtuosi such as the strange power of the *Echineus remora*. He also made short work of items like the sticks unnnaturally formed in such shapes as that of a St Andrew's Cross, straightforwardly rationalizing them with the words: "Tis probable, That these were bound together (as may be any other) when they were young, and with the Barque pared off, where contiguous; and so, by a kind of ingrafting, became coalescent'.[65] Hubert's curiosity about coconuts shaped like a mouth or a fish was quietly ignored, while Grew's hostility to the cult of rarity is further seen in his down-to-earth remarks on medicines, and how 'the greatest Rarity, if once experienced to be of good use, will soon become common'.[66]

There is also a contrast between Grew and Hubert concerning classification. Hubert had included in his catalogue sections such as '*Things of strange operation*';[67] Grew, on the other hand, was attempting a wholly natural arrangement, like other cataloguers of the time. Contrary to what

[64] Birch, *History*, iii, 333, 342, 475 – 6, 482 – 5 (the original manuscript of this survives as City of London Guildhall Library MS 1757, item 5); Grew, *Musaeum*, preface. See also below, 'Early Problems', pp. 275 – 6 and passim; Jeanne Bolam, 'The Botanical Works of Nehemiah Grew', *Notes and Records*, 27 (1973), 219 – 31.

[65] Grew, *Musaeum*, preface (for a comparable view, see John Ray, ed., *The Ornithology of Francis Willughby* (London, 1678), sig. A4) and pp. 32, 104 – 6, 184, 197 – 200. Cf. Hubert 1665 (n. 18), pp. 1, 24, 43.

[66] Grew, *Musaeum*, preface; Hubert, 1665 (n. 18), p. 46.

[67] Hubert, 1665, pp. 66 – 8.

has sometimes been claimed, he did not use the classification enshrined in Wilkins' *Essay*.[68] In view of the dissatisfaction which John Ray was by this time expressing about this, it would have been surprising if he had, though it is perhaps more surprising that he did not use Ray's classification either.[69] For instance, although Ray's edition of Willughby's *Ornithology* was available, Grew's arrangement of birds in the collection did not exactly follow that, his classification representing an eclectic blend of ideas derived from various treatises that he had studied.

On the whole, Grew's book was an inventory, which meant that his criteria of arrangement are rarely made explicit. But in one section he did feel the need to illustrate the taxonomy that he had employed, and this concerned the 600-odd shells in the collection. At the beginning of his section on this subject, he noted: 'According to the best Method I can at present think of, I shall here place them. And that it may be better judged, how far it is natural, or not, I shall afterwards digest them into Schemes'. The 'Schemes' take the form of seven analytical tables in which different types of shells are divided and subdivided according to their structural and decorative characteristics, and they fall into place among contemporary attempts to attain a sound classification of nature, illustrating clearly the relationship of Grew's book to the taxonomic efforts of scientists of the day.[70]

For all its merits, however, Grew's catalogue had its limitations, and these seem to a considerable extent to have been imposed on him by the collection for which he had been given responsibility. For one thing, the fact that he was cataloguing a finite collection cramped his style in terms of expounding classificatory principles. As he put it concerning the shells in the museum: 'The Reduction of all which to the Order of Nature, whoever shall go about [it], will find to be no little Task. Nor can it be perfectly done here, because as yet the Collection it self is not perfect'. In addition, though expressing in his preface the belief that nomenclature and classification should be closely related, Grew disavowed the intention 'actually to reform this matter' on

68 Dorothy Stimson, *Scientists and Amateurs* (London, 1948), p. 111; Slaughter, *Universal Languages* (n. 41), p. 175.

69 Raven, *John Ray* (n. 7), pp. 182f. and passim.

70 Grew, *Musaeum*, pp. 124, 150 – 3, and see plate 8. It is interesting that tables in a partial Latin translation of Grew's book, British Library MS Sloane 1927, differ slightly from those in the published edition: see esp. fol. 41v. For the context, see Martin Lister, *Historiae Animalium Angliae Tres Tractatus* (London, 1678), esp. pp. 110, 154, [203]. See also his *Historiae Conchyliorum Libri I-IV* (London, 1685 – 88), and Jeffrey Carr, 'The Biological Work of Martin Lister' (Leeds Ph.D. thesis, 1974), ch. 4.

Plate 8. Two of the 'schemes' arranging shells according to their structural and decorative characteristics included in Grew's Musaeum Regalis Societatis, *p. 151. Reproduced from Sherard 652 by courtesy of the Department of Plant Sciences, University of Oxford.*

Scheme 3.

- Conick, with the *Turbon*.
  - Long. With
    - The Right Lip Broad
      - Expanded and Turned out.
        - Armed with Fingers
          - On both Lips. Frog-Wilk. *Murex Coracoides.*
          - On the Right only. The Broad-Lip'd Wilk. *Aporrhais.*
        - Naked. The Marble-Wilk. *Murex Marm.*
      - Only expanded.
        - More
          - Even Lip'd. The Oriental-Wilk. *Murex Orientalis.*
          - A little angular.
        - Less. With the Whirle
          - Knobed.
          - Even.
    - Parallel Lips. With the Body
      - Belly'd With
        - Lips Even.
        - The left Furrowed.
      - Level. With the *Turbon.*
        - Knobed.
        - Spiked. *Murex Aculeatus.*
  - Low. With the Mouth.
    - Oval. The Wilk with plated Spikes.
    - Long. With the Body
      - Level. With the Whirle
        - Even.
        - Knobed
      - Belly'd. With the Rounds of the *Whirle*
        - Concave
        - Convex
      - Conick Snails. *Cochlea Conicae; vulgo Cylindricae.*

Scheme 4.

- Oval. With
  - Both Lips turned out. The Mouth
    - Even
    - Furrowed. With the Rounds of the *Whirle*
      - Un-interrupted. With the Body
        - Knobed.
        - Wrinkled.
      - Plated
    - *Persian*-Shells. *Conchae Persicae.*
  - One only
    - The Left. With a
      - Double Edge. With the Body
        - Oblong.
        - Short.
        - The Flat-Lip'd *Snail.*
      - Single Edge. The Mouth
        - Even. The four-Corner
          - Produced
            - Strait. *Cochlea Rugosa.*
            - Crooked. The Hook-Nose.
          - Reduced
        - With the *Whirle*
          - Low. The *Short-Nose.*
          - Immersed. The *Dipping-Snail*
        - Furrow'd. With the Mouth
          - Long. *Cochlea Cylindroide.*
          - Round. With the *Whirle*
            - Smooth
            - Spiked
            - *Cochl. Umb.*
    - The Right. With the Mouth
      - Even
      - Furrowed. With the *Turbon*
        - Smooth
        - Spiked
      - *Snailes.*

Scheme 5.

the grounds that he was not writing 'an Universal History of Nature'.[71] It might also be felt that his inventorizing format imposed shortcomings on the book: at times, Grew indulged in what can only be described as waffle in his attempt to find something to say about each object, while, despite his critical comments about Aldrovandi's classification, Grew's arrangement of the rarities had a good deal in common with that.[72] Indeed, the very task of cataloguing arguably diverted Grew from more valuable work that he might otherwise have done, such as continuing the investigations of plant and animal anatomy that he had begun earlier in the decade, and one might argue that the Royal Society's institutional needs here had a rather adverse effect on the intellectual development of one of those most closely associated with it.

What is more, in terms of the objects described, rather than the method employed in doing so, there is much more common ground between Grew's book and Hubert's, and it is almost as if the 'scientific' characteristics of Grew's catalogue to which I have referred were imposed on a collection inspired by quite other criteria. The fact that objects mentioned by Hubert could be identified with specimens described in Grew's book was first noted by Sir David Murray.[73] He only singled out a few instances of this, but fuller collation of the two works shows the overlap to be very extensive. Numerous items can be traced as going forward from one collection to the other, particularly fairly easily identifiable ones like exotic animals, birds and fish. This even included as number of objects which one might have considered out of place in the Society's scientific museum, such as the oddly-shaped sticks already referred to.

Some of Hubert's rarities seem to have disappeared (it is conceivable that the Royal Society acquired less than the whole collection, thus explaining the failure of such spectacular items as his Mummy to reappear).[74] In addition, Hubert's habit of referring to objects en masse at various points in his catalogue means that some specimens that appear in the Grew catalogue but not in the earlier one may have come from his collection. On the other hand, just from items whose donor is noted by Grew — quite apart from references in the Royal Society's minutes — it is clear that the Royal Society

[71] Grew, *Musaeum*, preface, p. [150].
[72] Ibid., passim; Wilma George, 'Alive or Dead: Zoological Collections in the Seventeenth Century', in Impey and McGregor, *Origins of Museums* (n. 2), p. 186.
[73] Murray, *Museums* (n. 2), pp. 132 – 3.
[74] The Royal Society had only one Mummy, which was presented by Henry Howard of Norfolk (Birch, *History*, ii, 202; cf. ibid., ii, 88; Grew, *Musaeum*, pp. 1 – 3); for Hubert's, see Hubert 1665 (n. 18), p. 1. For the possibility that a substantial number of rarities owned by Hubert passed into Sloane's collection, see Hawkins, *History of Music* (n. 18), iv, 379n, citing an MS collection on the history of London by William Oldys (this is then cited in *Biographia Britannica* (London, 1789), iv, 347, which adds the possibility that they reached Sloane via Courten). See also above, p. 133.

did add extensively to Hubert's collection, gaining benefactions especially from Fellows but also from merchants, travellers and others.[75]

Though numerical comparison is difficult and the ratio varies from section to section, the Royal Society's collection was between two and three times as large as Hubert's. In some sections, it was much more extensive: for instance, the section on sea-birds was much longer than Hubert's, with more native specimens represented, while the sections of seeds, mosses and earths were without parallel in the earlier collection, as were those of engines and instruments.[76] But despite this — and allowing for parts where Hubert's descriptions are very cursory — it is perhaps surprising how similar the physiognomy of the collection remained. Take, for instance, Grew's section on quadrupeds.[77] Some items are added, such as the skull and claws of a tiger, the skull of a hippopotamus (plate 7) and the skin of a rhinoceros. In other cases, donations meant that the Royal Society had multiple examples where Hubert had had only one or two: for instance, whereas Hubert had one rhinoceros horn, the Royal Society had three further specimens, one presented by the statesman, Sir Robert Southwell. But much remained the same — a sloth, a flying squirrel, armadillos, chameleons, crocodiles, lizards, salamanders, tortoises. In particular, what recurred was the collection's basic stress on non-indigenous specimens: with very few exceptions, exotic items outnumbered native ones. It thus retained strongly the character of a virtuoso cabinet, with its emphasis on the strange and the outlandish.

Indeed, in some respects it actually became more like a typical virtuoso cabinet than Hubert's had been. As has already been noted, Hubert's collection stood out — and perhaps seemed especially attractive to the Royal Society — because it was an assemblage exclusively of 'natural rarities', in contrast to the mixture of natural and artificial curiosities that was the norm in virtuoso collections of the day. As a result of the well-intentioned gifts of virtuoso Fellows and others, however, the Society's collection actually became more miscellaneous: by Grew's time it included such items as a box of 100 turned cups one inside the other, or two half-length figures in armour made up of parts of plants and insects inlaid in wax — both given by the virtuoso, Dudley Palmer — or a forest scene with animals 'all Cut in Papyr, in the compass of about three inches square'. It also gained a section of ethnographic curiosities — weapons, an Indian canoe, clothes, utensils like baskets, combs, and so on — which had not been paralleled at all in Hubert's.[78]

Hence the dependence of the repository on benefactors did much to define its character. Quite apart from the fact that it would have seemed churlish to refuse gifts and that these led to unnecessary duplication — as with the six saw-fish snouts that the Society ended up with, in addition to a

75 Grew, *Musaeum*, passim; Birch, *History*, passim.

76 Grew, *Musaeum*, part I, sect. 4; part II, sects 3, ch. 3, and 4; part III, sect 3, ch. 3; part IV, sect. 2.

77 Ibid., pp. 10 – 48. Hubert, 1665 (n. 18), pp. 2 – 7, 33 – 9.

78 Grew, *Musaeum*, pp. 378 – 9, and part IV, sects. 2 – 3, passim.

complete specimen — it is clear that it was principally odd or exotic objects which seemed to virtuosi to be worthy of a place in the collection. The 1666 announcement in *Philosophical Transactions* had specified the 'rare and curious' as what was required, and to a disproportionate extent it was the abnormal rather than the normal which materialized: two-headed calves rather than ordinary ones, African birds rather than British.[79] There was a preconceived expectation of what such a collection should contain, and in practice the valuation of ordinary items that we have surveyed among the Society's activists failed to work through to the mentality of well-intentioned donors.

It is also clear that it was the extraordinary rather than the ordinary which caught the attention of the visitors for whom the Royal Society, like Hubert previously, catered. On his visit to the collection in 1710, for instance, Z. C. von Uffenbach itemized as the 'things that pleased us most' unusual specimens, or specimens unusually well-preserved, rather than such things 'of a common sort' as were on show.[80] The same is true of English descriptions of the collection, such as that in the anonymous *British Curiosities of Nature and Art* (1713), which instanced such 'Animal Curiosities, as the Cameleon (that lives on the Air,) one Joint of the Vertebra of a Whale 30 l. Weight, &c.' Indeed, though based on Grew, the brief descriptions in the latter book are more reminiscent of Hubert's than his in their stress on the odd and the wonderful: for instance, 'The Webb of a Bermudas Spider, so strong as to snare a Bird', or 'The Tail of an Indian Cow, whose Hair is about a yard and quarter long: (This Creature is worshipped by the People, near the *Ganges*)'.[81]

At this point it does seem as if we have a conflict between what might be called 'scientific' and 'virtuoso' values, between the adulation of the exotic and the rare which characterized the virtuosi on the one hand, and, on the other, the aspiration to a comprehensive collection of objects, ordinary as well as extraordinary, seen in the public statements of the Society in the 1660s. The nature of the gifts which dictated the physiognomy of the repository thus illustrate clearly the way in which the Society's facilities were moulded — and perhaps trivialized — by the virtuosi who formed the rank and file of its support.

On the other hand, this is not the whole story, for, though the attitudes of the virtuosi may not have helped, there is reason to believe that the initial conception of a 'complete' collection was less realistic than had originally been thought, partly because of the limitations of the Royal Society's resources, and partly in its own right. The Willisel experiment exemplified the Royal Society's attempt to add to the collection in a systematic way, and the value of Willisel's work is illustrated by his success in bringing to light specimens in unexpected places which John Ray was able to incorporate in his definitive catalogue of English plants; Willisel was probably also responsible for the number of native examples which swelled some sections

79 *Phil. Trans.*, 1 (1666), 321; Grew, *Musaeum*, pp. 27, 84 – 6 and passim.
80 Quarrell and Mare, *London in 1710* (n. 2), pp. 99 – 101.
81 *British Curiosities* (n. 1), sig. A5, p. 44. See also the other works cited in n. 1.

of Grew's catalogue of the repository by comparison with Hubert's, for instance that of sea-birds.[82] But — quite apart from the fact that the Royal Society could only afford to employ Willisel briefly — it is worth pointing out that systematic accumulation of specimens of all the natural phenomena even of England would have resulted in a collection of enormous size, which would have placed a huge burden on a society with only limited resources.

The virtues of a 'complete' collection were especially questionable when it came to common species which could be equally easily studied in their natural habitat. Indeed, some had evidently pointed this out at the outset, as is suggested by William Balle's report to Henry Oldenburg in 1666 of his recollection of a discussion which presumably took place in 1664 over the proposal that 'all sorts of animals' should be included in the collection. 'As I remember', he wrote, 'when some were moving that all the curiosities of our owne land should bee first gotten others laughed att itt as too voluminous, to have ducks geese & hen &c.'[83] At that stage, this was clearly not a dominant view, as is shown by the efforts towards 'completing' the collection that followed. But, as with others of the Society's early grandiose schemes, like its hopes to improve technology, in the event difficulties came to light which illustrated a degree of naivety in the Society's original ambitions.[84]

It is thus interesting to find that, in practice, scientists as much as virtuosi consulted the repository for exotic specimens rather than common ones which were otherwise available. Of the items in the Society's repository cited by Walter Charleton in his *Onomasticon Zoicon* (1668), for instance, a disproportionate number were exotic species which he otherwise had no opportunity to examine, and from this point of view the collection certainly had a worthwhile role to play.[85] This is also illustrated by the fact that, insofar as Grew's selection of items for careful scrutiny was not random, he apparently devoted most attention to exotic objects which writers like Jonston had misdescribed because they had never actually seen. In addition, serious scholars as well as dilettantes could find a value in the monstrous: Bacon, after all, had stressed the need to study nature 'erring' or 'out of course' as a crucial way of penetrating her secrets, so it could be seen as genuinely important to preserve monstrosities and abnormalities as a resource for scholarly study.[86]

82 Ray to Lister, 17 July 1670, in John Ray, *Correspondence*, ed. E.Lankester (Ray Society, 1848), pp. 61 – 2. Cf. also Skippon to Ray, n.d., ibid., pp. 85 – 6. Grew, *Musaeum*, part I, sect. 4.

83 Balle to Oldenburg, 14 April 1666, *Oldenburg*, iii, 90. See also above, p. 125.

84 See Hunter, *Science and Society*, ch. 4.

85 See above, n. 17. Cf. e.g. Ray to Lister, 19 Dec. 1674, in Ray, *Correspondence* (n. 82), p. 112; Grew, *Musaeum*, passim.

86 Bacon, *Works*, iv, 168 – 9. See also Katharine Park and L. J. Daston, 'Unnatural Conceptions: the Study of Monsters in Sixteenth- and Seventeenth-century France and England', *Past and Present*, 92 (1981), 20 – 54.

But for common objects, and especially for live, animate ones, the value of a museum was less clear, and arguably the Society's initial ambition for a museum had been somewhat misconceived. As it turned out, its organizers had been wrong to presume that it was either necessary or desirable to have a complete series of dead specimens laid out in the same room in order to achieve a satisfactory taxonomy of nature. Though the work of Francis Willughby and John Ray was to exemplify the period's real achievement in building up an accurate classification of nature, it soon transpired that such systematization was best achieved not so much through the accumulation of specimens, as had originally been believed, as by books like theirs, the content of which was not co-extensive with any collection. Indeed, it is revealing that, insofar as Willughby and Ray collected themselves, they did so in a functional and temporary manner, buying fish, dissecting them on the spot, drawing them and then throwing them away. Even items which they brought home from their travels abroad appear to have been jettisoned once they had been classified.[87]

The position of the Royal Society's museum vis a vis this broader enterprise is well illustrated by the relationship of the collection with Willughby's systematic *History of Fishes*, posthumously edited by Ray and jointly published by the Royal Society and Bishop John Fell of Oxford in 1686. The repository was the source of the specimens engraved in some twelve of the excellent plates that were provided for the book by subscription; the remainder were mainly taken from life, from fresh examples of common varieties acquired from the slabs of London fishmongers.[88] Moreover, as C. E. Raven observed, the Royal Society's specimens have a distinctly 'shrunken and eyeless' look about them, suggesting that preserved exhibits were decidedly inferior to live ones where the latter were available. Hence it was only appropriate that it was from fresh specimens of common species that Willughby and Ray were able to build up their basic taxonomy — carp, tench and mackerel, herring, bream and mullet — and it was into this that they had to fit the exotics — globe fish, square fish or mailed fish — with which Grew's catalogue shows the Royal Society's collection to have been so profusely supplied.[89]

Hence in the early history of the repository, we can arguably see a significant process of education going on as to what the strengths and weaknesses of a museum *were* in relation to understanding the natural world. A museum had a real value, but ironically this was not altogether what had initially been thought. As with so many other facets of the Society's early institutional ambitions, a process of trial and error was at work in which it became clear what a scientific institution could usefully achieve and what it could not. Thus modifications were introduced in the light of experience, and it is ironic that — even if for the wrong reasons — it

87 George, 'Alive or Dead' (n. 71), p. 185, and a personal communication of 27 July 1983 from Dr George.

88 Francis Willughby, *De Historia Piscium Libri Quattuor*, ed. John Ray (Oxford, 1686), plates G9, I2, 7, 10, 20, 22 – 4, N13, O3 – 4, X11 and passim. See plate 6.

89 Raven, *John Ray* (n. 7), p. 357; Willughby, *De Historia Piscium* (n. 88), passim.

may have been the virtuosi rather than the scientists who had a better sense of what it was appropriate for a museum to contain.

If these reflections arise from the character of the repository at the time when it was reified in Grew's catalogue, its subsequent history further exemplifies the practical problems associated with administering a collection of this kind on the part of a voluntary institution. In terms of its content, gifts which are recorded in subsequent years continued to reinforce the 'monstrous' and exotic character of the collection.[90] As for its condition, the museum seems to have experienced a series of ups and downs, due to the fact that those who were supposed to look after the rarities rarely had as much time for the task as it required. In a letter of September 1681, John Evelyn told William London in Barbados that 'the Royal Society have lately put their Repository into an excellent method', and this was perhaps because the collection had been spruced up and rearranged while Grew was working on his catalogue.[91] But, though named 'praefectus Musei regalis Societatis' in 1682, Grew does not seem to have taken much interest in the museum after the publication of his catalogue: it is revealing that, though various possible appendages to that are mentioned, including an index and possibly an abbreviated version with the objects arranged 'in better order', none ever appeared in print.[92] Instead, responsibility for the collection devolved to the Society's more menial operators and clerks. It was apparently they who showed visitors round the museum, and they were also supposed to keep it in order, though their other commitments meant that this task was frequently not a priority for them. Moreover the difficulties that the Society had with individual employees — for instance, in the eighteenth century one was to abscond, while another was imprisoned for debt — deflected on the museum.[93] In addition, new acquisitions meant that a constant process of administration was required as the proper places for these had to be found, and catalogue entries made for them.[94]

This evidently explains the evidence that exists of neglect, perhaps best documented by the impressions of Uffenbach, who gave a critical account of his visit to the museum in 1710. As has already been noted, Uffenbach came to see the collection with expectations formed by Grew's catalogue, and he was shocked by what he saw. 'Hardly a thing is to be recognized, so wretched do they all look', he wrote, complaining how 'the finest instruments and other articles (which Grew describes), [are] not only in no

90 See esp. Royal Society MSS 413 – 4.
91 Evelyn to William London, 27 Sept. 1681, Evelyn, *Diary and Correspondence*, iii, 259.
92 Birch, *History*, iv, 171, 250, 557; Francis Aston to Martin Lister, 29 March 1683, Bodleian Library MS Lister 35, fol. 92; see also below, 'Early Problems', n. 74.
93 See A. D. C. Simpson, 'Newton's Telescope and the Cataloguing of the Royal Society's Repository', *Notes and Records*, 38 (1984), 187 – 214.
94 See esp. Royal Society MSS 413 – 4. See also Simpson, 'Newton's Telescope' (n. 93), nn. 50, 54.

sort of order or tidiness but covered with dust, filth and coal-smoke, and many of them broken and utterly ruined'.[95]

On the other hand, this neglect can be exaggerated on the basis of accounts like Uffenbach's, as it is for instance by Richard Altick, who writes the collection off as 'decrepit' virtually from the start.[96] In fact, the process was more cyclical than inexorable. Within months of Uffenbach's visit the Society moved to new quarters — it was possibly partly in the expectation of this that the collection had been neglected — and shortly afterwards a purpose-built, galleried room measuring forty foot by twenty-three foot was erected to hold the rarities, probably to a design by Sir Christopher Wren. Then, in 1712 a committee was appointed 'to take care of the due placing of the Curiosities in the New Repository', while in the 1720s, we find an attempt being made to reduce new acquisitions more or less to Grew's arrangement.[97]

By the early 1730s, however, the condition of the collection was again poor, as was revealed by the findings of a further committee set up to deal with it, which complained of the 'Disorder' and poor condition of the specimens and expressed the fear that unless steps were taken, 'the greatest part of the Repository will soon perish & become useless'. Surveying the state of affairs, they reported how the birds, for instance, 'are generally in a very bad Condition; many of them having lost their feathers & several having nothing remaining beside their head & feet'.[98] They therefore took various steps to improve the condition and care of the exhibits, including the installation of locked cases; they also arranged for a register of benefactors to be kept. In addition, in the 1730s an entirely new scheme of classification was attempted by Cromwell Mortimer, the Society's Secretary, a doctor and former student of Boerhaave, who had been assistant to Sloane: this still survives at the Royal Society and it would repay detailed study.[99]

But it was not long before things began to go wrong again. In 1752, the museum was said to be in 'a ruinous forlorn condition', and in the 1760s it again had to be taken in hand, reinventorized and rearranged.[100] By the late 1770s the state of affairs may well once more have been deteriorating. Though the lack of space in the Society's new quarters in Somerset House was ostensibly the reason for the offer of the collection to the British

[95] Quarrell and Mare, *London in 1710* (n. 2), pp. 97 – 8.

[96] R. T. Altick, *The Shows of London* (Cambridge, Mass., 1978), p. 14.

[97] See J. A. Bennett, 'Wren's Last Building?', *Notes and Records*, 27 (1972), 107 – 118; Simpson, 'Newton's Telescope' (n. 93), pp. 193 – 4 and n. 50; Royal Society MS 413.

[98] D.M. 5.105 (before 'lost', 'nothing left' has been deleted). See also Simpson, 'Newton's Telescope' (n. 93), pp. 194f., and Royal Society MSS 414, 416.

[99] Royal Society MS 415/2 – 5. Simpson ('Newton's Telescope' (n. 93), p. 196) is therefore incorrect in stating that it is not known to survive.

[100] Simpson, 'Newton's Telescope' (n. 93), pp. 199, 201, and Royal Society MSS 415/1, 417, 419.

Museum in 1779, it may well have been felt that a collection was something with which a voluntary society was less able to cope than had optimistically been presumed in the 1660s.[101]

Ironically, at this transitional stage, an institution like the Royal Society may have been less able to look after a collection of rarities properly than an enthusiastic individual like Sir Hans Sloane with a burning commitment to the enterprise. It is symptomatic that during the revamping of the museum in the 1730s, Sloane's collection was held up as an example to the Royal Society on one occasion.[102] Moreover, matters were made worse by the fact that an institutional collection stimulated high expectations in visitors which were easily disappointed, as in the case of Uffenbach, whereas private collections obviously stimulated fewer pre-existent expectations and were therefore likelier to please. It is, in fact, interesting that Uffenbach juxtaposed his critical comments about the Society's collection with precisely the sort of observations about the difficulties of keeping public institutions afloat which are so appropriate to the early Royal Society.[103]

Hence, in addition to the process of education which led to the abandonment of the idea of a comprehensive collection of natural things, the history of the repository also revealed a further problem that had not been apparent at the outset, namely that looking after a museum necessitated work-a-day administrative tasks that were simply beyond the resources of a voluntary Society with plenty of other commitments. So, though the Society's ambitions have been echoed by historians like Sir Henry Lyons, who have seen the formation of the collection as an almost natural corollary of the Society's new, institutional role in the science of the day, in fact their assessment is anachronistic in the light of the Society's actual circumstances in its early years.[104] Once again, therefore, we find ambitions having to be modified in the light of experience, as the Society first scaled its plans down to a collection of a practicable size, and then abandoned the collection altogether to a more specialized body in the form of the British Museum.

101 Simpson, 'Newton's Telescope' (n. 93), p. 202.

102 Royal Society MS 490 (CMB 63), minutes of meeting of 8 May 1733, cited in Simpson, 'Newton's Telescope' (n. 93), p. 194.

103 Quarrell and Mare, *London in 1710* (n. 2), p. 98. Cf. also Uffenbach's view of the Ashmolean, which he compared unfavourably with private cabinets such as that of Tobias Reymers of Luneburg, 'who is only a private person, [but] has certainly as many specimens again as one meets with here and far more important ones': W. H. and W. J. C. Quarrell, eds., *Oxford in 1710 from the Travels of Zacharias Conrad von Uffenbach* (Oxford, 1928), p. 26. For the collection of Reymers, see Murray, *Museums* (n. 2), i, 51. See also John Macky, *A Journey* (n. 1), i, 260.

104 Sir Henry Lyons, *The Royal Society, 1660 – 1940* (Cambridge, 1944), pp. 49, 63 – 4. Cf. *The Record*, pp. 33 – 5.

5

A 'COLLEGE' FOR THE ROYAL SOCIETY: THE ABORTIVE PLAN OF 1667 – 1668*

The Royal Society did not possess premises of its own until 1710. For most of the first fifty years after its foundation in 1660, the Society met at Gresham College, the educational institution in the City of London founded in the late Elizabethan period by Sir Thomas Gresham. By the Restoration the college had little vitality but plenty of space, and the Royal Society worked out a mutually advantageous relationship with its professors: the Society held its meetings and housed its facilities at Gresham, while the college gained some reflected prestige and any interested professors were allowed free membership of the Society. The only gap in the Society's sojourn at Gresham in these years occurred in the aftermath of the Great Fire of 1666, in which the Royal Exchange was destroyed. Gresham College was then commandeered for more urgent civic functions, and the Society, though not actually obliged to leave, felt it best to transfer its operations elsewhere. From the summer of 1667 the Society was instead given hospitality by Henry Howard, brother of the then Duke of Norfolk, at Arundel House in the Strand, the London house of the Howard family, returning to Gresham again only in 1673.[1]

If this is fairly well known, it is much less familiar that in 1667 – 68 a sustained effort was made to erect a purpose-built 'College' for the Royal Society.[2] Though this project proved abortive and the Society remained without a building of its own for another forty years, the episode is nevertheless of interest. For it illustrates an aspect of the Society's exalted ambitions in its early years, while significant conclusions can be drawn from the failure of the plan, which exemplifies the tension between the Society's aspirations and the actual difficulties that it encountered due to its dependence on purely voluntary support. This tension is important, and it has not been fully understood by modern scholars. A considerable amount

* In the preparation of this article I have been assisted by Mr H. M. Colvin, Professor H. J. Cook, Dr C. H. Josten, Mr A. V. Simcock, Dr David Sturdy and Mr Paul Wood.

[1] See Iam Adamson, 'The Royal Society and Gresham College, 1660 – 1711', *Notes and Records*, 33 (1978), 1 – 21.

[2] For existing accounts of the episode see especially C. R. Weld, *A History of the Royal Society* (London, 1848), i, 205 – 214, and Michael Hunter, 'The Social Basis and Changing Fortunes of an Early Scientific Institution: An Analysis of the Membership of the Royal Society, 1660 – 1685', *Notes and Records*, 31 (1976), 9 – 114, Appendix 2, pp. 60 – 62, 'The Plan for a College, 1667 – 1668'. The present paper supercedes the latter.

of recent scholarship has been devoted to chronicling the informal and rather precarious state of the early Royal Society, contrary to anachronistic presumptions that the Society was as effective as a modern scientific institution.[3] Though such arguments contain a good deal of truth, they have failed to do justice to the extent to which this informality characterized the Society in spite of institutional ambitions which *were* grandiose, but which never came to fruition for want of funds and support. The abortive college illustrates this clearly, since many believed at the time that the erection of such a building would have a galvanizing effect on the Society's institutional role and its scientific work. Yet the effort proved more than could be sustained, and the fact that the hoped-for contributions from outside the Society barely materialized illustrates the lack of support in Restoration society at large that the Society encountered.[4]

A long-term stimulus for the idea of a purpose-built scientific institute must be 'Solomon's House' in Francis Bacon's utopian sketch, *New Atlantis*, while a precursor of comparable date, J. V. Andreae's *Christianopolis* (1619), also included a collegiate institution devoted to the study of nature and the arts. A more immediate background is provided by the various plans for residential colleges devoted to intellectual pursuits that were put forward in the years around 1660. Indeed the notion of a 'college' may have been a deliberate allusion to such projected research institutions rather than to the more familiar model of colleges devoted to teaching found at Oxford, Cambridge and Gresham, since the Royal Society never anticipated their building serving functions other than of research and discussion.

One scheme was expounded by John Evelyn in a letter to Robert Boyle of 3 September 1659. Its aims were broader than scientific research, and he planned a quasi-monastic institution whose members were to live in 'apartments or cells' in the complex while the central facilities would comprise 'one handsome pavilion' together with separate smaller buildings housing a laboratory, a collection of rarities and the like. Evelyn's estimate of the cost was £400 for the pavilion, £150 for a chapel that was also projected, and £600 for the apartments, walls and outhousing.[5] A residential college was also projected by the Swede, Bengt Skytte, and the scale of this was comparable, while Abraham Cowley, the poet, outlined in his *Proposition for the Advancement of Experimental Philosophy* (1661) the idea of a 'Philosophicall Colledge', to be endowed to the tune of £4000 per annum, with a substantial layout of houses for the inmates built round three courtyards with corporate facilities such as a chapel and a library, not dissimilar to Evelyn's.[6]

3 See particularly Quentin Skinner, 'Thomas Hobbes and the Nature of the Early Royal Society', *Historical Journal*, 12 (1969), 217 – 239, and, for a general appraisal, Hunter, *Science and Society*, ch. 2.

4 For the background see ibid., esp. pp. 83 – 84, 139, 195 – 197.

5 John Evelyn, *Diary and Correspondence*, iii, 117 – 118.

6 Margery Purver, *The Royal Society: Concept and Creation* (London, 1967), pp. 226 – 232; A. Cowley, *A Proposition for the Advancement of Experimental Philosophy* (London, 1661), pp. 14 f.

In contrast to these, the Royal Society never planned a building in which more than a handful of staff would be resident (in central London this was hardly necessary), which meant that so extensive a layout was not required: in any case, the Society had to consider what it could afford with the resources at its disposal. But in terms of the concept of a research institute named a 'college', and even in some of the facilities to be housed in it, there is enough similarity to these schemes to suggest that the Royal Society owed something to them.

Nothing is heard of plans along such lines in the Royal Society's earliest years, perhaps largely because this was only one of several projects which depended on the adequate endowment which its founders sought at this time, hopefully from public funds: their hopes were unrealized except for the royal grant of the defunct theological college at Chelsea, mooted in 1664 and finally ratified in 1669.[7] But in a letter to Robert Boyle of 24 February 1666, the Society's Secretary, Henry Oldenburg, referred to plans to provide various facilities, though complaining of lack of support, and this may well already have presupposed the desirability of purpose-built premises. 'We are now undertaking severall good things', he wrote, 'as the Collecting a Repository, the setting up of a Chymicall Laboratory, a Mechanicall operatory, an Astronomicall Observatory, and an Optick Chamber; but the paucity of the Undertakers is such, that it must needs stick, unlesse more come in, and putt their shoulders to the work'.[8] Though the repository materialized, thanks to the generosity of the Society's Treasurer, Daniel Colwall, nothing more is heard of the other proposals, and hence their setting remains obscure.[9]

The first definite reference to the idea of a purpose-built college for the Society is to be found in the final pages of Thomas Sprat's *History of the Royal Society*, published in the late summer or early autumn of 1667,[10] a reference which is worth quoting in full since it shows (contrary to what has sometimes been thought)[11] that the notion of purpose-built premises was not seen as exclusive of the use of Chelsea College. 'The places of their *Residence* they have appointed to be two', Sprat writes: 'One a *College*, which they design to build in *London*, to serve for their *Meetings*, their *Laboratories*, their *Repository*, their *Library*, and the Lodgings for their *Curators*: The other the *College* at *Chelsey*, which the *King* has bestow'd on them; where they have a large Inclosure, to serve for all *Experiments* of *Gardning* and *Agriculture*: and by the neighbourhood of the *River* they have excellent opportunity of making all *Trials* that belong to the Water'.[12] This was echoed in an account

[7] See Birch, *History*, i, 391, and vols. i – ii, *passim*, and Hunter, *Science and Society*, pp. 38 – 40.
[8] *Oldenburg*, iii, 45.
[9] On the repository see Sir David Murray, *Museums, Their History and their Use* (Glasgow, 1904), i, 130 – 133, and above, ch. 4.
[10] Sprat, *History*, p. ix.
[11] See below, p. 177.
[12] Sprat, *History*, p. 434.

of the Society's plans given by the Italian visitor, Lorenzo Magalotti, who also saw the two sites as complementary, though wrongly believing that the development of both was dependent on royal grants.[13]

The college plan was announced to the Society itself when meetings resumed after the summer recess on 30 September 1667. John Wilkins — Oldenburg's co-Secretary and very much the leading figure in the Society's corporate initiatives in these years — moved that a committee be appointed 'for raising contributions among the members of the society, in order to build a college'. Wilkins was thereupon asked to bring a list of possible names to the next Council meeting.[14] This did not occur till 5 November, and though the committee was not yet nominated 'the business of voluntary subscriptions for contributing towards the carrying on of the ends of the institution of the Royal Society being considered of', it was agreed: 'That it was now a seasonable time for such subscriptions; and that they were to be made first by such of the council and of the society, as were both willing and able, and afterwards by such other well disposed persons not of the society, as should come in by the solicitation of a committee to be nominated by the council out of their own number, and out of the fellows of the society; which contributions should be employed in promoting the ends of the society, and particularly to the building of a college, as the most probable way of the society's establishment'.

A form of subscription was therefore drawn up as follows: 'We, whose names are underwritten, being satisfied of the great usefulness of the institution of the Royal Society, and how requisite it is for attaining the ends designed thereby, to build a college for their meetings, and to establish some revenue for discharging the expences necessary for trial of experiments, do heartily recommend it to the bounty of all generous and well-disposed persons for their assistance to a work of such public usefulness; and we do each of us, for ourselves, hereby promise to contribute to those good ends the respective sums subscribed by each of us, at four distinct quarterly payments, to be made to such persons, as shall be authorised under the seal of the Royal Society for the receit thereof; the first payment to begin at . . .'.

13 W. E. Knowles Middleton, ed. and trans., *Lorenzo Magalotti at the Court of Charles II: His 'Relazione d'Inghilterra' of 1668* (Waterloo, Ontario, 1980), p. 140. Magalotti states that the London building was for use in the winter 'because in that season it would be too uncomfortable to go [to] the one outside the city', hence implying that Chelsea would be the principal seat of the Society's activities. But he may well have misconstrued what he was told, as he is known to have done in other cases, and Sprat's testimony is to be preferred.

14 Birch, *History*, ii, 194. The Society had taken possession of the Chelsea site during the summer of 1667. The college plan is first mentioned in the minutes almost immediately after this was reported, and it is conceivable that the two were linked (as is suggested by J. A. Bennett, 'Wren's Last Building?', *Notes and Records*, 27 (1972), 116 n. 11), possibly due to the hope of financial profit from Chelsea (cf. the Halls' note in *Oldenburg*, iv, 437). However, it seems to me that this was at best a fairly marginal consideration.

This form was given to Sir Anthony Morgan and John (later Sir John) Hoskins to make it legally binding.[15]

Hence by November 1667 the project was definitely under way, and it is worth pausing to consider why it came to fruition at this point. It can hardly be coincidental that the plan is first heard of in 1667, when, as has already been explained, the Royal Society had left Gresham College in the aftermath of the Great Fire and taken up residence at Arundel House. Moreover the Society's organizers were possibly stimulated by the fact that their occupancy at Arundel House looked a little insecure within months of their arrival, though in the event the Society was to stay there for the next six years. As Oldenburg explained in a letter to Boyle of 1 November 1667, the Society was worried that Howard did not really have room for them at Arundel House, 'seing his Sons are come home, who need many rooms with their attendants'.[16] He added: 'Some of us are so sanguine, as to believe, we shall have a Colledge of our owne created before the end of next Summer. I know many Infidels in this case, of whom he professes himself to be one' — his scepticism presumably relating to the speed with which some hoped the scheme could be implemented, though it echoes earlier remarks in letters to Boyle, including that quoted above, in which Oldenburg voiced his well-founded misgivings about the dwindling support for the Society.[17] The link with the move from Gresham is clearly suggested by a remark of Oldenburg's in a later letter to Boyle, when he expressed his view that the college would 'fixe us (who are now lookt upon but as Wanderers, using precariously the lodgings of other Men) in a certain place, where we may meet, prepare and make our Experiments and Observations, lodge our Curators and Operators, have our Laboratory, Observatory and Operatory all together'.[18]

The background to the scheme is further illuminated by a proposal (known only from a later report but evidently dating from this time) that the Society should share premises with the London College of Physicians, whose building had been destroyed in the Great Fire. The idea that the Society should meet at the College of Physicians had been considered and rejected in 1660, but — presumably following the Society's removal from Gresham, since it was said that when the proposal arose 'the R[oyal] S[ociety] also wanted a convenience for their meetings' — it was suggested to the Society and the College by Daniel Colwall and Dr Christopher Merrett 'that a common place of meeting might be erected at the common charges of them both'. After initially being received with enthusiasm, we are told that this proposition was then rejected by the Council of the Royal Society, 'urging that both the name of the place and honour would be

[15] Birch, *History*, ii, 205 – 206.
[16] *Oldenburg*, iii, 505.
[17] For other instances, see *Oldenburg*, ii, 235, 320; iii, 155, 476.
[18] Oldenburg to Boyle, 21 January 1668, ibid., iv, 116.

wholy the Physitians'.[19] This abortive scheme may also have acted as a catalyst to the project of a building of the Society's own.

But, apart from the incentive provided by the short-term inconvenience of leaving Gresham College, there were also hopes that by this means the Society could achieve more than it had hitherto. In the background lay the foundation of the Académie des Sciences in Paris in 1666 and the expectation that Louis XIV's chief minister Colbert would provide lavish facilities for it. Rumours concerning the Académie and Colbert's intentions were in the air from the time of its foundation onwards, as is illustrated by comments in letters to and from Oldenburg, and, though the buildings for the Académie were not yet constructed, it was clearly felt that the English society needed to look to its laurels.[20]

Quite apart from such a wish to outdistance the Académie there seems to have been a more general sense that the erection of a special building would 'establish' the Society and help to offset the element of precariousness in its fortunes of which Oldenburg, for one, was aware even at this stage. In addition, there are hints that the Society would have liked if possible to have collected a sizeable enough endowment not only to erect a building but also to have a lump sum that could be invested to provide an income to supplement that derived from Fellows' subscriptions.[21] Oldenburg may have been alluding to either or both parts of the scheme in a letter to Boyle of 12 November 1667, when he told him that the Society 'are now continuing a way to doe something, that may more nearly concern their perpetual establishment'; he echoed this in a letter to an overseas correspondent in the following February, when he described the college as 'a means very probable to establish their Institution to perpetuity'.[22] His sentiments were thus close to those of the form of subscription agreed to on 5 November which has already been quoted.

At the next Council meeting, on 16 November, it was resolved that when Morgan and Hoskins had brought the form back 'it should then be offered promiscuously to able and willing persons, as well without as within the society', and for this purpose the elaborate preamble devised on 5 November was abandoned and replaced by the simple formula: 'I A. B. acknowledge to owe to the Royal Society the sum of . . . to be paid, &c.'[23] At the same meeting consideration was given to the question of a site. 'It

[19] Christopher Merrett, *A Short Reply to the Postscript &c. of H[enry] S[tubble]* (London, 1670), p. 2. I am grateful to Professor H. J. Cook for this reference. On the proposal of 1660 see Birch, *History*, i, 5.

[20] See Hunter, *Science and Society*, p. 40 and n. 31. On the Académie and its buildings see Roger Hahn, *The Anatomy of a Scientific Institution* (Berkeley and Los Angeles, 1971), ch. I.

[21] See particularly the form of subscription of 5 November quoted on p. 159 above.

[22] *Oldenburg*, iii, 592, and Oldenburg to Norwood, 10 February 1668, ibid., iv, 167. Compare also Oldenburg to Boyle, 14 January 1668, ibid., iv, 112, where he speaks of 'one of the most probable means of their establishment'.

[23] Birch, *History*, ii, 211 – 212.

being intimated, that the ground of the wardrobe was yet undisposed of, and that it was probable, that the King upon the motion of some of the council would not be unwilling to bestow it upon the society for the raising of a college upon it; the president [William, Viscount Brouncker] was desired by the council to take an opportunity to break this matter with his Majesty; which his lordship promised to do'.[24]

Nothing is heard of the college after this until the New Year: a Council meeting of 9 December was concerned with other matters.[25] But on 2 January, 'the business of voluntary contributions for building a college being moved again', it was decided after debate that those who offered to subscribe should not be obliged to do so until the total promised amounted to £1000. It was also agreed 'that those of the council, who had most interest to engage others both of the society, and without it, should be desired to employ the same in speaking to persons of both sorts, and to learn the sums of their intended contributions, thereby to make an estimate what the total was likely to amount to'.[26]

This was taken further at the next meeting, on 11 January: a committee was appointed to solicit contributions, the Society's membership list was examined and a selection was made of 'persons, who were thought both willing and able to contribute to the said building'.[27] These were then allocated to different members of the committee who were presumably thought to have an 'interest' with them, thus providing interesting evidence of acquaintance networks in the Society to supplement that available from the recorded names of Fellows' proposers which I have analysed in my monograph on the Society's early membership.[28] The courtier Sir Robert Moray was clearly seen as having useful contacts with the Scottish aristocracy while the influential role of Seth Ward, Bishop of Salisbury, in high circles is illustrated by the allocation to him not only of various bishops but also of statesmen like George Monck, Duke of Albemarle, or John, Lord Robartes, Lord Privy Seal. Robert Boyle and his protégé, Oldenburg, were requested to approach mainly figures with Irish connections, while John Evelyn and Thomas Henshaw were to dun Philip Packer and Sir Robert Paston respectively, with whom they are known to have been friendly from other sources.[29]

It is worth analysing this list in the light of the actual commitment of the Fellows involved to the Royal Society. Clearly it was compiled with high expectations of the most eminent and privileged section of the membership — aristocrats, statesmen and others. These figured in it to a disproportionate extent while Fellows of lower status — doctors, lesser courtiers, gentlemen and the like — were neglected, although many of them were regular weekly

24 Ibid., ii, 212. On the Wardrobe see William Kent, ed., *An Encyclopaedia of London* (London, 1937), p. 697.
25 Birch, *History*, ii, 224 – 5.
26 Ibid., ii, 234.
27 Ibid., ii, 238 – 9.
28 Loc. cit. See Hunter, *Royal Society*, esp. Appendix 2.
29 See Evelyn, *Diary*, ii, 548 n. 1; iii, 170, 186, and *passim*.

subscribers to the Society.[30] Yet, if those in the list are examined in the context of the returns which this and other appeals for support actually evoked, it transpires that the distinguished figures who appear here failed to respond either to this or to any other requests for help: only those who had evidently been selected for their loyalty to the Society rather than their eminence tended to make contributions.[31] This dichotomy therefore exposed with special clarity the Society's rather pathetic illusions about the intent of the aristocratic sponsors — the traditional patrons of intellectual activity — who had allowed their names to be added to the Society's roll. Its support in fact came from a more varied, but more committed, band of enthusiasts.

At the same meeting of 11 January, Robert Hooke, the Society's Curator of Experiments, was desired to bring to the next Council meeting a design 'for the building of the society's college': Hooke was an accomplished architect as well as a scientist.[32] Further light is shed on the discussion on this occasion by a letter from Oldenburg to Boyle of 14 January, which gives various details not recorded in the minutes. 'They guesse', he wrote, 'there will be 4 Classes of Contributors: Some of an hundred lb; some of 60. or 50, some of 40; some of 20. Our President [Brouncker] hath already declared for 100; and I think, the Bishop of Salisbury [Ward], for the like summ; Dr. Wilkins for 50; and Mr. Hayes, for 40, Sir Philip Carteret, for 50; and, if there be occasion, for more. We begin with the Councill, and proceed to the Society; that, when we goe on to beg of others not of our Body, they may not object, we would load others, and draw our own necks out of the yoak'. He noted how those nominated to solicit contributions were called the 'Committee of Beggars' — 'as they were pleas'd to terme it merrily' — and he remarked in a letter a week later how, though himself 'averse from and as much a Bungler in Begging, as any man', he was prepared to do it when 'Honor and publick Usefulnes' were involved, citing 'a maxime, I learned in my Logick, when a Boy, *qui amat finem, amat et media ad finem*', who wills the end wills the means[33].

In his letter of 14 January Oldenburg also mentions a crucial fact which the minutes strangely omit to note, which is that Henry Howard offered to compound his generosity to the Society either by contributing a substantial sum of money or by giving a piece of ground in the vicinity of Arundel

30 This evaluation is based on a collation of the list in Birch, *History*, ii, 238 – 9, with the data contained in Hunter, *Royal Society*, esp. the 'Catalogue of Fellows, 1660 – 1700', which tabulates the tens of regular subscribers not included in the list in the minutes.

31 Ward was particularly unlucky: none of those allocated to him helped the Society on this or any other occasion. Moray and Boyle also failed to raise any contributions, while Howard had a success rate of two out of six, Oldenburg of one out of four, Brouncker of three out of seven, and Wilkins of four out of eleven; the latter three were allocated mainly regular subscribers. None of those allocated individual Fellows had any luck, although all those involved were regular subscribers.

32 Birch, *History*, ii, 238. On Hooke as an architect see M. I. Batten, 'The Architecture of Dr Robert Hooke, F.R.S.', *Walpole Society*, 25 (1936 – 37), 83 – 113, and Margaret 'Espinasse, *Robert Hooke* (London, 1956), ch. 5.

33 Oldenburg to Boyle, 14, 21 January 1668, *Oldenburg*, iv, 112 – 113, 116.

House for the college to be built on. This offer of a site is also reported in other sources,[34] and it was presumably eagerly accepted by the Society, since John Evelyn records in his diary that on 24 January 'We went to stake out ground for the building a Colledge for the *R[oyal] Society* at *Arundel* house but did not finish it, which we shall repent of'.[35] Despite their earlier silence on this point, the Society's minutes make it clear that at least by 25 January Howard's gift was official, since at the Council meeting on that date it was decided that an acknowledgement of the donation should be inserted in a letter that Oldenburg had been ordered to draw up for Sir Robert Moray about soliciting contributions in Scotland.[36]

What can be said about the nature and location of the site promised for the college? Arundel House comprised the pre-Reformation palace of the Bishops of Bath and Wells, which had come into the possession of the Fitzalan family in 1549. In the earlier seventeenth century it had been the London residence of the great collector Thomas Howard, Earl of Arundel, Henry Howard's grandfather; he had extended the existing building with galleries for his collection of antique sculpture and paintings, though leaving much of the earlier structure intact. So it remained in 1668: Henry Howard had shown Evelyn a plan for redeveloping the site as early as 1659 and redevelopment was to occur in the 1670s and 1680s, but at this stage the house was still a rambling mansion of a variety of periods, with very extensive gardens amounting to some one and a half acres.[37]

The exact location of the piece of land offered to the society by Howard is unclear. The layout of Arundel House and its surroundings is shown by various contemporary maps and views, such as Wenceslaus Hollar's plan of the west-central district of London, the relevant section of which is reproduced here (plate 9). If one may presume that the building was to have had access without passing through Arundel House itself, it is perhaps likeliest that it was to have been situated in the western part of the extensive gardens of the house, which abutted Strand Bridge Lane (this is the thoroughfare shown on Hollar's plan which leads down to 'Strand Bridge' on the river, separating the gardens of Arundel House from those of Somerset House). When the staking out referred to by Evelyn as begun on

34 *Oldenburg*, iv, 113; Pepys, *Diary*, ix, 28 (Pepys's report under the date 17 January on information received from John Creed, a member of the Council); cf. Philip Skippon to Ray, 24 January 1668, *The Correspondence of John Ray*, ed. E. Lankester (Ray Society, 1848), p. 23.

35 Evelyn, *Diary*, iii, 505. On the completion of the staking out of the site see below, p. 174. If the cryptic aside at the end of Evelyn's diary entry is not a retrospective embellishment it may suggest that he felt that to have completed the process now would have expedited the legal transaction which was subsequently to prove a complicating factor in the affair.

36 Birch, *History*, ii, 242. The endorsement is mentioned in the minutes for this meeting although, ironically, the letter itself is not referred to in its own right until the next.

37 See C. L. Kingsford, 'Bath Inn or Arundel House', *Archaeologia*, 72 (1922), 243 – 277; J. M. Robinson, *The Dukes of Norfolk: A Quincentennial History* (Oxford, 1982), pp. 107, 123; Evelyn, *Diary*, iii, 234, 595 and ns.

Plate 9. Arundel House and its Surroundings: detail from Wenceslaus Hollar's plan of the west-central district of London, reproduced by courtesy of the Trustees of the British Museum.

24 January was completed in May, the dimensions of the Royal Society's site were given as 100 by 40 feet, and such a parcel of land could easily have come out of this section of the gardens. As we shall see, plans for the building were considered facing either towards or away from the river — which would again be quite feasible in this location — while another Fellow, Philip Skippon, described the site as 'towards the waterside' when reporting on the scheme in a letter of 24 January 1668 to the naturalist John Ray.[38]

Returning to the narrative of the project, at the meeting on 25 January further offers of help were reported — from Sir Samuel Tuke, the Earl of Devonshire and Thomas Povey (reported by Howard), from Edward Nelthorpe and Philip Skippon (reported by Wilkins) and from Robert Boyle, the Earl of Anglesey and his son, Lord Annesley (reported by Oldenburg). All these were Fellows except Anglesey, who had not yet joined the Society though his son had: he was to be elected in April, perhaps due to having been approached in this connection.[39] Again, further details of the proceedings are revealed by a letter written shortly afterwards from Oldenburg to Boyle, in which he notes that a drawing of the proposed structure (perhaps Hooke's) was shown and examined, stimulating Howard — who was in attendance — to give 'no small proof of his understanding and skill in that Art'. Oldenburg also disclosed that a form of subscription was to be drawn up in a manner designed to be 'obligatory in Law, and yet not severe in expression'.[40]

On 30 January, the revised form of subscription was produced, specifying the site at Arundel House and obliging the signatory to pay within a year. Two hundred copies were ordered to be printed, one hundred with the quarterly dates of contribution specified as beginning on the forthcoming Lady Day, and the other hundred without.[41] In addition, a draft was read of the letter drawn up by Oldenburg to be sent by the Council to Sir Robert Moray for soliciting contributions in Scotland (Moray was one of the Treasury Commissioners for Scotland). This letter restated how the college was the best way to 'establish' the Society, how it was hoped that

[38] See below, pp. 174, 176; Skippon to Ray, 24 January 1668, *The Correspondence of John Ray* (n. 34), p. 23. It is conceivably significant that Magalotti describes the site as 'near' rather than 'in' the garden of Arundel House (Middleton, *Lorenzo Magalotti* (n. 13), p. 140): this western section could be seen as a sort of annexe.

[39] Birch, *History*, ii, 242 – 3. On Angelsey see Hunter, *Royal Society*, catalogue entry 258. On the outcome of these offers see p. 171 and note 57.

[40] Oldenburg to Boyle, 28 January 1668, *Oldenburg*, iv, 120 – 1. On Boyle's own subscription, see, in addition to this letter, Boyle to Oldenburg, 1 February 1668, ibid., iv, 139 (which interestingly associates the delay in the arrival of his contribution with the ban on imports of Irish cattle ordered by the English parliament at this time) and Oldenburg to Boyle, 11 February 1668, ibid., iv, 169, in which Oldenburg acknowledged the arrival of the money, which he promised to keep and to inform Boyle of 'the time of the first actual payment made by any of the contributors'.

[41] Birch, *History*, ii, 243 – 4. At this meeting the first two bonds for subscriptions were actually signed, by Lord Brouncker and Sir James Hayes.

subscriptions would be forthcoming both from outside the Society and within, and how Howard had given a site and Fellows had contributed for the college, 'the raising of which they intend, God willing, to begin with this approaching spring; and if the design be seconded by chearful contributions, hope to finish by Michaelmas next'. The letter concluded by urging the Scottish peers who belonged to the Society to contribute, which 'cannot but redound to the immortal fame of the contributors'. At the same meeting, Sir Anthony Morgan was asked to discuss with Howard the question of the conveyancing of the land, 'so as to make it valid in law by collateral security', the first hint of legal complications that were to arise later, and this came up again at the subsequent meeting.[42]

The subscription forms had evidently been printed by 4 February, when Oldenburg enclosed one in a letter to Boyle, adding: 'If I had thought, there were any others at Oxford, that would cast in their benevolence, I would have sent more Copies. We hope, we shall very shortly lay our foundation' and thus the sooner subscriptions arrived, the better.[43] Oldenburg also wrote to correspondents such as Paul Rycaut, the English consul in Smyrna and an absentee Fellow, explaining about the project and the Society's hopes to begin it in the spring and 'to finish in a short time, if they be further seconded by cheerfull Contributions'. He had presumably also written to Henri Justel in Paris along similar lines, since Justel had rather ironically replied on 29 January: 'You must have a considerable endowment to build a College for your society, such as you tell me of'.[44] His scepticism would be shown to be justified.

The story then moves to the attempt to accumulate subscriptions. On 17 February Wilkins subscribed £50, as promised the previous month. He was also given permission by the Council to present half a dozen copies of Sprat's *History of the Royal Society* 'to some persons, from whom he expected contributions', an interesting illustration of the way in which Sprat's book was intended to be used to enlighten potential supporters of the Society.[45] In addition, Wilkins was asked to desire Dr William Croone to draw up a letter to be sent to the Duchess of Newcastle, whose visit to the Royal Society in 1667 had created such a stir, requesting her to contribute.[46] At subsequent meetings, the minutes record offers of contributions from various Fellows,[47] and other offers can be dated from manuscript sources, particularly 'A Liste of the Engagements, made by several Members of the R[oyal] Society, for

[42] Ibid., ii, 243 – 5, and 249. See also below, pp. 177 – 8.
[43] *Oldenburg*, iv, 145.
[44] Oldenburg to Rycaut, 30 January 1668, ibid., iv, 133; cf. Oldenburg to Norwood, 10 February 1668, ibid., iv, 167 – 8; Justel to Oldenburg, 29 January 1668, ibid., iv, 128 – 9.
[45] Birch *History*, ii, 249. On his promise see above, p. 163.
[46] Ibid., ii, 250. On the Duchess's visit, see Hunter, *Science and Society*, p. 72 and n. 39.
[47] For these, see below, nn. 103 – 110.

Contributions towards building the
COLLEGE.

contributing to the Building of a College upon the ground near Arundel-house given by the Honourable Henry Howard of Norfolk for that end', which gives the dates on which the 'Engagements' were signed.[48]

Information about sums promised — though not the date on which the offers were made — is also available from two other sources, a further list of 'Subscriptions' and a lavish volume bound in gilt-stamped vellum and inscribed 'Contributions towards Building the College', which still survives in the library of the Royal Society and is reproduced as Plate 10.[49] In this book, as Oldenburg had explained in his letter to Rycaut of 30 January, the names of benefactors were to be registered 'and thereby perpetuated to all posterity, as they shall well deserve, that doe assist according to their severall Abilitys, to render England the Glory of the Western World, by making it the Seat of the best knowledge, as well as it may be the Seat of the greatest Trade'.[50]

From these sources, it is possible to trace the progress of subscriptions, and the information they provide is summarized in the list at the end of this article. Two Fellows signed bonds in January, three in February, five in April, five in May, two in June and two, or perhaps three, in July.[51] A further six, or perhaps seven, offered contributions at dates that are unspecified, either because they did so after June/July 1668 or (more likely)

[48] Royal Society Miscellaneous Manuscripts, 11. 24. This list was evidently first compiled on 25 June 1668, when it was endorsed by Colwall to the effect that he had received from Oldenburg the nineteen 'Engagements' then listed in it, as ordered by the Council on 22 June (see below, p. 176). Oldenburg then entered three further subscriptions which were passed on to Colwall subsequently, on '26 July 1669' [sic], which could be an error for 1668.

[49] Misc. MSS 11. 25, and Royal Society MS 352. The former is entitled 'Subscriptions towards building a College' and endorsed: 'This is collected from the List of Engagements and the Catalogue of the Fellows', presumably an allusion to Misc. MSS 11. 24, and MS 352 respectively. A copy of the printed membership list for 1667 tipped in to the back of MS 352 records the contributions of Brouncker, Ward, Hayes and Howard only.

[50] *Oldenburg*, iv, 133. It should be noted here that I do not think that MS 352 is to be identified with the 'book in folio with vellum leaves, to contain the names of the benefactors to the society, together with the particulars of their respective donations' referred to in the minutes of the meeting on 13 April 1668 (Birch, *History*, ii, 265), which evidently never materialized.

[51] The uncertainty about numbers stems from the fact that Skippon's contribution is noted in the annexe to the list in Misc. MSS 11. 24, with no date, either because his bond was undated or because he signed it late in July 1668.

Plate 10. The special vellum-bound volume provided by the Royal Society as a record of subscriptions to the proposed College: Royal Society MS 352. Reproduced by permission of the President and Council of the Royal Society.

because they signed no bonds.[52] Some subscribers appear in two of the lists but not in the third, while the only other significant variation between them concerns John Evelyn, who is said to have offered £50 in the list of 'Engagements' but is down in the book of 'Contributions' as promising 50,000 bricks: £50 was perhaps their estimated value.[53]

In general, the contributors were predictable, comprising members of the Society who were both active and regular in their dues and others who were equally regular in paying but who are revealed by the minutes as somewhat less active. Only one was falling behind with his annual dues, the lawyer and office-holder, Henry Powle, though two others were somewhat irregular, the Denbighshire gentleman, Nicholas Bagenall, and Prince Rupert's secretary, James (later Sir James) Hayes; a fourth was not normally expected to pay any dues at all, Oldenburg himself.[54] One contributor, the financier Sir John Banks, offered assistance to the project on 24 July although not elected Fellow until 10 December 1668, while two men who offered financial assistance were not members at all. These were Sir Denis Gauden, who offered £40 on 23 March 1668, and Silas Taylor, who offered £10 on 24 March.

Taylor (alias Domville) was a former parliamentary captain and sequestrator who had acquired the post of Keeper of Naval Stores at Harwich in 1665. His scientific and antiquarian interests are well documented, and he contributed to the Royal Society's proceedings at various meetings in the 1660s although he never became a Fellow.[55] Gauden is not known to have had any association with the Royal Society apart from this: he was a prominent London merchant and alderman, Sheriff of the City in 1667 – 68, who was Surveyor of Victualling to the Navy. As a result he was a close contact of Pepys, who frequently refers to Gauden in his diary, and it was perhaps through Pepys that Gauden came into contact with the Society. But Gauden may have had other links with the circles from which Fellows were drawn: his brother was John Gauden, Bishop successively of Exeter and

[52] This is suggested by the fact that, although it was said that over £1000 had been promised by 4 May 1668 (Birch, *History*, ii, 275), the dated contributions up to that date listed in Misc. MSS 11. 24 amount to substantially less than that. It may also be relevant that those whose contributions are not dated include various people dunned in April 1668: see below, p. 171. The Fellows whose offers are undated are (apart from Skippon: see previous note): Moray, M. Wren, Powle, C. Wyche, Williamson and Stewart; all appear in MS 352 and Misc. MSS 11. 25, but not in ibid., 24.

[53] Evelyn, *Diary*, iii, 507, mentions the bricks but not the £50. In Misc MSS 11. 25, Evelyn is said to have offered both. Bagenall and Gauden appear in Misc. MSS 11. 24 – 25 but not in MS 352 (it may be significant that their names are separately listed in Misc. MSS 11. 25). See also previous note.

[54] On these Fellows, see Hunter, *Royal Society*, catalogue entries 27, 172, 111, and 33 respectively. Stewart (catalogue entry 238) was to prove only fairly regular; but he had joined only in October 1667.

[55] See ibid., p. 8, and *DNB*.

Worcester, who was briefly a Fellow before his death in 1662, while his wife was a correspondent of the Latitudinarian divine, Simon Patrick.[56]

If all the lists are collated, it appears that a total of £1285 was promised. In many ways, however, what is most striking is how few offers were made altogether, a point to which I shall return. Only a handful of those listed as likely contributors at the meeting on 11 January offered contributions, and the same is true of those mentioned on 25 January: in the event even the contribution of £50 or more by the naval officer, Sir Philip Carteret, which Oldenburg had announced to Boyle on 14 January as virtually certain, failed to materialize.[57] The Duchess of Newcastle, it may be added, did not subscribe, and neither did anyone from outside the Society apart from the two already mentioned. It is perhaps not surprising that, whereas in January Oldenburg had been optimistically assuring his correspondents that work would begin in the spring, by 10 February he was more cautiously talking about the summer.[58] On 13 April a further attempt was made to solicit contributions, which were clearly felt to be disappointing, and various Fellows — not all of them on the original 'Committee of Beggars' — were deputed to dun likely contributors: this was repeated on 20 April and 4 May, but even then by no means all those singled out for attention obliged.[59]

It could thus hardly be claimed that offers of assistance flooded in, but by 4 May 1668 it was reported in the minutes that over £1000 had been promised. The Society was now encouraged to commit itself quite seriously to the project, and Sir Anthony Morgan was asked to expedite the deed of conveyance for the site, which he promised to do 'as soon as the ground should be surveyed, to know the bounds allotted for the said building': as has already been seen, this surveying had not been completed in January.[60] The familiar message recurs in Oldenburg's correspondence at this time: as he told Hevelius on 11 May, 'Our Fellows are now of a mind to build a College for themselves, as a permanent accommodation for the Royal Society; the hope is, that when this is done their labours will meet with even greater success'.[61] Plans had already been requested from Christopher Wren

[56] J. R. Woodhead, *The Rulers of London, 1660 – 1689* (London, 1965), p. 75; Evelyn, *Diary*, iii, 510 n. 2, 512 n. 1. Gauden is included in the lists in Misc. MSS 11. 24 – 25, but not in MS 32, unlike Taylor, who appears in all three. On John Gauden and the Royal Society, see Hunter, *Royal Society*, catalogue entry 96.

[57] Offers of 25 January which failed to come to fruition involved the Earl of Anglesey, Lord Annesley, Tuke, Povey and Nelthorpe (see, however, below, p. 176). Of those suggested on 11 January (Birch *History*, ii, 238 – 239), only nine out of forty-three offered assistance.

[58] Oldenburg to Norwood, 10 February 1668, *Oldenburg*, iv, 167 – 168.

[59] Birch, *History*, ii, 265, 271, 275. Lord Berkeley, named at this time, had not been on the original committee. Of those named, M. Wren, Williamson and C. Wyche subscribed; Paston did not. There are also vaguer references to whole social and occupational groups among the membership who might be dunned.

[60] Ibid., ii, 275. See also ibid., ii, 284, 189.

[61] *Oldenburg*, iv, 395, 397 – 398. See also Oldenburg to Harpur, 22 May 1668, ibid., iv, 421, 424.

and Robert Hooke, while Howard had evidently himself also prepared a design, which 'he took delight to shew' Wren at Oxford at this stage.[62]

Hooke, who had initially been asked to prepare plans as early as 11 January, was asked on 11 May to bring in his 'draught' for the building and an estimate of its likely cost, while John Wilkins on 4 May and Sir Paul Neile on the 11th were asked to procure Wren's design.[63] On 18 May, it was agreed that the dimensions of the great meeting room of the college should be 28 feet broad, 42 feet long and 25 feet high, but no further discussion is recorded.[64] The precise dimensions of the site were mentioned at the next meeting, on 30 May, when it was recorded that Howard 'had set out the ground for building the society's college upon, viz. an hundred feet one way, and forty feet the other': the staking out of the site was thus now finally completed. In addition, Oldenburg was requested at this meeting to write to Wren at Oxford requesting him to wait on Howard about the proposed design.[65] This stimulated a reply from Wren, dated 7 June 1668, in which he outlined his design, a letter which is readily available in various printed editions.[66] Though Wren enclosed a drawing with the letter, this no longer survives: an elevation in a volume of architectural drawings in the British Library that has sometimes been thought to be of this building is almost certainly not by Wren, while there is absolutely no evidence to link it with this building, despite the fact that it has been reproduced in this connection.[67]

Wren's letter, however, gives a good idea of the scale of the building he conceived as appropriate, and the facilities that he would have liked to include. In its basement it was to contain a cellar, 'a Fair Elaboratory', workshops, kitchen and larder. At street level, a passage-hall was to lead from the front of the building to the back, opening to 'the great Staire' to the first floor and also to 'a fair Roome for a Library & Repository which may well be one roome placing the bookes after the moderne way in glasse presses, or if you will divide the roome with pillars it will the better support

62 Wren to Oldenburg, 7 June 1668, ibid., iv, 454.

63 See above, pp. 163, 166. Birch, *History*, ii, 275, 282.

64 Ibid., ii, 284.

65 Ibid., ii, 289.

66 Ibid., ii, 290 – 1; Weld, *History* (n. 2), i, 212 – 213; *Wren Society*, 13 (1936), 48 – 49; *Oldenburg*, iv, 454 – 5.

67 The elevation of British Library MS Sloane 5238, no. 66, is cited in *Wren Society*, 13 (1936), 49, and reproduced in ibid., 5 (1928), plate 27, and *Oldenburg*, iv, plate 4. It differs in various details, however, from the description given in Wren's letter, including its dimensions, the style of its roof, etc. It is probably an elevation for the facade of a private house (as is suggested in *Wren Society*, 5, 11), and is possibly by Hooke, like other drawings in this volume. I am most grateful to Mr H. M. Colvin for advice on this matter in a personal communication. See also E. F. Sekler, *Wren and his Place in European Architecture* (London, 1956), p. 146 n. 1; V. Fuerst, *The Architecture of Sir Christopher Wren* (London, 1956), p. 202 n. 551; and Batten, 'The Architecture' (n. 32), pp. 87 – 88.

the floor of the great roome above it, & soe place the presses for bookes in one part & the presses for rarities in the other'. A parlour for the housekeeper was also included. The first floor was occupied by the council room, by 'a litle closet for the Secretary' and by the great meeting room— '40 foot long & 2 stories high, divided from the Anti-chamber by a Skreen between columnes, soe that the whole length in case of an entertainment may be 55 foot'. On the second floor were rooms for the curators, one of them situated over the stairs and ante-room and looking down into the great meeting room, while in the attic was a 'passage gallery' the whole length of the building — useful 'for tryal of glasses & other experiments that require length' — together with further workshops and smaller chambers for the Society's humbler employees. There was also to be a platform on the roof for experiments, while 'in the middle rises a Cupolo for observations, & may be fitted likewise for an anatomy Theater'.

Wren's design seems thus to have had a somewhat grander meeting room than the Society had agreed they wanted on 18 May — two stories, as against 25 feet high, and 55, as against 42, feet long overall. In addition, the building as a whole was evidently 'somwhat greater then was proposed, as being 100 foot long & 30 foot broad' (thus taking up more of the proposed site), 'which length Mr Howard doth not scruple to allow you', as Wren assured Oldenburg: it is interesting to find that Howard had at first considered Wren's scheme too expensive, though he subsequently acquiesced. This theme of scale and expense recurs in the last paragraph of the letter, where Wren blithely remarks: 'As to the Charge of this fabrick, I confess it is my opinion that a fair building may easier be carried on by contributions with time, then a sordid one'. But he offered a compromise, of laying the foundations for the whole building but erecting only half of it for the time being, which 'will containe the necessary roomes, & soe you will leave your selves an opportunity of enlarging hereafter upon the same modell'. Wren folded the drawing that he enclosed to illustrate what should be built at present — presumably the end of the building containing the library and meeting room, since he suggested 'an extempore stairecase of deale boards & lath'.

Even Wren's compromise version, however, would have cost £2000, according to his estimate in the letter, while leaving a continuing commitment for further expenditure in the future. It is therefore perhaps not surprising that the Society turned instead to Hooke, who may have been readier to try to be economical. Despite Wren's view that 'a fair building' would attract contributions better than 'a sordid one', the Society's experience of the difficulty of raising funds at all doubtless suggested that the erection of a grand but half-finished structure would be an unwise gamble. But there was nothing surprising about the choice of Hooke, who was a versatile architect in his own right, though greatly overshadowed by Wren.[68] It appears that Howard's design also still remained in the offing, perhaps for reasons of tact towards the Society's host and benefactor,

[68] See above, n. 32: compare the opinion quoted by Bennett, 'Wren's Last Building?' (n. 14), p. 116 n. 8.

though it is possible that Howard was a skilful amateur architect himself, at a time when this was not uncommon.[69]

Hooke's design was produced at a meeting on 22 June at which Oldenburg was ordered to deliver the obligations that had been signed for contributions to the Treasurer, Daniel Colwall, so that he could 'begin to collect what is due by the said obligations'.[70] Hooke's design was 'agreed upon' and he 'was ordered to get a model of it made with one door, and to consider of the buying of the materials, and of contracting with workmen, to be paid by measure for so much a rod and square: as also to find out a person to be constantly present; and to see the workmen do their duty'. Subsequently, on 29 June, Hooke was asked to bring 'an estimate of the charge both of the materials and workmanship of the building', while he and Wilkins were desired to speak to Edward Nelthorpe, evidently a Fellow with relevant contacts, about the provision of timber.[71]

On 6 and 13 July, further consideration was given to the design. On the 6th, Hooke was ordered to make a draught of the building with the front towards the Thames (the frontage of the original design had presumably faced north) 'and to draw it with the windows'; he was also requested to estimate the cost of this, though unfortunately no drawings or specifications by Hooke survive to illustrate how his scheme would have compared with Wren's, either in its original or modified form.[72] Then, on the 13th, an alternative design with a river frontage by Henry Howard was considered, perhaps because Howard had broader plans for the whole Arundel House site into which the Society's building would have had to fit.[73]

A decision on this matter was deferred to the next meeting, but no such decision was ever taken. The Council did not meet again until 10 August, and at that meeting 'it was resolved, that the building of the college should be deferred till spring, and in the mean time good materials be provided'.[74] In fact, however, the project was never taken up again. In a letter of 13 July to Sir William Curtius in Frankfurt, Oldenburg circumspectly remarked that 'the building of a College is being looked into', but he made no subsequent reference to it in any of his letters; though Henri Justel, writing to Oldenburg on 23 September, thought that the project was still alive, he was out of touch, and after August the scheme disappears from the Society's records.[75]

69 Birch, *History*, ii, 305. Cf. Judith Hook, *The Baroque Age in England* (London, 1976), p. 63.

70 Birch, *History*, ii, 300. These were the 'Engagements' listed in Misc. MSS 11. 24. See note 48.

71 Birch, *History*, ii, 302.

72 Ibid., ii, 304.

73 Ibid., ii, 305. Howard was asked to bring to the next meeting 'his design of ordering the whole plot of his land'. On Howard's redevelopment plans for the site see above, p. 164.

74 Birch, *History*, ii, 313.

75 *Oldenburg*, iv, 537, 539; ibid., v, 63 – 64. On Curtius, see Evelyn, *Diary*, iii, 36 n. 1.

Why was the plan abandoned in this way? A minor and hitherto unnoticed complication may have been the delicacy of choosing between the designs of Hooke and Howard, as offered on 6 and 13 July: it might have been difficult to disappoint the latter, as the Society's benefactor, even if his plan was less suitable than Hooke's. But this seems unlikely to have been a major factor in the decision. A more serious suggestion, first made by C. R. Weld and subsequently repeated by other authors, was that the demise of the scheme could be linked with the Society's final completion of its ownership of Chelsea College on 8 April 1669 after years of negotiation, involving a new 'patent' to the Society giving it fresh corporate powers.[76] Yet there is no reason to believe that this affected the fortunes of the proposed building. The college was evidently intended to house the Society's main activities and facilities, and it is doubtful if it was ever considered desirable or feasible to transfer these out of London to Chelsea. In fact, as Sprat's opinion quoted earlier suggests, the two premises were seen as complementary, and the Chelsea site was in any case little used thereafter before being sold back to the Crown in 1682.[77]

A more plausible factor in the abandonment of the college, also noted by Weld, may have been the Society's concern about the security of tenure of the site.[78] Arundel House, together with other property owned by the Howard family, had been entailed by an Act of Parliament obtained by the Earl of Arundel in 1627 restricting his title and these lands to male descendants of the fourth Duke of Norfolk excluding the Suffolk family; this remained in force until 1956.[79] The Society had been aware of the need for safeguards of their tenure as early as January 1668, but, as the date for building approached, the matter was further discussed, Howard being said to be 'willing to enter into a bond of six thousand pounds for performance of covenants forthwith, and into another of six thousand pounds more, that he would within a twelvemonth either procure an act of parliament to enable him to make estates of the ground belonging to Arundel-house, notwithstanding the act of 3 Car. I that had intailed it; or other good and indefeasible title for the society; or else give them collateral security by conveying land to them'.[80]

That this was a major factor is suggested by the fact that at the meeting on 10 August when the decision to postpone the plan till the following spring was taken, Howard 'promised, that he would endeavour to procure an act

[76] Birch, *History*, ii, 363 – 371; Weld, *History* (n. 2), i, 214; *Wren Society*, 13 (1936), 49.

[77] See, however, the contrasting report of Magalotti, above, n. 13.

[78] Weld, *History* (n. 2), i, 214. See also Sir Henry Lyons, *The Royal Society, 1660 – 1940* (Cambridge, 1944), p. 66, and A. R. Hall's introduction to the Johnson Reprint Corporation edition of Birch's *History* (New York, 1968), p. xxi.

[79] Robinson, *The Dukes of Norfolk* (n. 37), pp. 115 – 116.

[80] Birch, *History*, ii, 243, 299 – 300; see also ibid., ii, 249, 265, 275, 284, 289.

of parliament for letting of leases against that time'.[81] Howard never obtained such an Act, though one may presume that he could have done, since Acts of Parliament were granted in 1671 and 1689 for redevelopment schemes for the site which *did* involve leases of part of it, including the gardens on which the college would have been built.[82] Is it conceivable that Howard — who, after all, had considered such redevelopment before the college was mooted — had had cold feet about his generous offer on reflection about the site's commercial value?

An alternative site could, however, presumably have been found if the Society had been able to afford it, though it is symptomatic of the Society's bad luck with its benefactors that even this would have been rather embarrassing after it had first gone along with Howard's offer. In fact, the principal reason for the abandonment of the project was evidently financial. After the steady flow of subscriptions between April and mid-June, by late July only two more subscriptions are definitely known to have been promised, and, even if some additional offers came in at this stage,[83] it was perhaps becoming increasingly obvious that sufficient funding was unlikely to be forthcoming for a building on the grand scale which the Society wanted.

Wren's design may have been too lavish, but the size of meeting room to which the Council agreed on 18 May would itself have involved sizeable expense, almost certainly more than the £1000 or so that the Society had by then collected. Even Wren's estimate for starting his building was twice this sum, while a further comparison may be made with a slightly smaller structure erected for a comparable purpose between 1679 and 1683, the Ashmolean Museum at Oxford, which cost £3970 to build (with a further £560 for the site).[84] It should be categorically stated here that, contrary to what has sometimes been thought, there is no evidence for any connection between the Ashmolean and Wren's design for the Royal Society.[85] But this

[81] Ibid., ii, 313.

[82] Kingsford, 'Bath Inn' (n. 37), p. 256.

[83] Some of the offers — totalling £300 — included in Misc. MSS 11. 25, and MS 352 but not in Misc. MSS 11. 24, may have come in at this stage. See also above, pp. 169 – 70.

[84] *The Life and Times of Anthony Wood*, ed. A. Clark, iv (Oxford Historical Society, 30, 1895), 78 – 79.

[85] See R. T. Gunther, ed., *Early Science in Oxford*, iii (Oxford, 1925), 293 – 295; C. H. Josten, *Elias Ashmole* (Oxford, 1966), i, 233 – 234, and iv, 1483 n. 1, and Martin Welch, 'The Foundation of the Ashmolean Museum', in *Tradescant's Rarities*, ed. Arthur MacGregor (Oxford, 1983), p. 47 (a compromise position). The only evidence concerning the design of the Ashmolean comes from Michael Burghers's contemporary engraving of the building, which states that the 'Architect' was the Oxford master mason, Thomas Wood (see H. M. Colvin, *A Biographical Dictionary of British Architects, 1600 – 1840* (London, 1978), p. 912). The Old Ashmolean is, like the projected college, a purpose-built scientific institute in the classical style, but claims that the Oxford building is markedly similar in layout to the design outlined by Wren in his letter to Oldenburg of 7 June 1668 cannot be sustained. The dimensions are quite different: the Old Ashmolean is only 60 feet long, whereas

comparison indicates the scale of expenditure necessary, a scale which the Society could never have afforded without a much higher level of response from Fellows and non-Fellows than was in evidence during the fund-raising period. The decision to defer the building and the failure to readopt the scheme is therefore hardly surprising. Indeed in many ways the most striking aspect of the episode was the disappointing response of Fellows to the request for funds. In the volume dedicated to 'Contributions towards Building the College' a space was left beside the name of every Fellow in the evident hope of widespread contribution, but only twenty-seven out of more than 200 Fellows promised anything, and the total promised was at most £1285.[86]

In the difficulties that it encountered in raising funds, the Royal Society may be paralleled by other London institutions which projected new premises in the late 1600s, including the College of Physicians, the College of Heralds and Sion College. In contrast to the Royal Society, each of these succeeded in carrying out their plans, but in each case the building costs ran into thousands of pounds, while voluntary contributions proved so disappointing that alternative sources of finance had to be found and the shortage of funds produced numerous delays and complications.[87] Matters were made worse by the fact that so much rebuilding activity was going on at once in the aftermath of the Great Fire, which had destroyed the premises which in these other cases were now replaced. 1668 was the year in which large-scale rebuilding activity got under way, and this too may have contributed to the demise of the Society's plans after the brief optimism of that year, tipping the balance by making finance and expertise scarcer than in normal circumstances.[88]

That the Royal Society's difficulties proved insuperable while those of the Physicians, Heralds and Sion College did not might be taken to illustrate the

---

Wren's proposed building was to be 100 feet, and while the Ashmolean's first floor room is 56 feet 2 inches long, Wren's was to be 55 feet only if the ante-chamber separated from it by columns and a screen was included, and was only 40 feet long without. In addition, there is a serious difference between the layout of rooms outlined by Wren and that at the Ashmolean. For whereas at the Ashmolean the first floor room was given over to the display of rarities, at the Royal Society the first floor was to be occupied by the meeting room, the repository and library being consigned to the entrance floor. It is true that a basement laboratory was a common feature of the designs, but beyond this the differences are more marked than the similarities.

86 Misc. MSS 11. 25. A pencilled total in MS 352 (£1075) differs, due partly to the omission already noted (n. 53) and partly to faulty arithmetic.

87 Sir George Clark, *A History of the Royal College of Physicians of London* (Oxford, 1964 – 6), i, 328 – 333; W. H. Godfrey, Sir Anthony Wagner and H. Stanford London, *The College of Arms, Queen Victoria Street* (London Survey Committee 16th Monograph, 1963), ch. 2; T. F. Reddaway, *The Rebuilding of London after the Great Fire* (London, 1940), pp. 259 – 260, 264 – 266.

88 Reddaway, *Rebuilding*, ch. 9.

relatively low valuation of natural philosophy as against other intellectual activities at the time. But matters were a little more complicated, for there seems to have been considerable difference of opinion among the Fellows as to whether a college was desirable at all. Oldenburg's enthusiasm is patent — despite his occasional pessimism about the likely outcome of the appeal for support that has already been noted — and his views were shared by Fellows like John Evelyn, who was convinced that failure to implement the plan would prove detrimental to the Society.[89] But the fact that neither all those on the committee for soliciting subscriptions nor all active members of the Council for the year bothered to contribute suggests a failure of many to agree wholeheartedly on the desirability of purpose-built premises like those planned, rather than simple parsimony.[90]

There are further indications of disagreement within the Society's ranks. The canny John Wallis, Savilian Professor of Geometry at Oxford and an experienced trimmer, wrote to Oldenburg when asked to contribute to the college and to solicit further contributions in Oxford: 'I wish your Building good successe; but cannot promise that I shal here be able to get you subscriptions toward it; nor do I presume to deliver my opinion of the design, (knowing so little of it,) but take for granted so many wise men see very good reasons to undertake it, & ways how to get it perfected'.[91] Perhaps the most interesting remark about the whole affair is Pepys's, who on 2 April noted in his diary how he and others were forced to subscribe, 'but', he continued, 'several I saw hang off, and I doubt it will spoil the Society — for it breeds faction and ill will, and becomes burdensome to some that cannot or would not do it'.[92] Hence, while I have listed the sums promised at the end of this article, conclusions about Fellows' failure to contribute should take account of these evident differences of opinion about the abortive project.

The attitude of those with reservations was probably comparable to that of the Fellows who, in 1710, resisted the move to Crane Court on the grounds that Gresham College offered the Society ample facilities and that premises of its own would involve unnecessary expense.[93] This makes it less surprising that plans for a building for the Society were never seriously revived in the interim — although there are occasional hints of the possibility of special premises, while the promised subscriptions, though never paid, were still being listed by Oldenburg among the Society's assets in 1674 – 75 (it is doubtful, however, if they were more realistic than the

[89] Evelyn, *Diary*, iii, 505.

[90] On the committee: John Hoskins and Sir John Lowther. Active Council members: Sir Paul Neile, Sir Anthony Morgan, Sir George Ent, Lord Brereton, William Erskine, Charles Howard, John Creed. On the unrealized offer of another Council member, Carteret, see above, pp. 163, 171. See Hunter, *Royal Society*, catalogue entries 92, 157, 5, 16, 31, 32, 79, 123, 155, 188.

[91] Wallis to Oldenburg, 8 February 1668, *Oldenburg*, iv, 159.

[92] Pepys, *Diary*, ix, 146.

[93] See *An Account of the late Proceedings in the Council of the Royal Society* (London, 1710), pp. 22f. and *passim*.

arrears of Fellows' annual dues alongside which they appeared).[94] The possibility of a royal grant of land for a building is mentioned in 1686, but this came to nothing, despite the fact that in 1682 the Society had acquired capital from the Crown in return for Chelsea College, which could have been used to fund such an enterprise.[95]

Similarly, nothing came to Robert Hooke's professed wish to devote his Estate to 'an handsome Fabrick for the *Societies* use, with a Library, Respositary [sic], Laboratory, and other Conveniencies for making Experiments', since he died intestate.[96] Instead, as already noted, the Society remained throughout the seventeenth century and into the eighteenth in borrowed premises, only finding its own when its tenure at Gresham came under threat from 1699 onwards. And though the possibilities considered at that stage included gaining a grant of land from Queen Anne for a building, or persuading the trustees responsible for Gresham to incorporate a building with meeting room, library, repository, laboratory and observatory for the Royal Society into their redevelopment scheme for the site, in the end nothing came of either idea.[97] Instead, the Society made do in 1710 with a house in Crane Court which was extended to provide room for the Society's repository — almost certainly to designs by Wren — but which was far more modest than the lavish structure planned in 1668.[98]

This, then, is the story of the Society's abortive plan to erect a college, which might have meant that London was graced by a purpose-built Baroque scientific research institute south of the Strand. In its failure, it reveals much about the Society's mode of operation in its early years, highlighting the vulnerability of its organizers to the 'general will' of the rank and file, and, more broadly, illustrating the Society's largely unrealized

94 R. K. Bluhm, 'Remarks on the Royal Society's Finances, 1660 – 1768', *Notes and Records*, 13 (1958), 94. The figure of £925 mentioned is the same as the total in the list of contributions up to 25 June 1668 only in Misc. MSS 11. 24, but this is probably coincidental: Oldenburg's figure is probably modified to allow for deaths in the interim. For a note by Robert Hooke suggesting that the possibility of using John Wilkins's bequest to the Society for a building was discussed in 1674, see Hooke, *Diary*, pp. 100 – 1. Beyond this, however, the references cited by Bennett, 'Wren's Last Building?' (n. 14), pp. 107 – 108, as concerning plans for a possible building at this time in fact refer to a revamping of the Society's organization: on this, see below, 'Towards Solomon's House'. Josten, *Elias Ashmole* (n. 85), i, 208, and iv, 1443, records a promise of £5 from Ashmole to Hooke in 1676 'towards building', but there is no evidence to link this with the Royal Society, as is there suggested.

95 Hunter, *Science and Society*, pp. 127 – 128. It is possibly in terms of hopes to acquire or construct a premises that one should interpret the comment of the Society's then Secretary, Francis Aston, in a letter to Martin Lister of 18 March 1682, that 'the Society is gathering in Arrears towards the making some purchase having sold Chelsy College for £1300': Bodleian Library, Oxford, MS Lister 35, fol. 64. On the arrears drive at this time, see Hunter, *Royal Society*, esp. ch. 3.

96 Hooke, *Posthumous Works*, p. xxvii; 'Espinasse, *Robert Hooke* (n. 32), p. 154.

97 Adamson, 'Royal Society' (n. 1), pp. 12 – 13 and *passim*; Wren's 'Proposals for building a House for the Royal Society', in Weld, *History* (n. 2), i, 363 – 4.

98 See Bennett, 'Wren's Last Building?' (n. 14); *An Account* (n. 93), *passim*.

— perhaps unrealistic — aspirations. Whether Oldenburg and others were right in seeing this as a crucial opportunity to 'establish' the Society, or whether more sense was spoken by those Fellows who saw premises as an unnecessary administrative burden, this is nevertheless an important episode in the Society's early history which deserves to be better known.

Contributions promised towards the 'College', in order of seniority within the Society, with dates (where known):

William, Viscount Brouncker (F1)[99]		£100	30 January[100]
Robert Boyle (F2)		£50	6 February[101]
Sir Robert Moray (F4)		£80	—
John Wilkins (F6)		£50	17 February[102]
Jonathan Goddard (F7)		£20	24 July
Thomas Henshaw (F15)		£20	7 May
Matthew Wren (F21)		£50	—
Henry Powle (F27)		£20	—
William Cavendish, Earl of Devonshire (F28)		£50	16 April
Seth Ward (F29)		£100	24 February[103]
Henry Oldenburg (F33)		£10	18 June
John Evelyn (F38)	50,000 bricks and/or	£50	2 April[104]
Daniel Colwall (F48)		£100	13 April[105]
Sir Cyril Wyche (F83)		£50	—
Theodore Haak (F93)		£10	2 April
Francis Willughby (F95)		£80	7 May
Joseph Williamson (F101)		£50	—
Sir Robert Southwell (F107)		£50	10 June[106]
James Hayes (F111)		£40	30 January[107]
Nicholas Bagenall (F172)		£20	10 May
Samuel Pepys (F187)		£40	2 April
Sir Theodore de Vaux (F197)		£25	21 May[108]

[99] These bracketed references denote entries in the 'Catalogue of Fellows, 1660 – 1700', in Hunter, *Royal Society*.
[100] Cf. Birch, *History*, ii, 244.
[101] See above, n. 40.
[102] Cf. Birch, *History*, ii, 249.
[103] Cf. ibid., ii, 252.
[104] See above, p. 170.
[105] Cf. Birch, *History*, ii, 265.
[106] Or perhaps 11. Cf. Birch, ii, 293.
[107] Cf. ibid., ii, 244.
[108] Or 28 May. Cf. ibid., ii, 286.

Henry Howard (F218)	The site	11 January
Philip Skippon (F227)	£20	—[109]
Sir Nicholas Stewart (F238)	£50	—
John Colwall (F262)	£80	21 May[110]
Sir John Banks (F272)	£20	24 July
Sir Denis Gauden	£40	23 March
Silas Taylor	£10	24 March

John Evelyn's Designs for his Proposed 'College', 1659

The proposal for a 'college' that John Evelyn made in a letter to Robert Boyle of 3 September 1659 is well-known, having often been reprinted since it was first published by Thomas Birch in his edition of *The Works of Robert Boyle* in 1744; it has frequently been discussed by those interested in the roots of the aspiration to institutionalize science which was to come to fruition with the foundation of the Royal Society.[111] In his letter, Evelyn gave details both of the components and layout of his proposed institution and of the regimen that he expected to exist there, while in a postscript he promised to bring Boyle 'a rude plot of the building, which will better fix the idea, and shew what symmetry it holds with this description' (in a slightly different version which Evelyn preserved in his letterbook, he promised 'the plot of this fabrique and designe, as I have heretofore scrabbled it out in perspective, and the Ichnography, the better to fix the Idea').[112]

[109] See above, n. 51.
[110] Cf. Birch, *History*, ii, 284.
[111] See above, pp. 7, 157, and, e.g., Charles Webster, *The Great Instauration* (London, 1975), p. 98.
[112] Boyle, *Works*, vi, 288 – 91 (in the folio edition of 1744, the letter occupies pp. 397 – 9 of vol. 5); Evelyn, *Diary and Correspondence*, iii, 116 – 20 (apparently derived from the above text, the only significant difference being 'one laboratory' for 'an elaboratory' on pp. 118/289)). For the copy in Evelyn's letterbook, see Christ Church, Oxford, Evelyn Collection, MS 39, no. 155. There are various verbal differences between this and the published version, but the only substantive one (apart from that noted in the text) is that in the MS version a further employee at common charge was added, namely 'an Artist that can be usefull in the Elaboratory, and turne his hand to any ingenious worke'. Possibly this reflects Boyle's influence; the cost of employing him does not appear separately among the expenses, as that for the chaplain does.

Plate 11 (over). John Evelyn's drawings of his proposed 'College', comprising a plan (above) and a perspective view (below). Reproduced from the originals now preserved in the Evelyn Collection at Christ Church, Oxford, by courtesy of the Trustees of the Will of Major Peter George Evelyn.

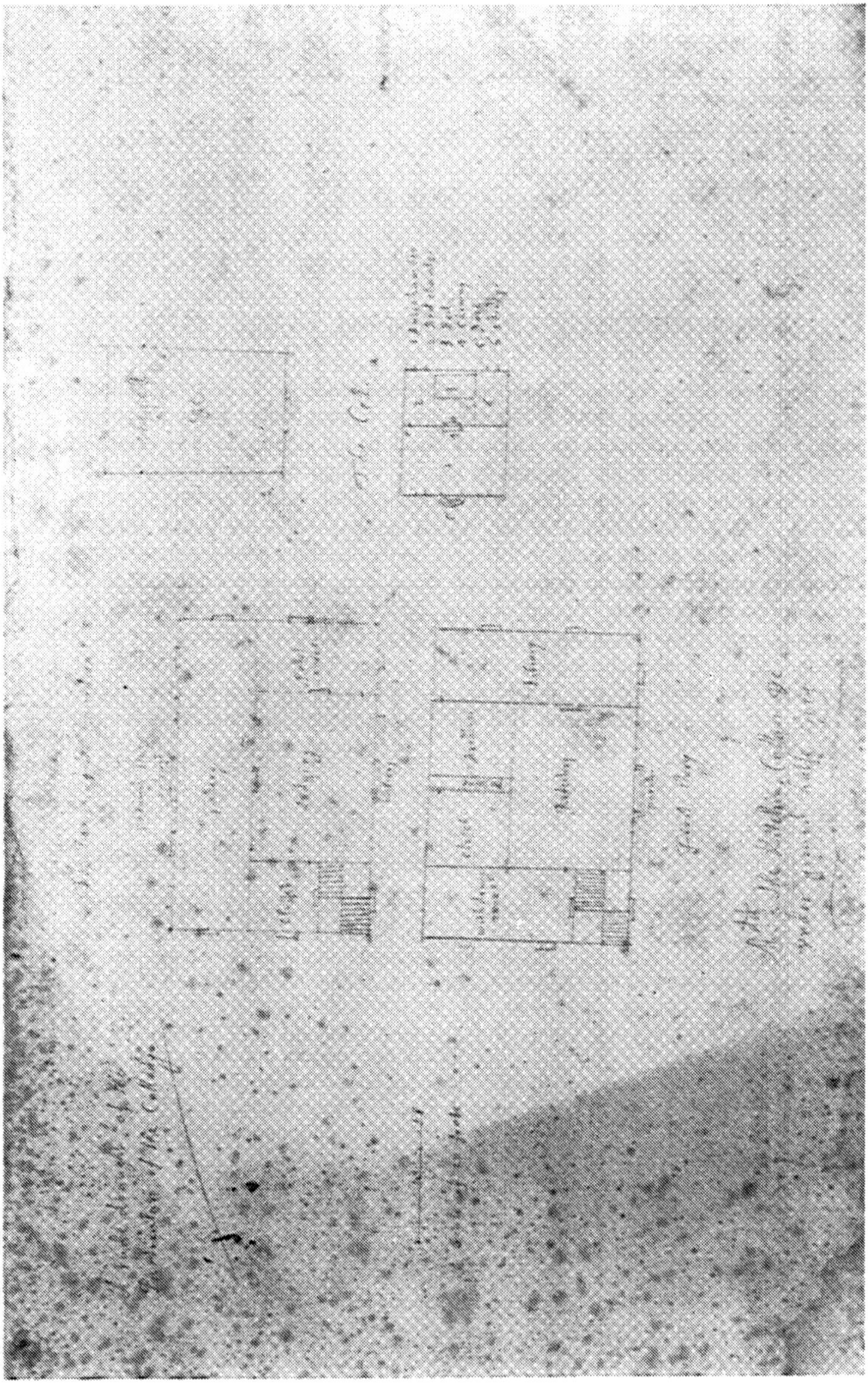

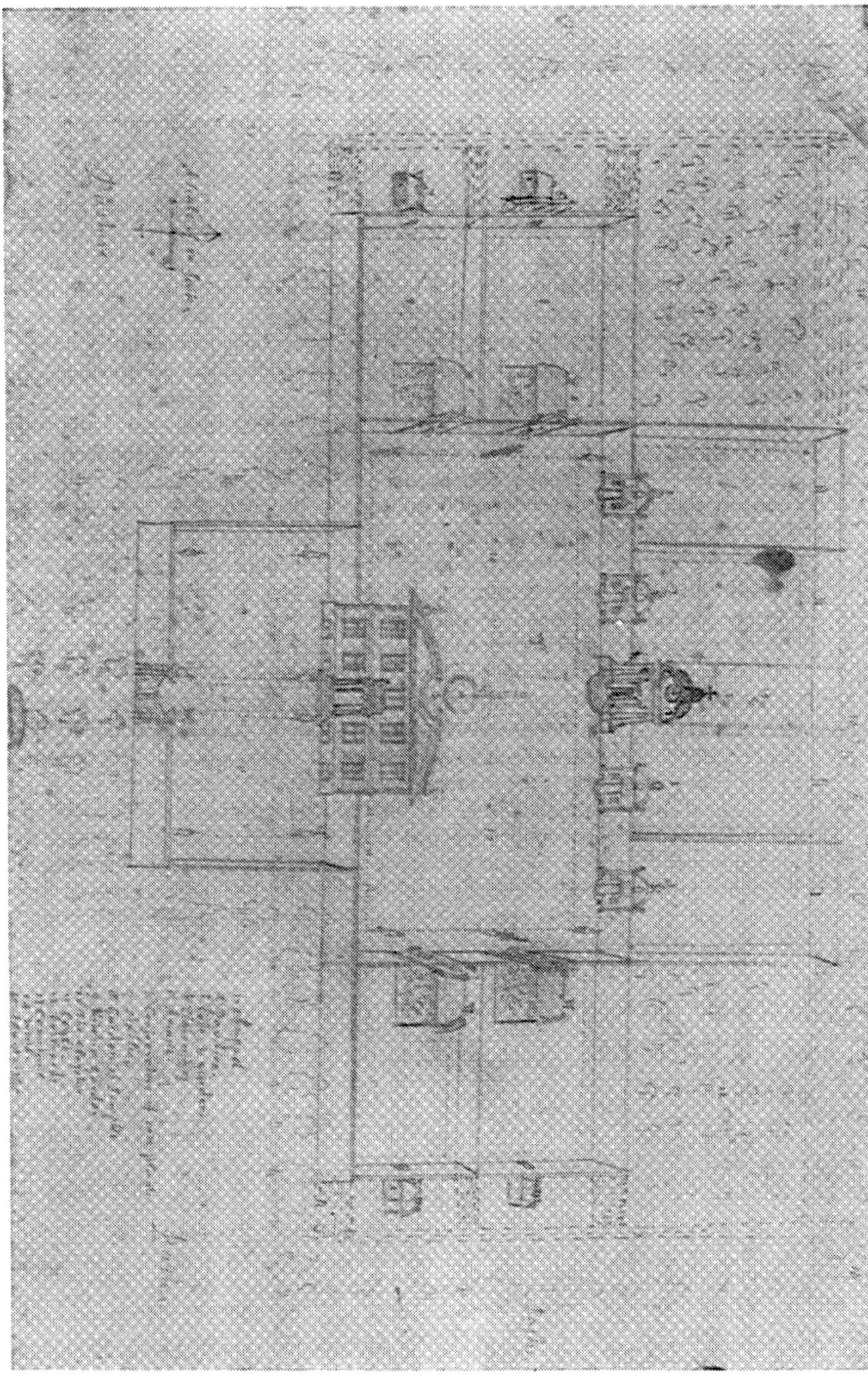

What has not hitherto been known is that this plan and 'Ichnography' still survive in the form of a pair of careful pencil drawings (with minor additions in ink) on one side of a folded sheet of paper. This is now preserved, together with other architectural drawings by Evelyn, in a paper folder in Box XI of loose papers entitled 'Evelyn Period Misc.' in the Evelyn Collection at Christ Church, Oxford. In view of their interest in providing a visual record of one of the schemes which form the background to the Royal Society's own 'College', it has seemed appropriate to reproduce them here, where they appear by courtesy of the Trustees of the Will of Major Peter George Evelyn.

The drawings — which are drawn to 'A Scale of 20 foote' — almost exactly reflect the scheme as outlined in the letter. The perspective drawing shows a 'Pavillion', the 'chappel', the six cells with their gardens, the laboratory, the 'Conservatory of rare plants', the stables and so on. All are laid out around a central courtyard with an adjoining forecourt, at the spot where the pasture and woodland met on Evelyn's thirty to forty acre site. A few features do not appear, such as the repository, though this was perhaps compensated for by the 'Archive' shown in the 'Pavillion' on the scaled plan. This plan gives 'a rude draught of the Particulars of the Colledge': it shows the layout of the rooms on both of the main floors of the pavilion (noting that the 'under ground halfe story' was to contain 'the Kitchin, Cellars &c.'), accompanied by plans of the chapel and of a cell.

The perspective well illustrates Evelyn's knowledge of the latest trends in architecture, particularly the pavilion, with its daring broken pediment and free-standing statues, and the chapel, with its attached columns, scrolled pediment and cupola: such features doubtless show the influence of Evelyn's observations of Baroque buildings during his travels in Italy. The depiction of the other buildings is too rudimentary for much to be learnt from it, but the general symmetrical layout, with a fountain at the centre, clearly reflects the influence of Renaissance planning. Like the verbal proposals that survive in connection with the Royal Society's 'College', Evelyn's scheme gives a tantalizing view of what enthusiasts for the new learning visualized as its appropriate architectural setting, otherwise only in evidence in England in the extant Old Ashmolean building at Oxford.

6

TOWARDS SOLOMON'S HOUSE: RIVAL STRATEGIES FOR REFORMING THE EARLY ROYAL SOCIETY*

Is there anything new to be said about the early Royal Society? Some readers may think it doubtful. The Society's high — perhaps undeservedly high — reputation in the international scientific community of its day; its association — albeit at times indirectly — with one of the most fertile periods in the history of English science; and its novel and influential institutional role, evolving into a 'type' of voluntary body devoted to the promotion of research which had been widely emulated since: all this has inspired a plethora of writings about the Society's 'nature' in its early years, aimed at elucidating how it originated, what it stood for, how it was constituted and just how much it achieved.[1]

We believe, however, that a fresh approach is possible, and in this paper we wish to consider, not so much what the early Royal Society was like, but what it might have been like. We will argue that a wider range of alternative strategies was considered in the Society's early years than is often thought; that these options — both intellectual and institutional — were more discussed than is generally realized; and that in the course of such discussion

* This paper was jointly written with Paul B. Wood, and I am very grateful to him for allowing me to reprint it here. The research for it was assisted by a grant from the Royal Society under their history of science research grant scheme. We are also grateful to the librarian of the Royal Society and his staff, especially Sally Grover, for their co-operation, to the Earl of Shelburne for permission to consult the microfilms of the Petty MSS on deposit at the Bodleian Library, Oxford, and to the following: Jenny Bray, Robert G. Frank, jr, Marie Boas Hall, John Henry and Wallace McLeod.

[1] For writings up to 1981 see Hunter, *Science and Society*. In addition see Lotte and Glenn Mulligan, 'Reconstructing Restoration Science: Styles of Leadership and Social Composition of the Early Royal Society', *Social Studies of Science*, 11 (1981), 327 – 64; Michael Hunter, 'Reconstructing Restoration Science: Problems and Pitfalls in Institutional History' (below, Appendix); *idem, Royal Society*; *idem*, 'Early Problems in Professionalizing Scientific Research: Nehemiah Grew (1641 – 1712) and the Royal Society, with an Unpublished Letter to Henry Oldenbury' (below, ch. 8), J. L. Heilbron, *Physics at the Royal Society during Newton's Presidency* (Los Angeles, 1983); Michael Hunter, 'A 'College' for the Royal Society: The Abortive Plan of 1667 – 1668' (above, ch. 5); *idem*, 'The Cabinet Institutionalized: The Royal Society's 'Repository' and its Background', in Oliver Impey and Arthur MacGregor, eds., *The Origins of Museums: The Cabinet of Curiosities in Sixteenth- and Seventeenth-century Europe* (Oxford, 1985), pp. 159 – 68 (see also above, ch. 4)

contemporaries deepened their understanding of the Society's nature and functions in a manner from which, retrospectively, we can learn much.

Our study is primarily based on an important series of documents, most of them surviving only in manuscript, which have hitherto been almost entirely overlooked. In them, such prominent early Fellows of the Royal Society as Robert Hooke, Henry Oldenburg and Sir William Petty considered — often at length — how the Society might be remodelled or reoriented so as better to serve the ends for which they believed it had been constituted. Perhaps not surprisingly, the authors involved were far from unanimous either in their assessment of the Society's proper aims or in the strategy they recommended to achieve them. Comparison of their ideas therefore throws much light on contemporary views of the nature of scientific activity and the proper role of a scientific institution.

This material is important both in its own right and for its implications. Above all, the range of opinions expressed by the different commentators helps us to get away from the rather one-dimensional picture of the Society's early history which is induced by reliance on the most easily accessible sources: either these were carefully prepared for public consumption, like Thomas Sprat's *History of the Royal Society* (1667) or Oldenburg's *Philosophical Transactions*, or they take the form of minutes as published by Thomas Birch in the eighteenth century, which tend to present an inscrutable consensus, documenting decisions reached but only hinting at the discussions which preceded them. Theodore Hoppen, it is true, has presented evidence of diversity of opinion in the Society on intellectual matters, while we also know of disagreements based on personal rivalries, as between Hooke and Oldenburg, or John Woodward and Hans Sloane.[2] But what we currently lack is any sense of rival strategies based on differing intellectual outlooks, making sense of discussions at meetings in terms of broad contrasts of attitude of the kind that are familiar in the Royal Society in the early nineteenth century. It is precisely this which these documents, collated with hints from other sources, can provide.

They also illustrate how divergent intellectual positions were reflected in contrasting attitudes towards the Royal Society as an institution. Hitherto, there has often been a tendency for intellectual and institutional approaches to the Society in its early years to remain separate, but the material deployed here demonstrates that, on the contrary, differing intellectual positions had immediate implications for commentators' views of the Society. Hence valuable evidence is provided to show how methodological, institutional and social issues interrelated.

This lengthy article is divided into four parts, of which the most significant is the second, in which we will scrutinize the views on scientific method and organization presented in the newly discovered documents, and draw out their significance. However, it is important to see how this

2 K. T. Hoppen, 'The Nature of the Early Royal Society', *British Journal for the History of Science*, 9 (1976), 1 – 24, 243 – 73; Margaret 'Espinasse, *Robert Hooke* (London, 1956), pp. 8 – 9, 63 – 70; J. M. Levine, *Dr Woodward's Shield: History, Science and Satire in Augustan England* (Berkeley and Los Angeles, 1977), ch. 5

broader heart-searching about the Society and its aims occurred at times in the institution's early years when more piecemeal measures were taken to deal with the problems that the Society faced. The first section of the article will therefore consider in detail the context in which these new documents originated, illustrating interconnections between their general strategies and the actual reforms that were implemented.

Then, in view of the significance of the material on which our study is based, we have considered it appropriate to include the complete text of three of the most interesting memoranda in full, one by Robert Hooke, one by William Neile and the third by an incisive author who unfortunately remains anonymous. These transcripts make up the third section of the paper, which is followed by two appendices which deal with questions of the authorship and dating of these papers.

I

The background to what we want to say is provided by what should now be familiar about the early Royal Society, and that is the debilitating tension between its ideals and its actual state in its early years.[3] Indeed, it was at moments when this tension reached a crisis that the most sustained heart-searching about how the Society could and should operate occurred.

The Royal Society was founded to promote a programme of intellectual reform by collaborative endeavour along the lines suggested earlier in the seventeenth century by Francis Bacon. To this end, it was intended that a systematic programme of information collecting and empirical enquiry should be implemented. The idea was that the Society should itself be the forum for the production as well as the discussion of scientific findings, through experiments actually carried out at meetings or by the efforts of experimenters employed on the Society's behalf. Thomas Sprat, for instance, referred to the Society's hope of employing '*Curators*' in the plural, and, though Robert Hooke was in fact the only one supported — in part through a benefaction to the Society from Sir John Cutler — there were from the first various plans to supplement his work, such as a proposal to employ the physician Walter Charleton for 'anatomical administrations' in 1664.[4] In addition, in 1664 committees were set up so that more expert and specialized groups could consider particular topics in detail.[5] A store of

[3] Hunter, *Science and Society*, ch. 2; R. G. Frank, jr, 'Institutional Structure and Scientific Activity in the Early Royal Society', *Proceedings of the 14th Congress of the History of Science (1974)* (Tokyo, 1975), iv, 82–101; Margaret 'Espinasse, 'The Decline and Fall of Restoration Science', *Past and Present*, 14 (1958), 71–89.

[4] Birch, *History*, i, 415; Sprat, *History*, p. 434. On Hooke's position see 'Espinasse, *Robert Hooke* (n. 2) pp. 4, 83.

[5] Birch, *History*, i, 402–3, 406–7. For the minutes of some of these committees see D.M. 5.62–68, and above, ch. 3.

equipment was another desideratum: in the Society's early years there were hopes to provide a museum, 'a Chymicall Laboratory, a Mechanicall operatory, an Astronomicall Observatory, and an Optic Chamber', possibly in a purpose-built 'college'.[6]

If the Society's founders thus had a vision of a high-powered scientific research institute, in fact matters did not work out as well as had been anticipated. Most important, the initial hopes for a large-scale endowment, preferably from public funds, never materialized, and as a result the possibilities for employing staff and providing facilities were limited. Instead, the Society was dependent on volunteers, both for finance and activity. After first considering limiting its membership to fifty-five, the Society then opened its doors to almost anyone who applied for membership, the only real restriction being social. To an extent, this was because of a genuine belief that all could contribute in some way to the growth of knowledge, a consideration which overlapped with the need for a quorum imposed by the Society's constitutional structure.[7] But it was also due to an awareness that money was potentially available from members' subscriptions: these, if paid fully, would have provided quite a healthy income, while there were also hopes for extra contributions in support of experimental work.[8] However, these members proved constantly disappointing, thwarting attempts to make the Society the powerhouse of corporate research to which its organizers aspired, and inspiring reform efforts which, even if temporarily successful, were never able to bring about fundamental change.

Already in the 1660s it was becoming apparent that the Society's activities would not be easy to sustain, and that the structure which the Society had adopted was seriously flawed. The problems associated with finance and attendance have already been surveyed elsewhere, as have the steps taken to deal with them at this time, which set a precedent for comparable measures later — notably the compiling of lists of those in arrears and the expulsion of a handful of extreme offenders.[9] Here it is important to note how concern about these matters overlapped with anxiety about the method of the Society's work, and with consideration as to how the Society's proceedings could be revitalized. Quite apart from the fact that the committees founded in 1664 quickly became defunct, the type of activity which the Society's voluntary structure tended to produce even at ordinary meetings left something to be desired.[10] On 4 December 1666 the question was raised as to 'how the experiments at the public meetings of the society might be best carried on; whether by a continuous series of experiments, taking in collateral ones, as they were offered, or by going in that

[6] Henry Oldenburg to Robert Boyle, 24 Feb. 1666, *Oldenburg*, iii, 45; Hunter, 'A 'College'' (above, ch. 5).

[7] Birch, *History*, i, 5; Hunter, *Royal Society*, esp. ch. 1.

[8] *Loc. cit.*; for contributions see for example Birch, *History*, i, 453, 472 – 3.

[9] Hunter, *Royal Society*, ch. 3.

[10] Hunter, *Science and Society*, pp. 42 – 43; Frank, 'Institutional Structure' (n. 3), pp. 94 – 6.

promiscuous way, which had hitherto obtained', while in 1668 a medal worth at least twenty shillings was offered to every Fellow who exhibited an experiment — an interesting precedent looking forward to the practice of the eighteenth and nineteenth centuries.[11] In the following year, when the question of improving the Society's experimental work was raised, Oldenburg suggested the revival of committees, mooting the idea of appointing two committees 'for directing of experiments' to meet at least once a month, one in the City of London and the other in Westminster. A further committee 'to consider of the most convenient and effectual way to carry on the business of the society' was appointed in June 1669, and the revival of committees — either for scientific or administrative purposes — was an idea that was to be a recurrent theme in attempts to ameliorate the Society's problems.[12]

Another idea was the extension of paid staff, and in 1667 – 69 various suggestions were considered for a further 'curator' to supplement the work of Hooke, involving such Fellows as Richard Lower, Walter Needham and John Ray, though nothing materialized.[13] In 1669 the Society did, however, appoint a peripatetic collector of plants and other specimens, Thomas Willisel, while in the same year the position of the Secretary, Oldenburg, was regularized by granting him a small salary for his assiduous work on the Society's behalf, in belated response to a rather desperate note on his part itemizing his duties and expostulating 'Whether such a person ought to be left un-assisted?'.[14]

From the 1660s we have only one manuscript memorandum, a piece with the title 'Proposalls humbly offered to better consideration' written by the mathematician William Neile. Though undated, a *terminus ante quem* for it is provided by Neile's death in 1670, and it probably dates from a year or two before that, when Neile was active in the Society.[15] Neile was one of the committee appointed in February 1669 to meet at Westminster to direct the Society's experiments,[16] and, since the paper deals with the issue of how experiments should be selected, it possibly gives a clue to the discussions that went on at those meetings. The 'Proposalls' reveal that Neile had deep misgivings about the rather narrow experimentalism espoused by some of his colleagues and celebrated in Sprat's *History*, which eschewed causal

[11] Birch, *History*, ii, 131 – 2, 265; Sir Henry Lyons, *The Royal Society 1660 – 1940* (Cambridge, 1944), pp. 172 – 3, 219 – 20, 247 – 8.

[12] Birch, *History*, ii, 344, 346, 378.

[13] *Ibid.*, ii, 178 – 9, 194, 206, 212, 355, 441.

[14] *Ibid.*, ii, 358, 371, 376, 378 – 9. Oldenburg had earlier been refused a salary as against piecemeal gratuities; see *ibid.*, ii, 273, and *Oldenburg*, iv, xxiv. Oldenburg's paper of spring 1668, printed in the *Correspondence*, is taken from British Library Additional Manuscript 4441, fol. 27. It was earlier printed in C. R. Weld, *A History of the Royal Society* (London, 1848), i, 135 – 6n., where it is wrongly dated to *c.*1663. *Cf.* H. G. Lyons, 'The Officers of the Society (1662 – 1860)', *Notes and Records*, 3 (1940 – 41), 121, and *idem, Royal Society* (n. 11), p. 46, where it is dated 1664.

[15] See below, Document A. For Neile's activity in the Society at this time see Birch, *History*, ii, *passim*, *Oldenburg*, iv – vi, *passim*, and Cl.P. iii (1) 48 – 53.

[16] Birch, *History*, ii, 346.

enquiries and confined itself almost exclusively to the compilation of facts through random experimentation.[17] By contrast, Neile insisted *inter alia* that experimentation was most effective when guided by theory, and, realizing that the experimental demonstrations exhibited at meetings had to amuse as well as instruct, he remarked acutely that 'experiments themselves are but a dry entertainment without the indagation of causes'. In addition, Neile suggested that committees should direct the work of the Society, collect information and advise the Society on its publication policy. He also felt that the library should be improved, perhaps sensing that the Society was falling behind in its intention of providing an infrastructure of resources for research. Neile's 'Proposalls' are printed in full at the end of this article.

If things had seemed less than perfect in the 1660s, matters got much worse in the early 1670s. Worries about arrears re-emerged in 1672, while the employment that year of Nehemiah Grew as a paid researcher on the Society's behalf was perhaps at least in part a response to declining activity.[18] But the mixed success of this venture illustrated the Society's problems, while the death of two of its leading and most energetic Fellows, John Wilkins in November 1672 and Sir Robert Moray in July 1673, engendered perhaps the most severe crisis in the Society's history in the whole of the seventeenth century. In the summer of 1673 Seth Ward considered 'the failing or Continuance of that Society . . . to be now brought to a Crisis',[19] and this stimulated a prolonged campaign of retrenchment which is the context of various of the manuscript memoranda with which this article is concerned.

The campaign began in August 1673 when Sir William Petty (who had recently returned from Ireland and was shocked at the state in which he found the Society) is recorded by Robert Hooke discussing possible moves to ameliorate the situation.[20] These deliberations were taken up by the Society's Council when meetings were reconvened after the summer recess in October that year, and over the next few months an attempt was made to reverse the decline in the Society's activity by persuading active Fellows to read papers or perform experiments at meetings, while the financial issue was tackled by listing and dunning those in arrears and encouraging members to sign a legally-obliging bond guaranteeing payment of subscriptions. At the same time, it was hoped to cut the dead wood of inactive and non-paying Fellows out of the Society, though it is clear that

[17] See Sprat, *History*, pp. 89 – 90, 100 – 9, 115 – 16. On the method expounded in Sprat see P. B. Wood, 'Methodology and Apologetics: Thomas Sprat's *History of the Royal Society*', *British Journal for the History of Science*, 13 (1980), 6 – 12.
[18] Birch, *History*, iii, 63, 70; Hunter, 'Early Problems' (below, ch. 8). For a complaint in 1671 that 'very many things were begun at the society, but very few of them prosecuted', see Birch, *History*, ii, 469.
[19] Seth Ward to Gilbert Sheldon, 19 Aug. 1673, Bodleian Library MS Tanner 42, fol. 29.
[20] L. G. Sharp. 'Sir William Petty and Some Aspects of Seventeenth-century Natural Philosophy' (Oxford D.Phil. thesis, 1977), p. 255; Hooke, *Diary*, p. 57.

opinions differed as to whether this should be achieved by expulsion or resignation.[21]

Various documents survive relating to this attempt to reinvigorate the Society, ranging in significance from dunning letters sent to those in arrears to full-scale reconsiderations of the Society's aims and methods, and, since there is a degree of overlap between the two, all of them will be itemized here. The most straightforward are a pair of letters drafted by the committee appointed in December 1673 to deal with the problem of arrears, which were sent to those most in debt to the Society in the first half of 1674.[22] Stress was laid on how money was essential 'to carry on the work of the Society, and particularly to defray the charges of their Experiments (in the vigorous prosecution whereof consists the life and honor of this Royal institution)'. In addition, the attractions of the fuller programme which it was hoped to provide were described, thus illustrating how the issues of finance and activity interrelated, in that more worthwhile activities at meetings were thought likely to encourage members to attend and to renew their subscriptions, while equally a larger revenue could assist in reimbursing demonstrators with their expenses and paying them a fee.

In the summer and autumn of 1674 a more elaborate 'declaration for restoring the Society' was prepared from drafts by Dr Jonathan Goddard and Sir William Petty, in which Fellows were urged to sign the form committing them to subscribe regularly in future. Printed copies of the 'bond' that was drawn up survive, but although 225 copies of the covering letter based on Goddard's and Petty's suggestions were printed and distributed,[23] none of these is extant, and the letter's tenor can therefore only be deduced from the existing drafts.

Goddard stayed fairly close to the point, reiterating how, thwarted by the lack of revenue which Fellows had promised to provide, the Society was prevented from carrying on its work and supporting its officers, so that, if all failed to oblige, the Society would collapse altogether, to the consequent 'dishonor' of those responsible. He disingenuously suggested that such considerations might not previously have occurred to the Society's debtors, noting that 'the charge cannot but be thought very reasonable and easy for a person in your condition'. He then provided a draft of a form promising regular payment of subscription in recognition of the benefits accruing from membership, suggesting that more care should be taken to ensure that all elected to the Society signed such a legally obliging bond on admission.[24]

21 Birch, *History*, iii, 94f.; Hunter, *Royal Society*, p. 38; Sharp, 'Sir William Petty', pp. 261 – 73, 401 – 7. On expulsion see further below, Section 2.

22 D. M. 5.8 and 1 (printed in R. K. Bluhm, 'Remarks on the Royal Society's Finances 1660 – 1768', *Notes and Records*, 13 (1958), 97). On the appointment of the committee see Birch, *History*, iii, 118 – 19; see also D. M. 5.11 (bis) and 12A, which evidently relate to the work of this committee, and Bowood House Petty Manuscripts H(8) 6, 8 – 9 and 11 – 13.

23 Birch, *History*, iii, 155; see also *ibid.*, 137 – 41, and, for references to replies to the declaration, 190 – 1. On the bonds see Hunter, *Royal Society*, pp. 96 – 97. See also D.M. 5.10 (see below, n. 27).

24 D.M. 5.13.

Petty's version was longer and more florid, elaborating on the reputation of the Society and the extent to which its work was endangered by shortage of funds, 'to the great discouragement of those, who had already actually been, and others who intended to bee Benefactors to the same', and especially the King. Alluding to the other part of the reform programme, he also explained how the Society had decided to select a limited number of Fellows who were prepared to contribute 'not onely by payment of their respective arreares, but also by advanceing other Summes, and most of all by their engagement Weekly to bring in some peece of the History of Nature or Art, to review, methodize & emprove what of that kind hath beene already registered, to fitt up an Apparatus for Several Illustrious experiments', and to supervise the work of the Society's curators. Those who refused to pay their arrears were requested to resign, and he intimated that in future elections the Society might become more selective.[25]

If these papers are merely administrative documents which amplify what is already known from such sources as Birch's *History*, what is much more interesting is a series of documents dating from this period which suggest how a higher level of finance might be used to extend the Society's activities, if not to alter radically its function. The concept of a 'a firm establishment of the said Society' — a reliable revenue either from subscriptions or other sources — seems to have been much in the minds of its organizers at this time.[26] Apart from seeking legal means to ensure that Fellows were reliable in their contributions, it was also hoped that capital might be acquired from various sources to provide an additional investment income and calculations were made of what level of funding would assure the level of work which could reasonably be aspired to.

Various strategems were considered for increasing the Society's funds, not all of them recorded in the minutes. One was of asking for an advance of a year's subscription from each Fellow, another of increasing the subscription rate from £2.12.0 per annum to £5.[27] A third was of capitalizing on the high status of the membership by establishing a sliding scale of membership fees according to social rank, with £5 for 'D[octo]rs

25 Bowood House Petty MSS H(8) 10, Transcribed in Sharp, 'Sir William Petty' (n. 20), pp. 399 – 400 (with the following errors: 'make' has been omitted between 'to' and 'pastime' in the tenth line from the bottom of p. 399 and 'those' has been mistranscribed as 'there' on p. 400, 1.5; Sharp does not note Petty's deletions). Petty's concern about potential benefactors may have been linked to the Society's hopes of Cutler at this time: see below. From the minutes (Birch, *History*, iii, 136 – 7) it sounds as if Petty was asked to get Fellows to agree to advance a year's salary all at once and to get them legally to oblige themselves to entertain the Society: these matters are mentioned in his version, but, as with Goddard, it is Fellows deeply in debt who are his main target.

26 Birch, *History*, iii, 100. Compare Hooke in Cl.P. xx.50, fols 85 and 98, wanting to make the Royal Society 'self subsisting'.

27 Birch, *History*, iii, 136 – 7; D. M. 5.10 ('A Forme propos'd of the Legal Obligation for paying the weekly Contributions to be subscribed by the Fellows of the R[oyal] S[ociety]', probably dating from summer 1674). It is perhaps worth noting that in D.M. 5.13 Goddard left the amount of the subscription blank.

&c', but with Dukes, Marquises and Archbishops paying £30, Earls £20, Viscounts £15, Barons £12, Baronets £10 and Knights £8.[28] But the Society had by no means given up the idea of achieving an endowment which, through investment, could increase its revenue beyond the level provided by Fellows' subscriptions. The possible sources of this capital sum were varied. The most definite was a bequest of £400 left to the Society by John Wilkins, which the Society knew it had been promised by October 1673 and which, after some delay, was actually invested in three fee-farm rents, yielding £24 per annum, in January 1675.[29] Other schemes mooted at this time, however, were not similarly brought to fruition.

One hitherto unnoticed fact is that in 1673 the Society appears to have solicited an extra endowment to the tune of £1000 from Sir John Cutler, who was quite active as a philanthropist in these years, endowing an anatomy theatre at the London College of Physicians in 1674 and taking an active interest in the rebuilding of St Paul's Cathedral in the late 1670s.[30] Details of this episode are known only from brief comments, mainly in Robert Hooke's diary and in a letter from Seth Ward to Archbishop Sheldon, which have hitherto been overlooked because Cutler was at the same time asked to be more regular in his highly erratic funding of Hooke; the sum involved is mentioned only in an unpublished memorandum by Henry Oldenburg.[31] This — together with still more shadowy hints in Hooke's diary of approaches to other merchants in the City — may possibly make sense of the otherwise puzzling reference in the Society's minutes, that it was encouraged in 1673 to move to Gresham College from Arundel House (where it had met since 1667) not least by 'the hopes, which they find grounds to entertain, of meeting with some considerable benefactors at that end of the city'.[32] These hopes were not realized in any case.

Further financial aspirations are recorded in a pair of memoranda concerning the Society's desired income and expenditure by Oldenburg,

28 D.M. 5.15. See below, n. 33.

29 Birch, *History*, iii, 93; see also iii, 70. H. G. Lyons, 'The Society's First Bequest', *Notes and Records*, 2 (1939), 44.

30 Sir George Clark, *A History of the Royal College of Physicians* (Oxford, 1964 – 66), i, 330 – 1; Jane Lang, *Rebuilding St Paul's after the Great Fire of London* (London, 1956), p. 98.

31 Seth Ward to Gilbert Sheldon, 19 Aug. 1673, Bodleian Library MS Tanner 42, fol. 29 (this letter suggests that it may have been because of perceived conflict with Cutler's impending benefaction to St Paul's that the Royal Society grant did not materialize); Cl.P. xx.50, fol. 91 (see below n. 39). The relevant references in Hooke's *Diary* may be divided up as follows: pp. 47, 56 almost certainly refer to the new endowment; pp. 49, 51 certainly refer to Hooke's arrears; pp. 53, 57, 60, 63, 67, 70 – 71, 74, 77, 78 could refer to either.

32 Birch, *History*, iii, 100; Hooke, *Diary*, pp. 40, 152. *Cf.* Ian Adamson, 'The Royal Society and Gresham College, 1660 – 1711', *Notes and Records*, 33 (1978 – 79), 2 and 16 n. 9.

one a more or less identical fair copy of the other.[33] Though undated, these clearly come from this period, probably from 1674.[34] Among the possible sources of funding listed by Oldenburg were arrears of subscriptions valued at half their actual total — presumably in the expectation of taking Fellows to court in pursuit of them, though this was not a step that the Society pursued at this stage. In addition, Oldenburg rather surprisingly included the sums promised by Fellows towards erecting a 'college' for the Society in 1668, an earlier project which Oldenburg and others had hoped would be instrumental in 'establishing' the Society, which had also proved abortive.[35]

Perhaps more interesting is Oldenburg's accompanying calculation of what level of funding would be desirable and feasible in which he suggested that the Society should have two paid secretaries with clerks (at that stage he was the only secretary), three curators with assistants, and a notional annual allowance for experimental equipment. In addition to Hooke, Grew was to be made a permanent fixture and a new curator for the History of Trades was to be added, with a young doctor in search of employment, Richard Blackburne, named as candidate.[36] All this would have cost £500 a year, and

[33] D.M. 5.14 – 15. The fair copy (item 15) is printed in Bluhm, 'Remarks on Finances' (n. 22), pp. 93 – 94. The chief differences between the draft and the fair copy are as follows. Firstly, the former has notes on the duties of the curators omitted from the latter: 'Dr Grew to have notice to bring in an Experiment every meeting. And Mr Hook undertakes [*sic:* replacing 'to agree with him', deleted] doe the like. And the account of the Experiments shall be brought in in writing and be read [', and left', deleted]'. Secondly, whereas D.M. 5.15 has arrears valued at £1100 and college subscriptions at £900, 5.14 gives the figures as £1000 each. Thirdly, the conjugate leaf of item 15 (not printed by Bluhm) has further notes originally written in pencil and then inked in, probably in a different hand, of which there is no equivalent in item 14. Here the idea of three secretaries and two curators, as against two and three respectively in Oldenburg's notes, is put forward; in addition, there are totals of Fellows in the classes 'Strangers', 'Travell[ers]' 'Exc[used]' and 'Doubtf[ul]', headed 'Not to be reckond upon', and calculations for a sliding scale of membership fees according to social rank (see above, pp. 192 – 3) with the total that this would give in terms of the number of Fellows from each class in the Society.

[34] This is suggested by the fact that the arrears are given as £2200, which is a figure at which they must have stood at some point in 1674 (see Bluhm, 'Remarks on Finances', p. 96). That these notes are later than October 1673 is clear from the fact that Oldenburg knew about Wilkins's bequest. The £20 from Sir John Cutler that appears may represent a realistic estimate of the extent of his expected funding for Hooke as Cutlerian Lecturer: Bluhm speculated that it might represent 'a projected further endowment by Cutler that never materialized' (*art. cit.*, p. 95), but it seems to bear no obvious relation — as interest or anything else — to the £1000 hoped for in 1673.

[35] See Hunter, 'A 'College'' (above, ch. 5), esp. p. 161. The figures given for college subscriptions in both D.M. 5.14 and 15 are less than the total actually promised in 1668, doubtless to allow for the death of some of those who had promised contributions in the interim: see *ibid.*, p. 179.

[36] For the background to this, see Hunter, 'Early Problems' (below, ch. 8), esp. pp. 271f.

Oldenburg then worked out (perhaps with the help of Hooke[37]) what level of endowment, invested at what rate of return, would be necessary in addition to Fellows' subscriptions to provide this, thus giving the figure which he then cast around for funds to fulfil.

A comparable undated paper survives among the manuscripts of Sir William Petty. Though hastily written and not altogether easy to interpret, this document appears to envisage the provision of three secretaries, one for correspondence and two for the Society's business; of two or perhaps three curators; and of equipment. Since Petty suggested paying these officers and their assistants a little less than Oldenburg, the total he considered necessary per annum — £440 — was slightly less. He also seems to have considered how this could be provided, though his financial jottings are impossible to follow in detail.[38]

So thought was going on about extending the Society's staff in a manner that was never actually realized, but, in addition, it is clear that deeper consideration was being given to the aims and nature of the Society. This juxtaposition is revealed most clearly by a further memorandum by Oldenburg bearing the date 1 November 1673, which survives in a volume of Hooke's papers at the Royal Society.[39] It evidently comprises the agenda for a meeting held at the home of the President, Lord Brouncker, on that date, at which, as we learn from Hooke's diary, he and others 'Discoursd and concluded severall things about [the] Royal Society'.[40] No minutes of it are preserved in the Society's archives, presumably because it was not a formal meeting.

This document illustrates clearly how the concern for financial stability in the early 1670s was linked to a more general aspiration to a 'permanent settlement' for the Society. It comprises a kind of analysis of the Society and its desiderata in the form of a series of headings and notes in Latin set out as a table. Short term considerations are indicated by the first section, of 'more especially pressing matters' relating to the Society's impending return to Gresham College. But the second part raised broader issues for consideration. One of these was the question of funding, the amount that the Society needed and the potential sources of it: Oldenburg listed the debts of Fellows, Cutler's putative donation and Wilkins's bequest, and itemized other possible benefactors from among the Society's ranks, ranging from the King to Lord Brouncker, the Treasurer, Daniel Colwall, and such

37 In D.M. 5.14 the words 'Which in houses amounts to £2860. in lands 3740' are added in what appears to be Hooke's hand.

38 Bowood House Petty MSS H (8) 5. Other officers are also referred to, but these were presumably voluntary. The fact that the same sum, of £40, is allowed for 'Contingents' in this and the Oldenburg papers just referred to could suggest that both are of similar date. A date earlier than the 1680s may be suggested by the fact that Petty here envisaged an expenditure of £440, whereas in 1687 he thought in terms of £500 (see below, p. 202). It should be noted that some of the financial calculations in this paper may not refer to the Royal Society.

39 Cl.P xx.50, fol. 91. It has been endorsed 'Scheme of the R[oyal] S[ociety] in Mr Oldenburgh's hand'.

40 Hooke, *Diary*, p. 67.

Fellows as Robert Boyle, Henry Howard of Norfolk, Sir William Petty and Seth Ward. In addition, he seems to have seen virtually the entire constitutional structure of the Society and its activity as up for discussion, for instance as to whether the charter needed to be supplemented or amended, whether the number of members should be fixed or indefinite, how many officers and staff the Society should have, and how often meetings should be held. Petty, too, combined financial calculations with a more analytical view of the Society and its functions, though there is a slightly whimsical element about Petty's.[41] He considered the Society's 'appearance & formall Institution' and its ends, listing its facilities and itemizing the number and characteristics of the members and officers, presumably as topics for discussion. Details in these papers will be further considered in the second part of this article, but what is important here is that they illustrate how, from the latter part of 1673, the reform efforts went with a broader reappraisal of the Society's activities.

Perhaps most interesting of all from this point of view is a long letter to the Council of the Royal Society reflecting on the current state of the Society and giving suggestions for improvement which a Fellow who signed himself 'A.B.' was stimulated to write by the discussions that were then in progress.[42] The letter is dated 19 October 1674, a date on which the Council met, and it has been endorsed in a different hand 'A proposal for the Advancement of the Royall Society'. The text of this highly important document is given in full at the end of this article; unfortunately its exact authorship cannot at present be established (see Note I), and hence it will hereinafter be referred to as simply by 'A.B.'.

The uncertain authorship of this letter is all the more regrettable since it well illustrates how the crisis that the Society experienced at this time stimulated profound heart-searching about the Society's very nature. 'A.B.' goes back to the roots of the Society's aspirations to reform knowledge. He expounds his views on the proper size and type of institution that this intellectual enterprise required, as well as on its activities and organization. He approved of the presentations at meetings that had been introduced as part of the reform effort; he wanted to increase the number of curators; and he gave his thoughts on how people were best motivated to be active in the Society and to support it financially. 'A.B.' also took up the theme of the Society's reputation which Goddard and Petty had alluded to in the drafts already discussed, seeing 'A stock of good & generall Reputation amongst men' and especially with the King as a priority, and giving some fascinating suggestions on how this might be achieved. In particular, he, like others, saw 'some good Stock or Estate of our own' as crucial, and he offered to contribute towards it himself on the grounds that — as he wryly put it — 'to whome doe men send their guifts, but to the rich?' On each topic 'A. B.' had trenchant, perceptive opinions, and his letter is one of the most fascinating

[41] For example, such headings as '1. Its Antiquity, Ever was & Ever will bee' and '4. Its Extent, for that All Trades are members of the Society': Bowood House Petty MSS H (8) 5.

[42] See Birch, *History*, iii, 136f., esp. pp. 138 – 9 for the Council meeting of 19 October 1674.

documents connected with the early Royal Society to have come to light in recent years.

After the autumn of 1674 evidence for such heart-searching ceases, though the reform programme continued to take its course for another year or more, involving the arrangement of a series of lectures by various Fellows through the winter season of 1674 – 75, the attempt to persuade Fellows to pay their arrears and sign the new bond guaranteeing subscriptions, and a token number of expulsions.[43] Thereafter, however, matters soon lapsed again, perhaps bearing out the views of the authors of the reform proposals just discussed of the need for more radical changes than had actually occurred. Soon there was renewed anxiety about the state of the Society: expressions of concern recurred in 1676, when fresh attempts were made to provide experiments for meetings.[44] Matters came to a head again, however, in the aftermath of the death of Oldenburg in 1677, when Lord Brouncker was ousted from his Presidency and a new regime was installed, including Robert Hooke, who now for the first time gained a place on the Council; further developments occurred under the aegis of Sir Christopher Wren, who became President in 1680. The Council minutes and such sources as the diaries of Hooke and John Evelyn give details of decisions taken and hints of discussions which went on, while evidence on the positions that were adopted in these debates is provided by manuscript memoranda, especially a number by Hooke which appear to date from this period. As earlier, worries about the content of the Society's work were juxtaposed with attention to its structure and finance, and concern about the arrears of inactive members culminated in the expulsion of some sixty Fellows between 1682 and 1685.[45]

On 19 December 1677 we learn from the minutes that it was ordered (among other things) 'that what experiments shall be undertaken by the curators shall be propounded a fortnight before the shewing thereof, that objections, answers, and confirmations may be timely thought of' and the 'design and usefulness' of the experiments explained. A special Council meeting 'about carrying on the affaires of the Society' had been held on 12 December, although no minutes for it have survived.[46]

A further bout of discussion about the Society's experimental policy occurred in the winter of 1679 – 80. In December 1679 it was decided to prosecute a single subject at a time as the 'main work', with one previously advertised experiment being shown at every meeting and members urged to go away and consult relevant authors. Over the next month the Council 'met several times to consider of the best ways of carrying on the design of

[43] See the accounts referred to in n. 21.
[44] Birch, *History*, iii, 309 – 10.
[45] Hunter, *Royal Society*, pp. 40 – 1, 99 – 100.
[46] Birch, *History*, iii, 365; Evelyn, *Diary*, iv, 125 – 6. See also Hooke, *Diary*, pp. 333 – 335.

the society', while a number of exchanges concerning experiments are recorded in the minutes. Thus Petty was to be found urging that all members should contribute 'such experiments or observations, as might prove of great and immediate use'. At a later meeting the President, Sir Joseph Williamson, argued that 'the best method of prosecuting experiments . . . was to proceed synthetically by first making the proposition what was designed to be proved, and then proceeding with the experiments to make the proof', whereas Petty claimed on the contrary that 'they would be more faithfully made and delivered, if they were not made to help out a theory, because that might prepossess and biass the experimenter'.[47] There was then a lull until 25 June, when a special Council meeting was summoned and it was ordered 'that the next experiments to be examin'd, should be my L: *Verulams*, & an account to be given of them to the publique from yeare to yeare 'til we had gon through them', a decision which reflects the Society's continuing Baconian commitment.[48]

Other proposals of this period revolved around the idea of reinstating committees. The discussions of December 1679 and January 1680 led to the appointment of a 'committee of experiments', while in January 1681 Wren, then newly elected President, proposed a series of committees to cover major areas of the Society's interests — anatomy, agriculture and cosmography — though none apparently materialized.[49] In addition, in 1683 extra staff were briefly appointed, thanks to a financial windfall in the form of the King's payment for the Society's relinquishment of its title to Chelsea College.[50]

There was also concern at this time about the Society's publication policy, a topic which had not come up in the early 1670s, presumably because, in the capable hands of Oldenburg, this had seemed to be the one area where all was well. Following Oldenburg's death, the *Philosophical Transactions* struggled through a bad period in 1677 – 79, with extensive space fillers from the Society's papers, before succumbing entirely.[51] As in the early 1670s, a crisis stimulated consideration of more novel desiderata as well as short-term measures. In December 1678, the idea was suggested of a journal which, in contrast to the *Philosophical Transactions*, might be restricted in circulation rather than freely distributed, an intriguing suggestion whose

[47] Birch, *History*, iii, 513 – 14; iv, 6 – 8, 20. There was also concern about the Society's records: *ibid.*, iii, 514 – 15, 518; iv, 9.
[48] Evelyn, *Diary*, iv, 205; John Evelyn to Samuel Pepys, 25 June 1680, in *H.M.C. MSS of J. Eliot Hodgkin* (London, 1897), p. 177. See also Birch, *History*, iv, 43, 46.
[49] Birch, *History*, iii, 516; iv, 7, 16, 20, 65.
[50] Hunter, *Royal Society*, pp. 41 – 42.
[51] *Philosophical Transactions*, 12 (1677 – 79), 923 – 1074. On the Society's publication policy see Hunter, *Science and Society*, pp. 51 – 54.

context will be elaborated below.[52] In 1679, Hooke began his *Philosophical Collections*, which, though different in style from Oldenburg's *Philosophical Transactions*, was not limited in this way. A further new suggestion came up in 1680 — during a gap in the issue of *Philosophical Collections* — this time for 'philosophical gazettes'. These were to comprise half a folio sheet and to cost not more than two pence, and it was thought that they 'would much propagate natural philosophy and the English language'. The proposed 'gazettes' would thus have contrasted with the rather bulkier and more expensive format of the *Philosophical Transactions*, though nothing came of this proposal.[53]

As with the early 1670s, it is possible to link some unpublished memoranda with the steps and discussions chronicled in the minutes. Oldenburg, of course, was now dead, and none of Petty's papers appears to date from this time,[54] but one figure whose profuse memoranda clearly do relate at least in part to this period was Robert Hooke. As Oldenburg's successor as Secretary, Hooke was newly influential in the Society, although the extent of his personal impact should not be overestimated since his was only one voice among many in the Society's inner counsels. Moreover, he was not always listened to with respect, as is apparent from the atmosphere of suspicion between Hooke and the rest of the Council noticeable in certain conciliar decisions and hints in his diary.[55]

It is clear from his diary, however, that Hooke was the source of the proposal 'about Experiments' which came up at the meeting of 19 December 1677, while the discussion of a 'philosophical gazette' recorded in the minutes early in 1680 is clearly linked to Hooke, who reported in a letter to the Somerset virtuoso, John Beale, on the discussion at the Royal Society as to the form that the publication should take — whether weekly, or fortnightly, monthly or quarterly and larger in size.[56] Hooke himself felt that a publication of between one and three sheets once a fortnight was best, 'and onely a small one for the use of the fellows of the Society once a Week which shall not be cummunicated [*sic*] to any person but to there Duty [?]

52 Birch, *History*, iii, 450 – 1. For further references to proposed publications at this time see *ibid.*, iii, 491, 501, 514, 518.

53 *Ibid.*, iv, 7, 16, 19.

54 Except for those relating to his arbitration in the dispute between Hooke and Sir John Cutler in 1682 – 83: see Bowood House Petty MSS H (8) 15 – 19. For the suggestion that Petty was discouraged by the limited outcome of his efforts in the early 1670s see Sharp, 'Sir William Petty' (n. 20), p. 271 and n. 3.

55 Hunter, 'Reconstructing Restoration Science' (below, Appendix), pp. 350 – 1.

56 Hooke, *Diary*, pp. 335, 337 – 8, 437 (where Hooke records 'much cavilling, caballing'), 439 – 40; Robert Hooke to John Beale, 6 March 1680, British Library MS Sloane 1039, fol. 171 (this copy or draft is in the hand of an amaneunsis, thus explaining its erratic spelling). For a further paper on publications by Hooke, of uncertain date, see Cl.P. xx.59.

which will guve [*sic*] a full account of our weekly proceudings [*sic*] and be sume satisfaction to our absent members and may encite them to act at a Distance and make good returens [*sic*] hither'. He then went on to refer to the Society's current interest in recording the variation of dipping needles, and this links the letter with an undated paper by Hooke in which the notion of a publication is also juxtaposed with a wish to investigate this and other phenomena.[57]

Beyond that, references in Hooke's diary show that he was giving thought at this time to how the Society should be run, and it is highly likely that at least some of a series of 'Proposalls for the Good of the Royal Society' that survive among his papers[58] date from these years, though it is possible that they belong to the early 1670s like the manuscript material already discussed, a time when it is clear that Hooke wrote papers on the running of the Society. Equally, it is not inconceivable that at least some may come from later years, like one document which is certainly of a later date and which will be discussed in a moment. Whatever their date, however, these papers undoubtedly illustrate the kind of view of the Society which Hooke would have brought to bear in discussions at this time, and they echo ideas — such as that of the restricted bulletin mentioned in the letter to Beale — which Hooke is definitely known to have entertained around 1680.

57 British Library MS Sloane 1039, fols 112 – 13. This is printed in Weld, *History* (n. 14), i, 148 – 9, and from thence in R. T. Gunther, ed., *Early Science in Oxford* (Oxford, 1923 – 67), vi, 123 – 4. Weld wrongly claimed that this document is to be found in British Library Additional Manuscript 4441. His transcription also contains a number of errors and omissions, of which the most important are the omission of the words 'and so hereafter from time to time' after 'this present' in the fourth line from the bottom of p. 148, the omission of 'such as the character of such new books or Discourses as are published here or elsewhere soe soon as they can procure them' after 'occur' in 1.4 on p. 149, the misreading of 'A theory' as 'Astronomy' in 1.19 on the same page, and the omission of the following incomplete new paragraph after the passage he quoted: 'Now that such observations may be made the more effectuall for the purposes aforesaid it is thought fit to communicate [replacing 'publish', deleted] those Instructions following for the better Directions of such as may not otherwise be soe well versed in the performance thereof.' On the dating of this manuscript see Note II. On dipping needles see Birch, *History*, iv, 18 – 19, 21 – 22, and Hooke to J. C. Sturmius, 6 April 1680, in Gunther, *Early Science*, vii, 552.

58 Cl.P. xx.50. Since this item comprises a number of separate papers, references within it are here given to the overall pencilled foliation of Cl.P. xx, of which this item occupies fols 85 – 98. On dating, including references to Hooke's *Diary*, see Note II. It is perhaps worth noting that most of the diary references of 1678 – 80 involve oral rather than written deliveries on Hooke's part: but since he was now on the Council this is perhaps not surprising.

These papers comprise the largest single group of memoranda dealt with in this article. Within them there is a substantial amount of overlap, though there is also some disagreement between different drafts.[59] Most are in Hooke's hand, though one is in the hand of an amanuensis with corrections by him.[60] In general, the papers may be divided into two types. The longest single item is an untitled paper densely written on both sides of two folio sheets, which has been transcribed at the end of this article. Somewhat comparable are another two sheets of detailed suggestions written on one side of the paper only, which may be incomplete parts of a longer document and which cover an overlapping but different series of points.[61] Then, there are a series of briefer summaries of Hooke's ideas, written concisely and sometimes almost approximating to tabular form: these were quite perceptively compared by R. T. Gunther, who printed one in full and another in part, with Hooke's 'algebras', such as that analysing 'the whole Businesse of Navigation' which he prepared for Samuel Pepys in 1686.[62] One of these summaries appears on the fly-sheet of the paper reproduced in full here, and, as will be seen, it summarizes the points in the longer paper more pithily. The others do much the same, often bringing out particular points more strongly and sometimes differing in detail. There are also a couple of papers of more miscellaneous notes on related topics, again overlapping with the longer papers and short summaries, while, in addition, bound with these papers of Hooke's is an anonymous paper which appears to present a sort of agenda for the Society's meetings, but in doing so raises issues about the Society's proper functions comparable to those dealt with in

[59] For instance, as to whether absent members were to receive bulletins fornightly (fol. 93v) or monthly (fols 95, 96); whether it was to cost two pence (fol. 93v) or 'not exceeding 1s' (fol. 97); whether putative members were to be attested by at least six existing members (fol. 87) or ten (fol. 94); whether admission should cost £5 (fol. 87) or £4 (fol. 94v); and as to how many staff the Society should employ, on which see below, pp. 211 – 2.

[60] Cl.P. xx.50, fol 97.

[61] *Ibid.*, fols 86 – 87. Points that arise here but not in the longer paper, fols 92 – 94, include the idea of incentives for curators and of a research and development fund discussed in Section 2, below. See also n. 123.

[62] Gunther, *Early Science* (n. 57), iv, xiii, from fol. 95. This section also appears in Hunter, *Science and Society*, p. 57, while the whole of fols 95 – 96 are transcribed in P. J. Pugliese, 'The Scientific Achievement of Robert Hooke: Method and Mechanics' (Harvard Ph.D. thesis, 1982), pp. 648 – 51 (a brief discussion of the papers as a whole appears in *ibid.*, pp. 38 – 40). In addition, part of fol. 85 and a short section out of context from fol. 86 were printed in Weld, *History* (n. 14), i, 138 – 40 (and dated, without any authority, to *c.* 1662), whence they were reprinted in Gunther, *Early Science*, vi, 121 – 2 (dated to 1663). For Hooke's paper for Pepys see Bodleian Library MS Rawlinson A 171, fols 245 – 6: see also Cl.P. xx.70 and Birch, *History*, iv, 468. It is worth noting here that fol. 98 is a heavily corrected draft for fol. 85.

the other papers described in this article.[63] It may well date from this time and be a contribution to the discussions which were going on.

In this group of manuscripts, Hooke gave his views on how the Society should best be run. In the fullest, printed below, he talked broadly about the Society's overall aims and organization, about the sort of members who should be recruited, about how efforts should be allocated to different parts of the enterprise, and about the role of curators: other papers overlap with this while amplifying some of the points within it. A theme which recurs in many of the papers — and perhaps suggests most strongly a date subsequent to the death of Oldenburg for them — is the question of keeping absent members informed of developments at the Society's meetings by correspondence or publications, concerning which alternative possibilities in terms of price and frequency are canvassed.[64]

For much of the 1680s and 1690s, there is relatively little evidence in the minutes of discussions about reforming the Society and improving its method.[65] There is also a dearth of manuscript memoranda apart from a paper by Petty entitled 'Things to bee done at the Royall Society', dated 1 December 1687 and evidently inspired by the fact that Petty had been elected to the Council at the anniversary meeting of that year for the first time since 1682.[66] This suggests an agenda for discussion reminiscent of those that Petty and Oldenburg drew up in the early 1670s, itemizing the Society's facilities, expenses and likely sources of revenue, and ways of matching the latter. It is novel in suggesting educational functions for the Society as a means of raising money: Petty proposed that courses in natural philosophy should be offered, with classes of twenty being charged a fee of £1 per month per student. The scheme may have grown out of Petty's

[63] Cl.P. xx.50, fols 88 (with information on, for instance, the regular performances required of members and honorary Fellows) and 97 (more explicit than other papers on secrecy). The anonymous paper is fol. 89. Its content is as follows: 'Symposia [deleted] / Assembly / Philosophicall [deleted] debate of causes / All busyed / Ordering Experiments / Curators or sollicitors by turne the contrary making others idle, drayning him & creating to[o] much dependance on him / Recompence piecemeale [substituted for 'by times', deleted] / And mulcts upon default latam sententiam & to performers immunity & reward / Registring remarkes & [? deleted] read first and the Reg[iste]r 1st next day / Letters opend & read / Answers dictated [substituted for 'directed', deleted] & sent ['& expected' deleted] / Registry & repository Kalendard & in publicke custody. / Accounts sent to absent wellwillers / Symposia'. There are no conclusive dating clues; the need for the repository to be calendared could suggest a date early in the 1670s; but the need to specify accounts for absent 'wellwillers' suggests a date after the death of Oldenburg. The paper by Oldenburg dated 1 Nov. 1673 already discussed is the next item in Cl.P. xx.50 (fol. 91), apart from an old cover sheet that has been bound in (fol. 90).

[64] *Ibid.*, fols 85, 95, 96, 97, and below, Document C. See also above, n. 59.

[65] We do, however, hear of 'Some proposals being read for regulating the business of the Society' in 1684: see Birch, *History*, iv, 246.

[66] Bowood House Petty MSS H (8) 14, transcribed in Sharp, 'Sir William Petty' (n. 20), p. 392. This transcription is correct, but it may be noted that the number of seven suggested for the council of management has apparently been substituted for five. On Petty's election to the Council see Birch, *History*, iv, 555.

increasing preoccupation with the education of his sons in the mid-1680s, though it also reflects his long-held belief in the pedagogical value of mathematics and science.[67] In the Society, the paper's background is possibly illustrated by expressions of concern about the Society's experiments and correspondence, about which a committee had met earlier in 1687.[68]

We then hear little until 1698 – 99, when a new reformist effort, almost on the scale of that of the early 1670s, occurred. As earlier, a purge of arrears was accompanied by attention to the Society's activities: the statutes relating to experiments were read and a paper drawn up 'concerning Promoteing and Encouraging the makeing Experiments', which was ordered to be printed in the *Philosophical Transactions*, though it never actually appeared there. On the suggestion of Sloane, it was decided to try to persuade Fellows to provide 'some Experiments or Observations in Natural knowledge' for meetings, encouraging them by offering to pay their expenses; a list of candidates for this was drawn up, a number of whom obliged.[69]

Though independent of it, these steps coincided with Louis XIV's reform of the Académie des Sciences in January 1699, and in March a committee was appointed at the Royal Society to consider the new statutes that the Académie was granted at this time 'and to make an Extract of what they think necessary to be Established for the Society'. Their report unfortunately does not survive, but the then Secretary, Hans Sloane, also wrote a letter on the subject to his co-Secretary Richard Waller (at this stage living in Hertfordshire), and Waller's reply is extant.[70] The reform of the Parisian Académie stimulated Waller to voice a sense of envy of the opportunities provided by its state endowment which others in the Royal Society had felt before him,[71] and he somewhat regretfully pointed out the limitations imposed on the Society in comparison with its French counterpart by its voluntary structure. But in enumerating items among the French institution's statutes that seemed to him relevant to the English one in spite of this,

[67] For a related draft see the Marquis of Lansdowne, ed., *The Petty Papers* (London, 1927), ii, 9. Lansdowne notes Petty's renewed interest in education in the mid-1680s: *ibid.*, ii, 3. For an earlier statement of his interest see [William Petty], *The Advice of W. P. to Mr Samuel Hartlib, for the Advancement of Some Particular Parts of Learning* (London, 1648).

[68] In February 1686 Council members were asked to come to the next meeting 'prepared to give their thoughts about the means of making experiments in a philosophical method', while the 1687 committee was intended 'to consider of experiments and the correspondents of the Society': Birch, *History*, iv, 455, 524.

[69] Royal Society Copy Council Minutes, ii, 141, 143 – 5, 148 – 9; Royal Society Copy Journal Book, ix, 118, 124; Hunter, *Royal Society*, pp. 45 – 46.

[70] Royal Society Copy Council Minutes, ii, 151; Richard Waller to Hans Sloane, 11 April 1699, British Library MS Sloane 4037, fol. 249. For the text of the regulations, which Waller refers to by number, see Leon Aucoc, *L'Institut de France: Lois, Statuts et Reglements* (Paris, 1889) pp. lxxxiv – xcii.

[71] See Hunter, *Science and Society*, p. 40.

he echoed the priorities of earlier reformers in seeing the need for a more streamlined and effective society in terms of membership and activities.

The theme of the contrast between the Royal Society and the Académie des Sciences is also at the heart of a further paper by Robert Hooke which evidently dates from between 1700 and 1702, entitled 'Proposals for the Advancement of the Royal Society'.[72] Though its title illustrates its affinity to the larger body of papers on the Society among Hooke's manuscripts, this is more down-to-earth and realistic than them, his chief suggestion being that committees should be reinstated, as in the Society's early years. Hooke grudgingly acknowledged the role even of members who came to meetings 'only as to a Play to amuse themselves for an hour or so' as long as they supported the Society by paying their subscriptions. He also commented on certain classes of Fellows from whom contributions to the Society's work might be expected, including 'Mathematicians', 'Gentlemen who enter themselves for diversion only' and doctors, though he criticized some of the latter for joining with the ulterior motive of promoting their practice and losing interest in the Society once this had been achieved.[73]

Though this article is mainly concerned with seventeenth-century material, it is worth noting that comparable concern about the Society continued into the eighteenth century. Thus under Newton a predictable attack on arrears was accompanied by the suggestion of specialist committees dealing with different areas of the Society's activity 'for the more Effectual promoting of the Ends of the institution of this Society', together with the notion of employing additional expert staff specializing in five branches of scientific knowledge. The latter idea is expounded in a 'Scheme for Establishing the Royal Society' introduced by more general comments on the Society's aims, for which Newton scholars have often mistakenly presumed originality.[74] In fact, as will by now be apparent, it closely parallels the recurrent suggestions of the late seventeenth century which have been surveyed here.

A further discussion of the state of the Society at much the same time is to be found in an exchange of letters among the friends of Martin Lister in 1709 – 10, in which the Académie des Sciences was admired and the 'defects' of the Royal Society deprecated. Lister was mainly preoccupied by the factionalism then rife in the Society, which he saw as self-destructive, suggesting that offices should rotate annually and seeing an effective secretary like Fontenelle as crucial. His correspondent, Thomas Smith, however, showed a defeatism lacking in the authors of the papers which form the main subject of this article in his conclusion that 'it would bee a

[72] D.M. 5.12, partially printed in Bluhm, 'Remarks on Finances' (n. 22), p. 92. On its identification and dating see Note II.

[73] On the tendency for doctors to drift away from the Society see Hunter, 'Early Problems' (below, ch. 8), p. 278; Frank, 'Institutional Structure' (n. 3), p. 97.

[74] David Brewster, *Memoirs of Sir Isaac Newton* (Edinburgh, 1855), i, 102 – 4; F. E. Manuel, *A Portrait of Isaac Newton* (Cambridge, Mass., 1968), pp. 444 – 5 and ch. 13; R. S. Westfall, *Never at Rest: a Biography of Isaac Newton* (Cambridge, 1980), p. 685 and ch. 13 *passim*.

presumption in any single person to find fault, much more to point out a method, how they may bee supplyed, & all disorders reformed'.[75]

II

Hence from the late seventeenth century we have a whole series of papers inspired by a sense of dissatisfaction with the Royal Society as currently constituted, often dating from episodes in the Society's history when the need for change was intense. As we shall see, a number of common themes runs through them concerning the best means of rectifying this situation, but, in addition, there are marked disagreements between the authors involved, and it is not least for this reason that the documents are so significant.

To some extent, these divergences stem from the different purposes that the papers were serving. Some deal primarily with the Society's organization, others with the content of its work, though there is an overlap between the two. More important, the documents differ according to how idealistic the authors were being in the suggestions that they put forward. While most aimed at strategies that the Society could implement within the resources that it could reasonably expect to mobilize, some — and especially the main group of Hooke's papers — were more Utopian in their suggestions, and are the more revealing for that.

Other disagreements stem from the authors' differing responses to the Society's problems in its early years. Indeed, one of the most important features of the papers is the extent to which they comprise attempts to confront real problems that the Society had encountered and with which their authors were familiar, rather than being theoretical schemes unfounded in any practical experience. The authors draw attention to matters which particularly concerned them, often providing telling evidence about the Society's structure which is not available from other sources, while their differing perceptions of the Society's actual or potential role in Restoration society also affected their view of what was feasible.

Most interesting of all are the disagreements and differences of emphasis that stemmed from contrasting intellectual positions within the Society. Insofar as such tension has been recognized in the past, it has been taken to represent a division of interest between serious scientists and virtuosi, and it is true that the trivializing effect of the amateurs who made up the rank and file of the Society's membership is in evidence in certain episodes in the Society's history in these years.

75 Thomas Smith to Martin Lister, 25 Feb. 1710, Bodleian Library MS Lister 37, fol. 179. See also Smith to Lister, 16 July 1709, *ibid.*, fol. 154, and Lister to Smith, 6 July 1709 and 15 Feb. 1710, Bodleian Library MS Smith 52, fols. 3 and 16.

Thus while some hoped that the Society's 'repository' or museum should be built up into a comprehensive collection of objects which would be a novel and serious research tool, this failed to make headway against a virtuoso valuation of the exotic and the rare.[76] A tension between more-or-less single-minded attitudes is also in evidence in views on the library that the Society was given by Henry Howard of Norfolk in 1667. Some, like Hooke, felt that the Society's library should 'consist of all books of Arts and Naturall History and of noe other whatsoever' and would have disposed of non-scientific works, but this step was resisted by others who evidently relished Howard's books for 'their rarity and the reputation they carry along with them upon that account'.[77]

It is also clear that the Society's voluntary structure meant that the virtuoso membership on which it depended for income from subscriptions could have a deleterious effect even on the reform efforts that occurred, so that instead of research the Society's priority became the amusement of Fellows. In the reform of 1673 – 75, for instance, part of the Council's object in providing 'experimental entertainments' was to make the meetings 'more considerable' and thereby attract more subscriptions.[78] Moreover the kind of spectacular and novel experiments that appealed most to the virtuosi were not necessarily the most scientifically significant ones.[79]

On the other hand, it would be simplistic to see the Royal Society's problems as exclusively due to a war between scientists and non-scientists, as was the tendency in such older historians of the institution as Sir Henry Lyons.[80] Rather, the documents on which this article is based bear out hints that may be gleaned elsewhere to reveal differences of strategy even among those most seriously and actively committed to the Society's enterprise, which were related to differing areas of interest and priorities in research. Disagreements of this kind have hitherto been almost entirely overlooked, but they are significant both in their own right and for the ramifications they had in terms of contrasting views of how the Society should be constituted and organized.

All agreed that to achieve its avowed aim of improving natural knowledge, the Society needed to perform experiments, collect observations, maintain a correspondence network, and collate the writings of naturalists both ancient and modern. But there was disagreement on the relative importance to be assigned to these activities, which seems to be

76 See Hunter, 'The Cabinet' (n. 1), pp. 162 – 6. See also Hunter, *Science and Society*, pp. 65 – 69 and above, ch. 4.

77 Robert Hooke to Francis Bernard, 7 April 1674, Bodleian Library MS Smith 45, fol. 105; Birch, *History*, ii, 352. See also Hunter, *Science and Society*, p. 146.

78 Birch, *History*, iii, 137. For comparable comments on later occasions, see *ibid.*, iii, 516 – 17; iv, 187 – 8.

79 See Frank, 'Institutional Structure' (n. 3), pp. 91 – 4, and Heilbron, *Physics at the Royal Society* (n. 1), ch. 1. For a highly suggestive study of related themes see Simon Schaffer, 'Natural Philosophy and Public Spectacle in the Eighteenth Century', *History of Science*, 21 (1983), 1 – 43.

80 Lyons, *Royal Society* (n.11), *passim*. For comments on this see Hunter, *Royal Society*, pp. 2, 49.

particularly associated with a contrast between those primarily involved with the physical sciences and those whose main interest was in the life sciences, the former taking a more single-minded stance while the latter adopted a more tolerent and pluralist one.

Indeed, a certain amount of mutual jealousy between these groups is evidenced by our sources. 'A.B.' was at pains to defuse any tension between experimenters and naturalists — 'an Artist or Experimenter, is not to be taken for a maker of gimbals, nor an observer of Nature for a wonder-monger', he wrote. Comparable evidence survives from the 1680s, in a comment by the Irish intellectual, William Molyneux, whose own commitment to the physical sciences helps to explain his slightly caricatured view of biological research: writing to Edmond Halley in 1686, he reported how he had heard that 'there was a party arising in the Society, that were for rejecting all kinds of useful knowledge except ranking and filing of shells, insects, fishes, birds, &c. under their several species and classes; and this they termed *Natural History*, and *Investigating Nature*, never attending to the uses and properties of these things for the advantage of mankind, and reckoning chemistry, astronomy, mathematics, and mechanics, as rubs in their course after nature'.[81]

Such contrasting priorities are to be discerned in the differing strategies suggested by the authors of the documents dealt with in this article. On the one hand there were those like Petty and Hooke who believed that the Society should principally concern itself with a coherent programme of experimentation.[82] On the other, 'A.B.' — and it is significant that, insofar, as he can be identified, he was apparently a physiologist — attached as much significance to 'Observations' as to experiments, holding that 'so far as we have any new matter of fact before us, we have really advanced further then we were before'. Moreover there are hints that within the ranks of the Society there were others less discerning still, who tended to regard meticulously planned experiments as being on a par with random, and often uncritical, observations.[83]

These divergent assessments reflect a tension in the Baconian corpus which shaped the emergent styles of Baconianism to be found in the Society. Thus some stressed those of Bacon's methodological principles which proclaimed the value of systematic experimentation as opposed to casual observation. In his *Novum Organum*, Bacon insisted that our experience had to be enlarged 'in accordance with a fixed law, in regular

[81] See below, Document B. Molyneux to Halley, 8 April 1686, in Birch, *History*, iv, 476. *Cf.* J. G. Simms, *William Molneux of Dublin*, ed. P. H. Kelly (Blackrock 1982), p. 56. On Molyneux see also K. T. Hoppen, *The Common Scientist in the Seventeenth Century* (London, 1970), esp. ch. 5.

[82] Petty's view of the primacy of experimentation is stated explicitly in his 'Advertisements for the Dublin Society', reproduced by Hoppen, *op. cit.*, pp. 202 – 3. His activities in the reform efforts of the early 1670s are consistent with this view: see Birch, *History*, iii, esp. 102, 119. For Hooke, see below, Document C. Compare William Neile, below, Document A.

[83] See below n. 85 and Document B. On 'A.B.''s identity see Note I.

order', underlining the need for 'experiments of *light*' designed to reveal the true causes of natural phenomena, and such passages provided a charter for the sophisticated methodology espoused by authors like Hooke.[84]

But in addition, and particularly through the example provided by his own accumulation of an 'indigested heap of particulars' in his *Sylva Sylvarum*, Bacon also sanctioned an alternative tradition comprising the random compilation of observations and the occasional rudimentary experiment. It is symptomatic that the retrenchment at the Royal Society in 1680 took the form of renewed attention to *Sylva Sylvarum*, and there were certainly those in the Society's ranks who took their inspiration from this facet of Bacon's legacy, a position to which public expression was given by Thomas Sprat in his *History of the Royal Society*.[85] Though more sophisticated, there is a definite echo of this in the view of 'A.B.' that as much information as possible was needed, on the grounds that no observation or experiment, however apparently trivial, should be 'sleighted', 'if truly made'.[86]

Significant methodological differences existed even among the group which advocated increased emphasis on experimental demonstrations in the proceedings of the Society. While some of Petty's remarks imply that the programme of experimentation which he proposed was structured by the metaphysic of the mechanical philosophy, we have seen that he also maintained that experiments should not be made in order to prove some preconceived theory, as this would 'prepossess and biass the experimenter'. In this he was echoed in an anonymous memorandum on the Royal Society published in the nineteenth century by C. R. Weld, which claimed that 'till there be a sufficient collection made of experiments, histories, and observations, there are no debates to be held at the weekly meetings of the Society, concerning any hypothesis or principle of philosophy, nor any discourses made for explicating any phenomena, except by speciall appointment of the Society or allowance of the President'.[87]

By contrast, William Neile thought that it was more fruitful to first formulate causal hypotheses and then proceed to test them experimentally, and Hooke also recommended that the Society adopt this procedure, writing that 'WHOSOEVER therefore doth rightly make Experiments . . . doth consider what Circumstances and Effects, in that Experiment, will be material and instructive in that Enquiry, whether for the confirming or destroying of any preconceived Notion, or for the Limitation and Bounding thereof'.[88] To Hooke, this method was perfectly compatible with that

[84] Bacon, *Works*, iv, 95. *Cf.* P. B. Wood, 'Francis Bacon and the "Experimentall Philosophy"': A Study in Seventeenth-century Methodology' (London M.Phil. thesis, 1978) pp. 39 – 45.

[85] Bacon, *Works*, ii, 335; Birch, *History*, iv, 43, 46; Michael Hunter, *John Aubrey and the Realm of Learning* (London, 1975), ch. 2; see also above, n. 17.

[86] See below, Document B.

[87] Birch, *History*, iv, 8. *Cf. ibid.*, iii, 102. Weld, *History* (n. 14), i, 147: on this paper see further below, n. 211.

[88] See below, Document A; William Derham, ed., *Philosophical Experiments and Observations of the Late Eminent Dr Robert Hooke* (London, 1726; London, 1967), p. 26.

prescribed by Bacon, since, following Boyle, he assimilated the testing of hypotheses to Bacon's form of eliminative induction.[89] Moreover for both Hooke and Neile the principles of matter and motion and hypotheses derived from them were to guide projected experimental enquiries, with the search for causes forming a central element in the Society's work. As Neile put it, 'it is very likely the right setling of [the principles of matter and motion] will give much light to experiments'.[90]

This is seen clearly in the context in which William Neile's paper originated. Prompted by his desire to secure and elaborate the foundations of the mechanical philosophy, Hooke proposed in October 1668 that 'experiments of motion . . . be prosecuted, thereby to state at last the nature and laws of motion', and subsequently the Society embarked on a programme of experimentation which elicited theoretical contributions from Neile, Wren, William Croone, Christiaan Huygens and John Wallis.[91] Neile's correspondence with Henry Oldenburg during the course of these discussions is particularly revealing both epistemologically and methodologically, for in his letters he suggests that experiments are only fully intelligible within a theoretical framework, and he displays a sense of the complexity of the logical relations between theory and experiment not to be found among his more narrowly Baconian colleagues.[92] While Neile acknowledged that experiments were necessary to confirm theoretical principles, his stress on the importance of reason and rational assessment of such principles contrasts with the naive empiricism expounded in Sprat's *History*. Neile himself was critical of this form of empiricism, complaining to Oldenburg that 'of late people are growne so distrustful of reason that they take her to bee like a lyer who so often say's that which is false that though he may speake true yet no body ought to relye upon it barely for his word'. Reflecting his own confidence in the power of reason, Neile emphasized the need to discover fundamental principles such as those governing matter and motion, since in his view 'to know a thing barely by experiment is good for use but it is not science or philosophye'.[93] We see, therefore, that there was a broad spectrum of methodological opinion within the Society, ranging from Hooke's and Neile's combination of mechanism, hypotheticalism and inductivism, to the less theoretically-oriented Baconianism of 'A.B.' and others, and this diversity resulted in contrasting proposals for institutional reform.

89 Wood, 'Francis Bacon' (n. 84), pp. 222 – 3; Robert Boyle, *Certain Physiological Essays*, in Boyle, *Works*, i, 303.
90 See below, Document A.
91 See Birch, *History*, ii, 315, 318, 320, 328, 332 – 3, 335, 337 – 9, 340, 341, 342, 344, 347, 350, 352 – 4, 358, 361, 362, 377, 381 – 2, 387 – 8, 392, 397.
92 See especially William Neile to Henry Oldenburg, 7 May 1669, in *Oldenburg*, v, 517 – 18. On Neile's position see also J. A. Bennet, *The Mathematical Science of Christopher Wren* (Cambridge, 1982), pp. 118 – 19, 122.
93 William Neile to Henry Oldenburg, 13 May, 7 May 1669, in *Oldenburg*, v, 544, 518. See also Neile's other letters in *ibid., passim*, and below, Document A.

How, then, were the activities of the Society to be organized? All commentators seem to have been happy with the basic constitutional structure of the Society, which they aspired to supplement rather than replace. Thus the desirability of having a President who acted not only as titular head but also as 'moderator' of the Society's proceedings was propounded by both 'A.B.' and Hooke, while Petty, in his recommendations for the Philosophical Society founded in Dublin in 1683, also suggested that the President should play a part in planning and organizing research.[94] It is no coincidence that for a substantial part of the late seventeenth century the Society did have an active president who was knowledgeable in scientific matters and well qualified to supervise research, notably during the regimes of Lord Brouncker from 1662 to 1677 and Wren from 1680 to 1682, while a similar state of affairs was to exist under Newton.

As for the Council, again all believed that it served a useful function, but more reservations seem to have been felt about the manner in which it actually operated. 'A.B.', who interestingly described the Council as the Society's 'Parliament', thought it essential that attendance should be maintained so that its managerial role could be carried out effectively. Evidently dissatisfied with what purely voluntary effort was achieving from this point of view, he suggested a system of 'Rewards' — in the nature of gratuities rather than salaries — which would have been awarded to the Council members and officers: indeed, though feeling the need to justify this suggestion, he even recommended that the President should be gratified in this way.[95]

Petty similarly suggested the possibility of remuneration, while he was also unhappy about the Council's size, which he considered inefficient and would have liked to streamline, reducing it from its existing size of twenty-one to seven or even five people. Moreover, whereas most were happy to see the Council combining administrative and scientific functions, Hooke in one of his drafts advocated greater specialization, proposing that 'a select committee' be elected annually to oversee the Society's finances and to revise the statutes when necessary. Significantly, the members of this committee were to be exempt from other duties so that they could devote themselves entirely to their managerial role.[96]

A related strategem recommended with almost predictable regularity both in these papers and in the actual steps that were taken to reform the Society was the formation of committees. These were intended either to deal with specialist areas of the Society's work — as with the initial group of committees appointed in 1664, or the administrative committees appointed

[94] See below, Document B; Cl.P. xx.50, fol. 86; Hoppen, *Common Scientist* (n. 81), p. 203. See also the anonymous memorandum quoted on p. 208, above.

[95] See below, Document B.

[96] Bowood House Petty MSS H (8) 14, printed in Sharp, 'Sir William Petty' (n. 20), p. 392 (but see above, n. 66). For Hooke see below, Document C; but for another draft which implies that the Council combined administrative and scientific functions see Cl.P. xx.50, fol. 85.

at various intervals — or to act as a 'standing Committee' which would deal with theoretical issues and supervise the Society's research programme as a whole, as in Neile's suggestion.[97] In either case, an aspiration to greater efficiency is in evidence, and it is symptomatic that the committee system became firmly entrenched at the Académie des Sciences in Paris during the late seventeenth century.[98]

Above all, the key to improving the Society's performance was seen by almost all commentators to lie in adding to the size of its paid staff. Even here, the lack of consensus in the Society is illustrated by the fact that the notion of specialist employees appears to have been opposed in one anonymous memorandum, which argued that it was better for curatorships to rotate than to be vested in a specific individual, 'the contrary making others idle, drayning him & creating to[o] much dependance on him'.[99]

Such views were clearly in a distinct minority among those concerned with the Society's mode of operation, however, and in every other case, commentators were agreed on the desirability of having paid specialists. On the other hand there are some significant discrepancies among the suggestions that were made, related in part to contrasting scientific priorities and in part to the extent to which people limited themselves to what they believed the Society could actually afford.

None seems to have thought that the Society could do with fewer than five staff with assistants, as against the two which it employed for most of this period.[100] Oldenburg in the early 1670s would have liked to have had two paid secretaries and three curators — 'one to be Physico-Mathematical . . . the second, Anatomico-Chymicall: The third, for the History of Trades', and he expected the first two to demonstrate experiments at every meeting of the Society.[101] 'A.B.' aspired to five curators in addition to a secretary. Three of these would have dealt with experimental matters, being respectively allocated the fields of 'Mathematicks, Anatomy, & Chymistry'. The other two would have been devoted to information-collecting, perhaps overlapping with the functions of Oldenburg's secretaries, though the one given responsibility for gathering information from men as against books might also have partaken of some of the functions of Oldenburg's curator for the History of Trades, whose task would presumably mainly have been to collect oral data about industrial and technical processes.[102]

While Petty's calculations appear to involve three curators and three secretaries, the demarcation of their responsibilities not being made clear, it is revealing that some anonymous notes on staffing in the Royal Society's archives envisage three secretaries and only two curators.[103] Thus the

97 See above, *passim*, and below, Document A.
98 Roger Hahn, *The Anatomy of a Scientific Institution: The Paris Academy of Sciences 1666 – 1803* (Berkeley and Los Angeles, 1971), pp. 21 – 22.
99 Cl.P. xx.50, fol. 89. See above, n. 63.
100 See, however, Hooke as referred to below, p. 212.
101 D.M. 5.14 – 15 and see above, n. 33.
102 See below, Document B. On the History of Trades see Hunter, *Science and Society*, ch. 4.
103 Bowood House Petty MSS H (8) 5; see above, n. 33.

priorities of Oldenburg and 'A.B.' in the allocation of resources for collecting information and doing research were here reversed, evidently implying a somewhat different valuation of the Society's respective roles on the part of its author.

The most idealistic author was Hooke. Though in one of his memoranda he limited himself — like other commentators — to what the Society might realistically hope to afford, insisting on 'two Secretarys & two Curators at least', in the paper printed at the end of this article he was more ambitious, suggesting that six curators be allocated to information-collecting, while for experiments he aspired to '8.10.12.20 or more . . . as the stock increaseth'.[104] Here again the relevance of these writers' scientific priorities to their institutional proposals is made clear, since it is revealing how, in the event of more funds becoming available — and in contrast to the anonymous author just discussed — the commitment to empirical and especially experimental research would increasingly have outdistanced the task of collecting information in Hooke's scheme.

In addition, as perhaps befitted the Society's Curator of Experiments, Hooke described the curators' responsibilities in more detail than anyone else, recommending an elaborate set of procedures governing the demonstration of experiments. The curator was expected first to give some account of the nature of the experiment he intended to perform, and then, before demonstrating it at a regular meeting, he was to run through the experiment on a set day in the presence of interested Fellows and others given permission to attend. At the meeting itself, the curator was expected to discuss the experiment and its theoretical implications with a member assigned to 'propound difficultys or objections that may occurr', and the main points of these discussions were to be minuted. Each curator would have been continually involved in this process, as Hooke expected them to provide a minimum of one experimental demonstration per meeting.[105] Curators were also to keep careful records of their enquiries which were to be read to the Society, and they were to publish reports of experiments or inventions which the Society thought should be publicly announced. Elsewhere Hooke suggested that they be obliged to deliver a discourse based on experiments to the Society once a quarter. In addition to these duties, curators were to serve on *ad hoc* committees established to vet experiments submitted to the Society by outsiders.[106] Hooke thus recognized that one function of the Society was to validate the work of others, but he gave predecence to the Society's role as an institution devoted to the

[104] Cl.P. xx.50, fol. 85 (it is revealing that 'at least' is an insertion); see below, Document C: '20' may have been deleted, perhaps due to Hooke's realizing that his plans were getting out of hand.

[105] Cl.P. xx.50, fols 86 – 87. See also Hooke, *Philosophical Experiments* (n. 88), pp. 26 – 28, where Hooke describes this process in some detail.

[106] Cl.P. xx.50, fols 86 – 87, and below, Document C. Petty too recognized that the society validated the work of others: see Bowood House Petty MSS H (8) 10, transcribed in Sharp, 'Sir William Petty' (n. 20), p. 399. On this function in the Académie des Sciences, see Hahn, *Anatomy*, (n. 98), pp. 21 – 24.

pursuit of original sustained research, and he defined the curator's office accordingly.

Just where Hooke thought the finance for his veritable band of research workers would come from is only partially clear. Oldenburg's and Petty's proposals, as we have seen, were worked out in terms of calculations of what the Society would be able to afford if Fellows' subscriptions were supplemented by income from the investment of relatively modest amounts of capital. 'A.B.' was rather less realistic — in view of the Society's experience of the difficulty of attracting endowment, particularly from the King — in hoping that benefactors might come up with the funds required.[107] Hooke's response to this problem may be partly reflected in two proposals made in his notes which are not paralleled in other authors. Whereas commentators like Oldenburg and Petty visualized the curators being paid a fixed stipend of £40 or £50 per annum, a suggestion that appears in some of Hooke's papers, elsewhere he argued 'That the Curators salary be but small and that there be other incouragements given according to desert upon each new invention or Discovery either in money plate medalls or other gratuitys. but that his salary & gratuitys shall not annually exceed £50'.[108]

In addition, Hooke outlined in one of his memoranda the novel and ingenious idea of trying to tap the expertise of the Society's Fellows and staff by setting up a research and development scheme from which the Society would benefit. He proposed 'That a certain number of the Society be appointed to manage the prosecution of any new invention soe as to bring it into use & make it profitable for the Society and the inventor. [and] That the moyety of all the profitts that shall arise upon the said prosecutions after the deduction of 6 per cent for the stock shall be made, shall accrue and belong to the Society'.[109]

Elsewhere in his notes Hooke ignored such ideas, and presumed — in contrast to those who aspired to benefactions — that the Society would be wholly dependent on 'the bounty of its own members', that rank and file of the Society who did in fact provide the bulk of its revenue throughout the period under study. Indeed, evidently because of this need for finance, Hooke reluctantly made allowance for membership on the part of those whose 'occasions will not permitt' them to participate fully in the Society's work, in contrast to the normal requirements which he would have imposed on Fellows, which will be considered shortly. Such 'Extraordinary' Fellows were, however, to pay a double subscription.[110]

What, therefore, of the membership at large? No-one, it should be noted, seems to have aspired to a wholly professional body like the Académie des Sciences, partly because of realism about ever acquiring funds on a scale that

107 See below, Document B.

108 Cl.P. xx.50, fol. 86. For a slightly different suggestion see below, Document C. Compare the system operated under Newton: Heilbron, *Physics at the Royal Society* (n. 1), pp. 18 – 19.

109 Cl.P. xx.50, fol. 86.

110 *Ibid.*, fols 85, 88, and see below, Document C.

would have made this possible, and partly because of a belief in the genuine fruitfulness of juxtaposing the work of full-time researchers with the activity of part-time enthusiasts at meetings, along the lines expounded by Hooke. But though the basic desirability of a voluntary structure was agreed, here again significant disagreements are in evidence.

'A.B.' had a fairly tolerant view of the membership. He held that Fellows should *not* be obliged to participate in the proceedings of the Society. While they might voluntarily contribute in some way, their main role as he saw it was to provide moral and monetary support for the officers and paid staff. Furthermore, he did not regard philosophical ability as a prerequisite for membership. Since he saw the Society as dedicated primarily to the compilation of histories of nature and art, this required a combination of talents, not all of them scientific. Hence virtually anyone sympathetic to such aims could play a useful part, even those who 'may have no genius to progress in philosophy, who yet may have witt enough in their own way'. The pursuit of a Baconian natural history programme implied that the Society should be broadly based and open to all — 'As for persons, we cannot be too many employd, since we have all Nature to turn us in' — and it is hardly surprising that 'A.B.' gave no consideration to the question of assessing the appropriateness of applicants for membership.[111]

Hooke's view was radically different. He accepted the same Baconian premise that natural knowledge was to be advanced through co-operative endeavour, but his contrasting view of the corps which would be involved in the reformation of the sciences is epitomized in his comparison of it to the Spanish conquerors of Mexico — 'a Cortesian army well Disciplined and regulated though their number be but small'.[112] In Hooke's ideal scheme, all Fellows should be actively involved in the Society's research programme, and in the paper printed at the end of this article he envisaged their being allocated to the different facets of the Society's programme in the same proportion as the curators, with 10% devoted to the examination of books, 20% to the maintenance of its correspondence and 70% to making experiments and improving 'Artificiall inventions and productions'. Moreover he would have made mandatory the attempt to persuade active members to present the results of their research at meetings which had been part of the reform effort of 1673 – 75, insisting that all Fellows deliver a discourse before the Society either quarterly or annually, while each Fellow was also expected to cultivate and report on his philosophical correspondence.[113]

Here we return to broader principles, for the demands which Hooke thought should be imposed on the whole body of Fellows were evidently inspired not only by the exigencies of the Royal Society, but also by his conception of the consummate natural historian which he outlined *c.*1668 in his '*General Scheme, or Idea* Of the Present State of Natural Philosophy'. He there described the ideal naturalist as one who was unbiased, yet

[111] See below, Document B.
[112] See below, Document C.
[113] *Loc. cit.* and Cl.P. xx.50, fols 87, 96. On the 1670s see above, nn. 22 and 43.

knowledgeable in mathematics and mechanics, cognizant of the rival systems of philosophy ancient and modern, and ready to perceive the theoretical ramifications of empirical research. Far from being an omnivorous collector of random observations and recondite facts, Hooke's natural historian set himself well-defined problems to investigate and exercised considerable discrimination in compiling his information. As Hooke put it, 'there must be Judgement in the Historian to discern what will be material and useful in general, and what will be more especially adapted for the Inquiry whatever he designs'.[114] Hence the histories written by the model naturalist are not mere catalogues; instead, they address theoretical questions and present an organized summary of the relevant phenomena. For Hooke, the natural historian's task was thus an intellectually demanding one, requiring a sophisticated method and a high degree of competence in the sciences.

If each Fellow was to be required to make a serious contribution to the Society's programme in this way, this meant that the membership would have to be more select, and — evidently reacting against the ease with which people were currently elected to the Society — Hooke would have imposed severe restrictions on the membership, looking forward to the practice of more modern times.[115] None, he thought, should be admitted 'but such as shall be eminently deserving for his extraordinary skill and zeal for promoting the ends of the Society', something which was to be attested by six or ten Fellows or 'by some publique Action or publication'. He even wished to insert a promise not to allow anyone undeserving to be elected into the oath that new members were to take.[116] Regarding foreign members, Hooke was even more exacting, and here, it must be admitted, existing arrangements were almost more haphazard. To gain election, he required savants from abroad to submit two or more discourses for approval. If elected, they were to send quarterly 'some such discourse discovery experiment or Invention as shall [be ap]proved worthy to be accepted', to act as scientific intelligencers, and to supply information requested by the Society.[117]

It was not only concerning election that Hooke made rigorous requirements of his putative Fellows. He was equally dictatorial in other

[114] Hooke, *Posthumous Works*, pp. 19 – 20: on the dating of Hooke's scheme see Pugliese, 'Scientific Achievements of Robert Hooke' (n. 62), pp. 9 – 10. It is possibly significant that a manuscript lecture which overlaps with this material is to be found juxtaposed with the papers on the Royal Society analysed here, Cl.P. xx.50a. For a discussion of Hooke's conception of natural histories see Wood, 'Francis Bacon' (n. 84), pp. 216 – 22, and Pugliese, 'Scientific Achievement', pp. 9 – 31.

[115] See Maurice Crosland, 'Explicit Qualifications as a Criterion for Membership of the Royal Society: A Historical Review', *Notes and Records*, 37 (1982 – 83), 167 – 87.

[116] See below, Document C, and Cl.P. xx.50, fol. 87.

[117] See below, Document C. On existing arrangements see Hunter, *Royal Society*, appendix 10.

respects. In financial terms, his response to the Society's experience of Fellows' unreliability with their dues was to insist that non-payment of subscriptions should lead to automatic suspension.[118] In addition, Fellows were to take an oath to follow the Society's instructions as to what they might or might not divulge of what passed at meetings, while in the event of their leaving the Society, they were even to be forbidden ever to divulge why they had resigned.[119]

The difficulty which Hooke had to face was how to make people accept these almost draconian requirements, particularly when — as Richard Waller observed in contrasting the Royal Society with the Académie des Sciences — 'our society consisting wholy of volunteers the members will be unwilling to be too much confined and obliged to strictness in some particulars'.[120] Hence at this point Hooke's ideas once again overlap with the related thoughts of other potential reformers, their differing views providing a revealing insight into early reactions to that imponderable which all organizers of voluntary bodies have faced since, namely how to motivate members to undertake administrative duties and to participate in meetings. This problem was particularly intense for the Royal Society in its early years because the Society remained peripheral to the scientific activity even of many of its members, while at the same time it tended to provide more services more widely than it was recompensed for. As a result, activities encouraged by the Society flourished in the country at large while its London meetings faltered for want of support.[121]

The problem of what was practicable as against what might seem theoretically desirable clearly worried Hooke, since he himself wondered, in a passage in the memorandum printed at the end of this article which he subsequently erased, whether restrictive ordinances needed to be limited so that 'the strictness and Rigour of these first ingagements may not frighten and deter such person[s] as are every ways proper & desirable to be members of this Society from listing and entring themselves thereof'. More to the point, as he realized himself, the Society simply lacked the coercive power to enforce the strict conditions he aspired to. Noting that 'Punishment is impracticable', he therefore turned instead to considering the positive means by which people could be persuaded to take part in the Society's activities, noting that 'The Benefit of every Member thereof in this Way is the Soule & life of the Society'.[122]

One idea which appears fleetingly but is not developed is the notion that participation in the Society's meetings gave enthusiasts 'An opportunity to

118 See below, Document C, and Cl.P. xx.50, fol. 87.

119 See below, Document C. The last suggestion was presumably linked to the factionalism from which the Society suffered: see Hunter, *Science and Society*, pp. 44 – 45.

120 Richard Waller to Hans Sloane, 11 April 1699, British Library MS Sloane 4037, fol. 249. *Cf.* Hooke's own view in D.M. 5.12, printed in Bluhm, 'Remarks on Finances' (n. 22), p. 92.

121 Hunter, *Science and Society*, pp. 45, 56 – 57.

122 See below, n. 174; Cl.P. xx.50, fols 97, 85.

Shew their Abilitys with advantage to Get a character of the Society' — thereby showing an aspiration to give the Society the role in accrediting scientists which is what modern scholars have often presumed the early Royal Society to have had, but which in fact it largely lacked. 'A.B.' likewise clearly saw as a worthwhile reward 'the Societys acceptance of what is done': he was uncertain, however, whether people would respond to this and it was for this reason that he suggested the gratuities that have already been referred to.[123]

More important was Hooke's stress on the tangible 'benefits' which the Society could offer to its members, something echoed by such other commentators on the Society as Petty.[124] Hooke made clear that his object in itemizing these was to make membership of the Society so essential to any serious enthusiast that its removal would be a real threat. Noting that 'The Power of the Society is only alluring not otherwise coercive, and therefore the more allurements the more power it hath and the greater will be the terror of the only punishment it hath namely, Expulsion or Ejection', he listed in detail the 'Allurements' which the Society could offer — its facilities such as its library, museum and records, and the opportunities for information and advice available through it. Moreover he was adamant that these should be available only to members: 'there must bee somewhat to bee had by those that meete & are regular members which others must want.'[125] Even the Society's museum — one of the minor tourist attractions of London — was to be denied to non-members in one of his drafts, and he also repeatedly insisted on the need for secrecy. 'Nothing considerable in that kind can bee obtaind without secrecy because els others not qualifyed as abovesaid will share of the benefit.'[126]

Throughout Hooke's papers there is an obsession with secrecy. In one, he considered how it would work in detail, regarding matters reported orally at meetings as relatively easy to safeguard, 'memory alone not being able to carry away exact particulars'; the Society's archives, on the other hand, were to be kept under strict supervision, while even the reports of proceedings which were sent out to absent Fellows were to be kept secret for twelve months by their recipients.[127] Though this may have owed something to Hooke's bad personal experiences with ideas which he believed that he had

[123] See below, Documents B and C. On the Society's validating role see above, p. 212: see also Hooke in Cl.P. xx.50, fol. 86, where he suggests 'That any person that hath found out any new & usefull invention which upon his discovery thereof to the Society shall be approved of as such, shall Receive from the Society a publique attestation thereof under their common seal. and be further Gratifyed by a medall picture or some mark of honor & respect suitable to his invention'.

[124] Bowood House Petty MSS H (8) 5. The 'benefit and satisfaction' of membership is also referred to by Goddard in D.M. 5.13.

[125] Cl.P. xx.50, fols 95, 97. *Cf. ibid.*, fol. 85, and below, Document C.

[126] *Ibid.*, fol. 97. See below, Document C, and *ibid.*, fols 85, 86 – 7, 95, 96. In fol. 86, Hooke was a little less strict about the museum: non-members were allowed to see it with the Society's permission. On visitors to the museum, see R. T. Altick, *The Shows of London* (Cambridge, Mass., 1978), p. 14.

[127] Cl.P. xx.50, fol. 97, and see below, Document C.

originated and others plagiarized, it owed more to his feeling that only in this way would it be possible to assure a committed membership, in that 'the Benefit Receivd be greater then the expense and trouble will purchase elswhere'.[128]

How widely were such ideas shared in the Society? 'A.B.' agreed that a policy of secrecy would attract members and make tradesmen and others more willing to divulge their discoveries, while similar views were echoed at the Council meeting of 15 October 1674. The presence of non-members at meetings was said to be 'troublesome and prejudicial', while, to prevent members from being shy in divulging their findings, a statute was prepared 'injoining secresy to the members of the Society in such matters, as shall be . . . declared to be kept secret'.[129] A further echo is to be found in the idea which, as we have seen, was suggested to the Council in December 1678 — a journal which, in contrast to all the Society's publications up to that time, 'shall not be sent or sold to any person but members of the Society, and to such as correspond with Mr Hooke by the Society's directions, and make considerable returns to him for the Society's use', though this was never brought to fruition.[130]

If such suggestions had been followed up, the Society's institutional requirements would thus paradoxically have overridden the commitment to the free dissemination of knowledge which had originally inspired it and it would have reverted to a 'closed' organizational form reminiscent of Bacon's Solomon's House, with its secretive aura and its hand-picked staff. Moreover, though such ideas clearly arose primarily in reaction to the specific problems arising from the Society's voluntary structure, it is worth noting that they were also symptomatic of a deeper ambivalence about the need for secrecy in the science of the time, to be found in the Accademia del Cimento, the Académie des Sciences and in the thought of many seventeenth-century naturalists.[131]

On the other hand, it is significant, not only that the notion of a restricted bulletin suggested in 1678 never materialized, but that during the same period the minutes also record a diametrically opposite suggestion — that of the cheap broadside to 'propagate natural philosophy' which has already

[128] *Ibid.*, fol. 85.

[129] See below, Document B; Birch, *History*, iii, 137 – 9. Richard Waller was later hostile to the presence of non-members at meetings for a slightly different reason — 'some forrainers possibly comeing only to spye what is a doing and will never contribute any to us of worth': Waller to Hans Sloane, 11 April 1699, British Library MS Sloane 4037, fol. 249.

[130] Birch, *History*, iii, 450 – 1.

[131] Hahn, *Anatomy* (n. 98), p. 16; W. E. Knowles Middleton, *The Experimenters: A Study of the Accademia del Cimento* (Baltimore, 1971), pp. 300 – 1; Hunter, *Science and Society*, p. 100, and the references cited there. Hahn is perhaps being overly Mertonian when he claims that scientific academies in the late seventeenth century were 'open' and that intellectuals freely communicated their discoveries: see Hahn, *Anatomy*, pp. 3, 43 – 44. Compare Merton on the norm of 'communism': R. K. Merton, *The Sociology of Science*, ed. N. W. Storer (Chicago, 1973), p. 274.

been referred to.[132] Though this was not implemented either, it would have represented a reversion to and intensification of the Society's normal policy of freely promoting the new science as widely as possible, a policy which clearly had advocates in the Society's ranks despite the institutional problems that it caused.

Hence we can again discern differences of strategy among the Society's active membership, reflecting contrasting responses to the Society's actual experience and differing views on the best means of promoting the new science in contemporary society. Such disagreement is perhaps most clearly in evidence in connection with a more controversial matter, that step which Hooke named as the Society's 'only punishment' for those who abused their position in its ranks, 'namely, Expulsion or Ejection'.

Provision had in fact been made for expulsion in the Society's statutes, and the step was taken against a few members who showed total lack of interest in the Society's activities in the 1660s and 1670s. Then, as we have seen, the policy was applied on a large scale in the early 1680s, and it is interesting that this was juxtaposed with the implementation of another of Hooke's suggestions, of imposing stricter entry requirements on candidates for election. For in January 1682 it was decided that candidates should be scrutinized to see if they were 'so qualified, as in probability to be useful to the society' and approved by the Council.[133] This was evidently intended to prevent a recurrence of the indulgent membership policy of the Society's early years which had made expulsions so necessary, though it is doubtful if — after the first few years — the quality of recruits was much altered by this step. Hence in 1698 – 99 a handful of fresh expulsions were made, while in the latter year we find Richard Waller echoing Hooke in wishing that membership should be more exclusive, perhaps related to known achievement in the field.[134]

Though expulsions were thus made in an attempt to streamline the membership, it is clear that opinions within the Society were divided on the desirability of this stratagem, partly on practical grounds and partly on broader ones. Even at the time when expulsions were first made, in 1666, there is evidence that opinions differed as to whether the French writer, Samuel Sorbière, should be treated thus.[135] In 1669, John Beale warned against ejecting deliquent Fellows on the grounds that it might cause unnecessary difficulties, and this view was shared by 'A.B.', who saw the question of expulsion as 'a ticklish poynt'.[136]

Caution was in evidence when expulsion was mooted at Council in 1673, and 'It was resolved, that if any fellow should be ejected on this occasion, the cause of ejection be recorded', while opinions were evidently still divided in 1674. This is suggested by the fact that, though the threat of

[132] See above, p. 199.
[133] Birch, *History*, iv, 158; Hunter, *Royal Society*, p. 41 and appendix 8.
[134] *Loc. cit.*; Richard Waller to Hans Sloane, 11 April 1699, British Library MS Sloane 4037, fol. 249 (see above, n. 70).
[135] Birch, *History*, ii, 124, 127.
[136] Hunter, *Royal Society*, p. 101; see below, Document B.

expulsion appeared in the letter concerning arrears sent out in April that year, it was not directly alluded to either in another letter on the same subject compiled early in 1674 or by Goddard or Petty in their drafts for the Society's 'declaration' prepared that summer. Instead, Petty merely suggested 'That all Such Members of the Society as resolve not to pay their arreares . . . would forebeare comeing to the meetings of the said Society . . . Further Intimateing unto all such as desire to quitt the Society, that their leave shall bee readily granted them according to the statutes'.[137]

In the case of 'A.B.', it is easy to see how this reluctance to countenance expulsion dovetailed with his generally tolerant attitude concerning membership, but for him and others the more important issue was clearly practical. The point was not whether a streamlined society purged of the dross of inactive members was theoretically desirable, but whether such a policy, if implemented, would do more harm than good. 'A.B.' felt that expulsion was too final a step, preventing those who 'shall bethink themselves of reforming' from doing so, and there is some evidence that this was the case, in that some members who were expelled in the 1680s did indeed have to be reinstated and subsequently proved useful to the Society.[138] More important was the view of 'A.B' that the Society already had enough enemies without making more, which echoed Beale's concern about the 'miss-understandings' that might result. Even Hooke's concern that those who left might speak ill of the Society — against which he would have legislated with doubtful effectiveness — bears witness to a similar anxiety.[139]

Here, therefore, one might discern a further conflict between what seemed desirable for the effective running of the Society and what was advisable to protect its public image at a time when the new science was by no means as widely accepted as is often assumed.[140] Moreover this state of affairs meant that even within the meetings of the Society there was a place for a didactic function as well as a research one, for educating semi-informed enthusiasts in a manner from which the single-minded pursuit of scientific productivity to which Hooke aspired might have detracted. An awareness of this even on Hooke's part is perhaps in evidence in the concept of 'Pleasant Entertainment' in the attractions of the Society that he listed.[141] While this had the potential danger of trivializing reform efforts, there was no necessity why this should be the case, as seen, for instance, in the worthwhile programme of research in which Nehemiah Grew was

137 Birch, *History*, iii, 119; Bowood House Petty MSS H (8) 10, printed in Sharp, 'Sir William Petty' (n. 20), p. 400, and see above, Section 1.

138 See below, Document B; Hunter, *Royal Society*, p. 100.

139 See below, Documents B and C.

140 See Hunter, *Science and Society*, *passim*. The preoccupation with gaining wider support in the latter part of the letter of 'A.B.', below, Document B, further illustrates this.

141 Cl.P. xx.50, fol. 95. Compare Neile's view, below, Document A.

subsidized in the 1670s not least for the sake of the demonstrations which he was able to give the Society on the strength of it.[142]

Indeed, as this episode revealed, the most constructive approach to the problems involved in running the Royal Society in the late seventeenth century was an ability to work with the voluntary membership on which the Society depended for financial support, attendance and activity. For it is arguably even doubtful whether there were enough serious scientists around in the late seventeenth century to sustain a society of the exclusive kind to which Hooke aspired in his main group of papers, and an example of the true range of people who sustained the Society is revealed by the list of those who responded to the Society's request for 'lectures' in 1674 – 75, which included veritable dilettantes.[143] In these circumstances, the more pluralist view of the Society espoused by 'A.B.' was more realistic than the rigorous research institute to which others aspired, and it is revealing that Hooke oscillated between the idealism of his main group of papers and the rather frustrated acceptance of the status quo reflected in his later 'Proposals for the Advancement of the Royal Society'.[144] Quite apart from the contrasting intellectual positions among the commentators, therefore, there were also differing degrees of practicability in the actual circumstances of the late seventeenth century.

Only in the early nineteenth century, by which time the establishment and popularization of science had made great strides forward, was it feasible for the Royal Society to be systematically remodelled along the lines to which Hooke and his like had aspired. Indeed, there are significant continuities between the abortive plans of the late seventeenth century and the events that took place a century and a half later. A couple of parallels were pointed out by Sir Henry Lyons forty years ago, but the new material utilized in this article illustrates just how many precedents for the reforms of the nineteenth century are in evidence in the strategies discussed earlier, such as tightening up the criteria for election, reforming the Society's finances, organizing specialist committees and improving the Society's proceedings.[145] Hence these papers provide a background to later reform efforts hitherto largely ignored, illustrating that, though the actual structure

142 Hunter, 'Early Problems' (below), esp. pp. 273 – 4. Compare Heilbron, *Physics at the Royal Society* (n. 1), pp. 20 – 25.

143 Sharp, 'Sir William Petty' (n. 20), pp. 401 – 7; Birch, *History*, iii, 143 – 271, *passim*.

144 See above, p. 204.

145 Lyons, *Royal Society* (n. 11), pp. 243 – 4, 257. On the reforms which took place at this time see, in addition to *ibid.*, ch. 7: D. P. Miller, 'The Royal Society of London 1800 – 1835: A Study of the Cultural Politics of Scientific Organization' (University of Pennsylvania Ph.D. thesis, 1981); *idem*, 'Between Hostile Camps: Sir Humphrey Davy's Presidency of the Royal Society of London, 1820 – 1827', *British Journal for the History of Science*, 16 (1983), 1 – 47; Crosland, 'Explicit Qualifications' (n. 115), pp. 179 – 83; and M. B. Hall, *All Scientists Now: The Royal Society in the Nineteenth Century* (Cambridge, 1984), ch. 3.

and context of the early Royal Society imposed constraints on the extent to which change was feasible, a wider range of strategies was considered than is often realized, while the discussion of such alternatives was related to serious disagreements among activists which were not so unlike those of the early nineteenth century.

Finally, the documents printed here — William Neile's 'Proposalls', the letter to Council of 'A.B.' and the papers of Robert Hooke — are important for the light that their combined attention to intellectual and institutional considerations sheds on the evolution of the definition of the natural philosopher's role in the early modern period. As Robert Westman and Robert Frank have recently indicated, methodological precepts define social roles through the prescription of legitimate procedures and practices. Thus the instrumentalism propounded in Osiander's preface to Copernicus's *De Revolutionibus* proscribed the combination of physical speculation and mathematical model-building to be found in Copernicus's text, while Harvey's investigations signalled a change from the learned to the research-orientated physician.[146] Similarly, the different styles of Baconianism evident in the early Royal Society embodied divergent roles. The vulgar Baconians confined the philosopher to the co-operative collection of empirical data. Theory was to be eschewed, but utility was not. Fellows such as Hooke accepted these co-operative and utilitarian aspirations, but insisted on the validity of causal inquiry. Hence William Neile wrote that 'it seems a litle belowe the name and dignity of Philosophers to sitt still with the bare registring of effects without an inquiry into their causes'.[147] These conflicting norms inevitably resulted in contrasting institutional ideals. Those preoccupied with the compilation of natural histories were prepared to enlist anyone willing to contribute however marginally to this endeavour, whereas those seeking to advance theoretical understanding wanted to restrict membership of the Society to men of proven 'philosophical' ability, and they also emphasized the production rather than the validation of natural knowledge. Consequently the debates over the rival reform proposals discussed here were of more than local relevance, for they touched on the very nature of scientific organization and practice in the late seventeenth century.[148]

[146] R. S. Westman, 'The Astronomer's Role in the Sixteenth Century: A Preliminary Study', *History of Science*, 18 (1980), 105 – 47; R. G. Frank, jr, 'The Physician as Virtuoso in Seventeenth-century England', in Barbara Shapiro and R. G. Frank jr, *English Scientific Virtuosi in the Sixteenth and Seventeenth Centuries* (Los Angeles, 1979), pp. 59 – 114.

[147] See below, Document A; on the different styles of Baconianism see Wood, 'Francis Bacon' (n. 84), *passim*.

[148] Compare here Hahn, *Anatomy* (n. 98), pp. 31 – 32, 36f.

DOCUMENTS AND NOTES

Document A: William Neile

Royal Society Domestic Manuscripts 5.11 (first item)[149]

fol. 1

Proposalls humbly offered
to better consideration

The businesse of the society is to make experiments in order to which when an experiment is to be made let there be set downe in writing

1st The subject or heads of the experiment that is to say let it be drawne into a Probleme

2dly (If there be occasion) let the manner or processe of the operation be specified

3dly What is cheifly to be attended in the operation

4thly The scope or designe of the experiment

5. Lastly the successe or event with any new emergent suggestions.

[150]Finally let it be referred to a Committee to bring in a possible cause or causes, of the experiment which may be registred according as the society thinks fitt.

now to say something to this last clause why the giving of causes should not be neglected / **fol. 1v**

First one of the cheife ends of the society is to advance knowledge. To this if it be objected that that duty cannot be better dischardged then by making the experiments I answer That that society of men which is fittest to make experiments and to benefit the world that way, is also fittest to judge of the

149 Endorsed 'Proposals about the R. Soc:' on fol. 5v. Fols. 4v and 5r are blank. The third item in D.M. 5.11 is a partial transcript of Neile's paper by Oldenburg entitled 'Proposals concerning Experiments'. The attribution to Neile is based on the similarity of the handwriting of D.M. 5.11 (first item) to other examples of Neile's script: see, for instance, Cl.P. iii (1) 49. The reference in the document printed here to 'the nature of body and motion in generall' also links it to the material by Neile in Cl.P. iii referred to in n. 15 above. In our transcription, the punctuation has been regularized and 'u' and 'i' have been modernized as 'v' and 'j' where appropriate. Here and in the other documents deletions have been noted, but not insertions.

150 Before 'Finally' the words 'Now to say s. . .' have been deleted.

experiments when made, and not only to afford the world the benefit of their fancy and invention and of their manuall industry but also to steere their judgment and reason in the indagation of the causes of those experiments which they themselves make:

2dly, The experiments themselves are but a dry entertainment without the indagation of causes

3dly The bringing in of causes will very much advantage the carrying on the work of experiments For as the experiments doe hint out light in themselves towards the investigation of their causes so the forming of causes will also exhibit light towards the discoverie of new and unthought of effects, so that whereas you have one experiment now perchance you may have 2, or 3, then if this way be taken./ **fol. 2**

And I conceive the cheife error of former Philosophers was the excluding or at least too much neglecting of experiments and over much captivating their reason to one sect, from whence it does not followe that they should make no use at all of their reason and imploy nothing but their senses in taking notice of matter of fact[151]

4^{thly} This will no way impede the main businesse of making experiments, since the investigation of the particular causes of things is referred to committees, and need not be at all debated by the society in generall

Lastly it seems a litle belowe the name and dignity of Philosophers to sitt still with the bare registring of effects without an inquiry into their causes. Which if it be referred till they have more experiments to enlighten them in the discovery of truth. First they know not till they try whether they have not already enough to afford them a competent light towards the discovery of knowledge at least in some reasonable measure

2dly The other is a great way off when perchance The present members of the society may be dead and so both they and the world may come to lose the benefit (they might otherwise have) of their owne sagacity / **fol. 2**^{v}

Finally when all is done they can but doe their endeavor towards the finding out of truth, for they have no assurance of any certain successe by their experiments therefore they may as well venter to grope in the darke nowe as doe it a great while hence after tedious expectation.

To conclude there are some things, as the nature of body and motion in generall, which perchance are as fitt now to be referred to the consideration of a committee as ever they will be after never so many experiments. and it

[151] 'Captivating' has been substituted for 'emancipating', deleted. Earlier in the sentence, 'over' is substituted for 'too', deleted, while, later, the letter 'e' has been erased before 'reason', as has the letter 'd' before 'tedious' in the first sentence on fol. 2v, and the letters 'sp' before 'suplyed' in the second sentence on fol. 3.

is very likely the right setling of those principles will give much light to experiments.

Also let it be referred to a standing Committee to suggest newe experiments least otherwise the society should sometimes want worke, for that that is every bodies businesse is no bodies businesse.

Let it also be referred to this Committee to make a Catalogue of all former experiments made by whomsoever, and also of all the Phænomena of nature though never so vulgar or obvious that may be thought any way to conduce to the bettering of Phisiology, of which the causes to be deduced in [the] manner aforesaid / **fol. 3**

Also to collect all passages out of authors that may be worthy to be referred to the consideration of a committee. As where the author hints at the cause of any naturall effect, let them consider what is deficient in those places, how it may be suplyed, where the author's ratiocination is invalid[,] what need there is of further disquisition, how it may be improved to more advantage, and if out of this may be collected a handsome Theorie of that particular, I think it may properlie enough be offered to the society in order to their aprobation for the registery of it.

Let a Committee be likewise appointed to abstract from time to time out of the Registry such things as may be profitable to be printed first offered to the society for aprobation not as indubitable axiomes but as things worthy the notice and inspection of the publick.

Further more let there be every yeare a summe of money laid aside out of the society's stocke to be laid out at the discretion of the Council in books whose cheife scope is naturall Philosophy or mathematicks for the use of the society.

Document B: 'A.B.'

Royal Society Miscellaneous Manuscripts 4.72[152]

fol. 1

Most honoured Sirs,

[152] On the verso of the second leaf, this item has been endorsed in another hand 'A proposal for the Advancement of the Royall Society'. There is also a twentieth century pencil note saying '?Is this Dr Andrew Birch', evidently on the basis of the Society's membership lists, in which Birch is the only member with the initials A.B.

The present posture of this Society, & of its affairs, beeing so uneasy, & so unsetled, & upon both accounts so very unhappy, as it is: I think there is no true Friend among us unconcerned, to endeavour the reduction of of [*sic*] it to a better. Wherefore, although I presume not, as a Member of the Councel, & unasked, to give my advice: yet take my self obliged, as a Member of the Society, to suggest my thoughts. And what I have to say, I shall propose in that plaine Method, which every discreet man, in the ordering of his affairs, is wont to use. Which is; First, to ask himself, & see distinctly what he hath to do. Secondly, by whome his Work must be done, by himself, by others, or by both. And thirdly, what those things are which may assist & further himself & others in the doing of it.

As to our Work, we are all well agreed, or should be so, That it is not to whiten the walls of an old house, but to build a new one; or at least, to enlarge the old, & really to mend its faults. And that for this, Materials must be fetched from[153] from Observation & Experiment, lying in the two Worlds of Nature & Art. This I think is our foundation maxime; which, if it be throughly [*sic*] weighd, will assuredly produce & settle these two Conclusions in our minds. First, that the Observations of Nature, & the Performances of Art or Experiment, are both duly to be valued. That an Artist or Experimenter, is not to be taken for a maker of gimbals, nor an observer of Nature for a wonder-monger. That the one is to be accepted as the Author or advancer of an Art of his own; the other, as a finder out of those Arts which are divine. In short, That the one may be truly styled a Master of Nature; the other, one of God Almightys Scholars. So that here, to make comparison, as it is narrowmindedness in a single person, who would measure all the world suo modulo ac pede; so in a society, is morover factious, & will certainly tend to its ruine. Secondly, we shall hence conclude, That no Observation or Experiment, if truly made, is therefore to be sleighted, because it may seeme to be but mean. For so far as we have any new matter of fact before us, we have really advanced further then we were before. And many things which seeme trivial in themselvs, may be a foundation for that which is of greater moment; which no man but will be sensible of, who considers the original & gradual advancement of all the greatest Arts & Inventions in[154] the world. We shall conclude, that the meanest observation of what is really existent in nature, is more valuable than the most illustrious, if ungrounded, phancy: for something, may be some foundation. That a man may be the author of a good Invention, though neither a perpetual motion, nor the philosopers [*sic*] stone: or of a good observation, though not of all the Countrys of the World in the moone. Even as Aristotle was once thought to be; who though he pretended to no such great things, but only the making some observations about Animals; for his encouragement & assistance therein, had both the acceptance & the treasure of his Prince. In short, we shall, I say, conclude, That though a whole pile of bricks is not an house, yet that every batt of a brick, & every lump of lime is concurrent to make it up. And that the

153 'Natural History, that is' has here been deleted; 'from' remains undeleted.
154 'in' has been substituted for 'of', deleted.

thinking, which one time or an other usually ends in saying otherwise, will not only arrest our Work by confounding our Language, but by weakening our very foundation, make way for its fall. For no man will be very industrious about that, which either himself, or any other concerned in what he doth, shall take for a trifle. And so many amongst us as want these principles, or have not throughly [*sic*] Learnt them, should be taught them. And thus much for our Work.

As for the persons, we cannot be too many employd, since we have all Nature to turn us in.[155] But though all may be desired & encouraged, to doe somewhat; yet it may not be convenient for all to be obliged. For first some may have no genius to prog[ress] in philosophy, who yet may have witt enough in their own way, & may be ready enough to assist with their mony & other ways those that will. So that we have no more reason to expect, that all should be philosophers or Curators, then that all should be Benefactors; or that every Love-Brother of the Company, should know the Trade. Neither have all Leasure, who have a genius. Nor yet are all those who have both, borne to Labor; which yet is equally necessary. For what Horace said of a Poet, he might have done as well of a Philosopher, — Ego nec studium sine divite vena, Nec rude quid prosit video ingenium —.[156] It is not so easy a task to philosophize now, as heretofore; when it was enough for a man to say so much after Aristotle or his Followers, as to shew he understood what he had read. But now the question is not what saith another [,] for I can tell that my self; but say you? & how doe you prove what you say? & here experiments & observations must be made; which lye not in a bottle of Ink, but spread all over the world; & for which therefore every man will not be bound to travaile. Wherefore it seemeth best, / **fol. 1**^{v} That though all may employ themselves, yet that some only be intrusted.

Now the Persons to be intrusted are, Lecturers, Curators, one or more Secretarys, Councels, & a President. The four last are allready established upon mature deliberation by our Laws. For Lecturers, it is that which hath been lately, & I think very rationally, propounded. As beeing a thing very plausible; not without good example, but conformable to the received Custome of Universitys; & assureing us allways of meeting with somewhat, either[157] a Lecture, or discourse upon it, when we come. To these Nature & Art, I presume will be committed more at large; whereof some smaller parcels only, may suffice to be managed. But for the Curators, it is intended, that some larger Provinces should be more especially assigned to them. And of these it would be wished there were three kinds; vizt, for Books, for Men,

155 Illegible deletion of 2 – 3 words before 'But'. In the following sentence, 'love' has been cancelled following 'every' (at the end of a line).

156 A quotation from Horace's *Ars Poetica*, lines 409 – 10. The relevant passage reads in English: 'Often it is asked whether a praiseworthy poem be due to Nature or to art. For my part I do not see of what avail is either study, when not enriched by Nature's vein, or native wit, if untrained; so truly does each claim the other's aid, and make with it a friendly league': Horace, *Satires, Epistles and Ars Poetica*, trans. H. Rushton Fairclough (London, 1961), p. 485.

157 'of' has been deleted after 'either'.

& for Things. First a Curator for Books; whose business should be, to make a diligent search & particular collection, of all that is, or may hereafter be published, of the History of Nature or Art, by any person, at any time, in any Country or Language whatsoever. Secondly a Curator for Men; from whome he should make the like Collection, as the former from Books: enquiring of persons of all Professions, Trades, & Occupations; both at home & abroad. As the former would give us chiefly the History of Nature, so this of Art; whereof it might Lead us into an unknown world. Thirdly, & principally, Curators for Things. Of which it seemeth necessary there should be three at least, answerable to those three more ample fields of Philosophy, scilicet: Mathematicks[,] Anatomy, & Chymistry. That to the last be committed the Chymical History of Minerals & Vegetables. To the second, the Anatomical History of Vegetables & Animals; & the Chymical History of Animals; which I particularly propose, as that which hath never yet been done; & would certainly be to good purpose, as I conclude from some new experiments I have already made. And to the first, Astronomy, Meteorology, Mechanicks &c. Only with this caution, that they be not over strictly confined; for this will not be to assigne them their Province, but their Goale. As for the Secretary his business is, To take weekly Journals; To make Registers; & to keep a Philosophical Correspondence. As for a Councel, though it be so called, yet is rather the Societys Parliament, by all the Fellows whereof, & not the President alone they are made choyce of. I therefore passionately desire it may be rememberd, That they profess themselvs to be entrusted with the generall care of all our business. And that one necessary thing they are to doe, Is to meet, when they are summond: otherwise things will certainly fall to the ground. So that unless our Charter Law had specifyd away [*sic*] for their haveing Proxys, the bad example of Parliaments, will here be worse followd. And lastly, the President is our Sun, giving us light, heat & influence. Whose business therefore is, To preside the Meetings of the Society. To see that all things are carryd on according to our Charter Laws, or the Decrees of the Councel. To view both the Lectures, & what ever els is produced at our Meetings, beforehand; or to nominate by whome to be viewd: or at least it will be convenient, that he be before acquainted with the Subject matter thereof. For though he be not constituted a Moderator, yet it oftentimes falleth out, that so he is; & is both an Interpreter, & a Judge as often as he pleaseth. And lastly his business is, To acquaint, or nominate others, to acquaint the King with what is performed amongst us, so often as either Himself, or a Councel shall see convenient. For since the King is our Founder, I think he should understand what we doe. To the above said Officers, our Charter hath also added a Treasurer; which did suppose, we might one time or other grow rich: & upon that supposition did well constitute him with the rest. And thus much for the Persons by whome our Work is to be carryd on.

The next question is, By what means shall these persons be animated & assisted in their Work? To which the answer, in short, is, By Rewards. That the one will not be carryd on without the other, is an experiment as old as the World. Doe any say, This is not our business, but only our diversion; then is it full time for every man to depart to his business: for if this be not, nor hath been so, we have been at play long enough. Or is it said againe, That it is both our business, & encouragement is also given; viz[t], the Societys acceptance of what is done. I answer, This certainly is a fair Reward, & the meriting thereof worthy the ambition of any person, whose expectations are not sordid, & can distinguish betwixt a Pearl, & a barly corne. For Honour is a kind of orient thing, especially when it comes polished out of the hands of those, that know when to give it. But then it must be provided, that the persons concernd, be made to understand this acceptance: if so much be done, acceptance will be very cogent. But if not, it is in danger of loosing its effect, scilicet the incouragement of those who are concernd: because the better & more desirable any thing is, the more apt men will be to fear their missing of it: there ought therefore to be some reall Testimony to assure it. The speaking whereof puts me in mind of a passage told us, betwixt one of our Queens (I have at present forgot which) & her Ladys: who asking her, when they should see her great belly? She replyd, That the King indeed seemed to be very Loveing, but it was not, good morrow forsooth, & a kiss in the morning, that would make great bellys. /

fol. 2

And as it is necessary, that Rewards be allowd & setled; so may it be advisable, they should be extended unto all. And that therefore, not only the Lecturers, Curators, Secretary, & Treasurer be rewarded: but that also the President himself be presented, at some certaine time annually, as suppose upon St Andrews day, with some token of our thankfullness for his care of our affairs. I know that Noble person, who at this time is our President, takes his very beeing so, as an higher obligation upon himself, then any we can superadd. But what is that to us? unless it be thus much, that it makes our gratitude, to be so much the more our duty. And lastly, that those of the Council likewise should be annually gratifyd, as either by a Treatment, or some other way as shall be judgd convenient. For though it be never so little, yet if it be somewhat, it will have some meaning; & will make some obligation: for all business, in all other Societys is a dry & stiff thing, without a little oyl to make it limber: & no generous mind will endure to receive any thing for doeing nothing.

It will be said, That to doe all this, a good Revenue is necessary: which must be granted. And for a good Revenue, we must have good Benefactors; & for good Benefactors, we must have a good Reputation; & for a good Reputation, we must have some good Stock or Estate of our own. First, for a good Revenue, we must have good Benefactors. And if we cannot doe without them, I know no reason we should be shye of applying to them. To expect what men may possibly doe of themselves, is to trust to meer chance. For it is very unlikely, they should be guilty of such a contradiction, as to be our suiters, that they may become our Benefactors: we[158] must therefore ask them. For which also we have as good example, as reason; not only that of all men liveing upon the like occasion, but of the King himself, who vouchsafeth to ask his subjects for what he needeth: since he doth it not to be pursed up, but layd out for the common good: which is our case. And

[158] 'we' has been inserted to replace a word rendered illegible by deletion.

somewhat better, beeing perhaps more sure of obteining that allways in part, for which we lay it out. And if we should have the ill luck sometimes to be denyd, there is reason, we should be contented as well as he.

Yet to be allways denyd, will doe us no good. To secure ourselvs from which, we must take the best courses we can. One whereof will be, To make sure of the King in the first place, for more or less. For every man will say, why should he concerne himself more than the King, who is our Founder? Another, & a principal one, will be,[159] A stock of good & generall Reputation amongst men. And first of all, to take care, that the King himself, if he cannot be our Benefactour, yet at least he have a good opinion of us. Regis ad exemplum — is as true in this, as any other case. If he speak sleightly of us, who will not learne to doe the like? the Court will learn of him, & the City & Country of the Court; & so we shall teach all men to despise us. And at last perhaps shall be ready to despise ourselves: like those miserable Creatures, who beeing hooted at by all the Towne for witches, begin to think themselvs to be no other. It is therefore necessary, that the King should as really be our Patron, as our Founder. Which I doubt not, but he would heartily be, if this one thing were taken care of; that is, That he be made acquainted with what we are doing & with what we have already done. For were the Journal, & other Books perusd; & a Catalogue or Collection discreetly made & shewd; it would appear, That there hath been enough done in a few years, ever since our first foundation, not done before in many centurys, to keep us from contempt. Especially beeing performd by a number of persons, who have done more then their modesty would let them promise; & who have had so little to doe it with all. Which being considerd, we have not much reason to doubt of his Majestys gracious acceptance. And the less, because he professeth expressly, the large extent thereof in our Charter, scilicet by encouraging all those who shall, not only novam philophiam [*sic*] extundere, but veterem expotire.

The good will & opinion of all others, as well as of the King, should likewise be sought, & care taken, that no man be engaged to speak, or to think against us. As knowing, how great a matter, a little fire kindleth: how strongly envy & malice will spread, though originally managed by the most inconsiderable hand. That 'tis commonly one ill nature which disturbeth the whole parish; & but a small club of male contents that are able to embroyle a Kingdome. This makes me look upon the Ejection of Members as a ticklish poynt: least after we have already too many Enemys, we should make our selvs more: & those worse then the most we have as yet: beeing now only inclind, but then obliged, as they will think, & dared too, to doe us all the mischief they can. Which also they will be still the more ready to doe, because desperate, & without hopes of beeing received into our number againe; unless they will leave their snubing, & promise to be good boys; which men will never doe: nor, if they would, were their word to be taken. It might therefore, I conceive, doe better if some midle way were taken, as by reservedness, & not sommoning them to our Meetings, or some other

159 'A Stocke sure' has been deleted before 'A stock'. In the penultimate sentence of this paragraph 'beeing' has been altered to 'being'.

token of / **fol. 2**^{v} neglect & just displeasure, which may have some weight with them; that so if they shall bethink themselves of reforming, they may be left in a capable state, for their so doing. Whereto I add, that it is also necessary we should have a fair & cleer correspondence amongst our selvs; & that our business we have in hand, & the severall steps we take therein, be generally made known, & understood. But for the winning the good opinion of most other men, not of our number, we must doe it by takeing a contrary course: not by communicateing, unles what we publish, but by beeing reservd. For to be open, will be to expose every thing, either to be catched up[160] by some, or to be droll'd upon by others, or sleighted by all. Wheras, on the contrary, a secret is wont to be admired, what ever it be. And men doe usually give their mony to see the show, & not when they have seene it. Besides, that it will make Tradesmen & others more free & open, when they understand they may dare to trust their secrets with us.

Lastly, a principall means to get a good repute, & consequently good Benefactors; will be to have some good Stock or Estate of our own. In order to which, for my own part, I shall not only assure the weekly payment of what we have allready engaged for; but proportionably to my small estate, shall further contribute & ten pounds I shall most readily lay down upon the board; provided that others will doe the like; riseing from ten to fifty pounds, or perhaps an hundred; which some upon a general, unanimous & hearty contribution, may be willing to doe; especially if they have the King for their example, as hath allready been suggested. Which contribution, if it were made, would purchas us somewhat which we might call our own, & trust unto; And which, by giveing us some repute, would quickly bring us in more. For to whome doe men send their guifts, but to the rich? On the contrary, what is it that makes a society contemptible, as well as a single person, let him otherwise be never so valuable, but because poore? So certaine it is, that our beeing poore, doth necessarily keep us so. Whereas had we once somewhat considerable of our own; men would then look upon us, & what perhaps they will now give upon asking, would then give without it. For these things are often measurd less by strict reason, then humour & opinion; wherewith it is the part of wise men still to accord, that they may make use thereof to their purpose.

I conclude, with my humble submission, in all things here said, to your better judgments, & profess my self

most honoured Sirs,
Oct:19. 74 Your most cordial Friend &
most humble Servant,
A.B.

[160] Illegible deletion of two words following 'up'. In the subscription to the letter, 'y' has been deleted after 'my self'.

Document C: Robert Hooke

Royal Society, Classified Papers xx.50, fols 92 – 94[161]

fol. 94

The Designe of the Royall Society being the Improvement of Naturall knowledge all ways and meanes that tend thereunto ought to be made use of in the prosecution thereof. [162]Naturall knowledge then being the thing sought for. we are to consider by what meanes it may be soonest easiest and most certainely attaind.

These meanes we shall the sooner find if we consider where tis to be had to wit in three places. first in bookes, 2dly in men. 3^{ly} in the things themselves. and these three point us out the search of bookes. the converse & correspondence with men[,] the Experimenting and Examining the things themselves. [163]under each of these there is a multitude of businesse to be done but the first hath the Least [and is] the most easily attained, the 2^{d} hath a great Deal and requires much en[deavour] and Industry; and the 3^{d} is infinite and the difficultest of all.

The work then being soe very great, It follows that it cannot be done by the indeavours [of a] single person nor Indeed by the separate indeavours of thousands both these having been experimentally proved ever since the beginning of the world. but tis most probably to be done by the joynt unanimous and Regulated Labour of a multitude. this shews the necessity of a Society.

There hath not been wanting in all ages and places great numbers of men whose genius and constitution [164]hath inclined them to delight in the inquiry into the nature and causes of things, and from those inquirys to produce somewhat of use to themselves or mankind. but their Indeavours having

[161] This item has been wrongly bound in the volume, so that the leaf which should come first (fol. 94) has been placed last. Fol. 93 is a fly-sheet which is blank on its recto (except where a few words at the top of fol. 92v encroach across the fold). Apart from Hooke's notes on the verso of fol. 93, it is also endorsed 'Proposals for advancement of the R[oyal] S[ociety]'. Words in brackets in the transcription have been obscured by repairs to the manuscript; 'u' has been modernized to 'v' where appropriate throughout. Hooke seems to have made alterations both during and after writing the paper, and his original words are often indecipherable. In one case (see n. 172) his emendations significantly change his sense, contradicting what he says at the end of fol. 92v about members' obligation to honour changes in the Society's statutes made after their oath of admission.

[162] One word has been deleted before 'Naturall': 'erecting'?

[163] 'And these' deleted before 'under'. Six words later, 'is' has been substituted for a word rendered illegible by deletion: 'are'?.

[164] Illegible deletion of two words before 'hath'.

been only single and scar[c]e ever united improved or regulated by art have ended only in some small inconsiderable product hardly worth naming. But though [165]mankind have been thinking these 6000 years and should be soe six hundred thousand more yet they are and would be much whereabouts they were at first, [w]holly unfit & unable to[166] conquer the difficultys of naturall knowled[ge]. But this newfound world must be conquerd by a Cortesian army well Disciplined and regulated though their number be but small.

It Remaines then to consider of the uniting and regulating the indeavours of such men who are naturally inclined to these inquirys, into an effective corporation or Society or body.[167] to Effect this care must be taken to make it desirable by all that shall be members thereof first from its suitableness[168] to their inclinations. soe That the Labour thereof may be pleasant delightfull and naturall to them. Secondly upon the account of Interest & benefit [t]hat the labour and trouble susteind therein may be abundantly recompensed by the benefits that doe [a]ccrue thereby more then what can be any other ways attaind by the like [169]expense of money or Labour.

To effect the first It must be made an inviolable order or Law that noe person whatsoever upon any account or respect be admitted into the number thereof that shall not, by ten persons of the Society or by some publique Action or publication, be known to be naturally & zealously inclind to the prosecution of the Designe of the Society.[170]

Secondly that noe person whatsoever shall be admitted a member thereof that shall not solemnly take an othe upon the holy evangelists to yeald during the time of his continuance of the same a perfect obedience and submission to the Laws & constitutions of the said Society already made at the time of his admission[171] but as to such as shall be thought fitt to be made by the major part of the whole Society after his admission he shall be free to

165 Largely illegible deletion of 3 – 4 words preceding 'mankind'.
166 These words replace a long and largely illegible deletion following 'first'.
167 'or Society' is written above the line over 'or body', which it was presumably intended to replace, though these words are not in fact deleted. It is followed by a long, partly illegible deletion, evidently the result of Hooke reconsidering the way in which he expressed himself in the course of composition. In the previous line, a short deletion ('th'?) precedes 'to'. In the following line, the words 'suitable [first: inserted] and' have been deleted before 'desirable'.
168 Hooke originally wrote 'suitable' and changed this to 'suitableness'. In the next sentence, 'and' has been deleted before 'delightfull'.
169 Illegible deletion before 'expense': 'indeavour'?
170 In this sentence, 'now'[?] has been deleted before 'It', 'to' before 'be', and 'be' before 'by', while an illegible deletion of 1 – 2 words precedes 'or'.
171 'but shall also [?]' deleted before 'but as to such'.

submitt, or not to submitt thereunto.[172] and that he shall not either directly or indirectly disclose any matter or thing whatsoever that by the constitutions of the Society are made secrets or not to be discovered — That he shall during the [173]time of being a member zealously & streanuously indeavour to the utmost of his Power and ability indeavour [*sic*] to promote the good and interest of the Society in such ways as shall be judgd most requisite and proper by the major part thereof. and that he shall not upon his Leaving thereof either directly or indirectly doe or speak anything that shall tend to the detriment or defamation thereof, nor shall he at any time [d]eclare or make known the Reason of his Leaving thereof. / **fol. 94**^{v}

[174]The Laws then of the Society should be very few but those inviolably kept and Observed by all the members of the Society.

[175]that noe member shall give his vote for the election of any perso[n] that he doth not verily beleive will be a faithfull industrious & zealous prosecutor of the Designe o[f] the Society. *This is to be a part of the oath.*

That noe person upon any account or Respect whatever shall be admitted a member of this Society till he hath solemn[ly taken the] oath, and paid down the sum of 4^{lb} for admission. and [176]either before or after his [admission] hath given some signall specimens of his good will and abilitys for the good of the Society.

That noe member[177] whatsoever shall be permitted to come into or sitt in the society, upon any day after the Electio[n] Day that shall not upon the said Day pay into the Tr[easury] the summe of [178]4 pounds for the year insuing.

172 There is a single line through 'shall be . . . made by the', presumably in error, as it is required by the sense in addition to the inserted 'after his . . . submitt thereunto'; before this, 'that' is deleted, but 'or', which precedes 'that', has been accidentally left undeleted. The result of these alterations is to contradict Hooke's original view that members should submit not only to the existing statutes but also to changes to them made while they were Fellows.
173 One word deleted before 'time'. In the following line, an insertion above the line over 'Power and ability' has been deleted.
174 Before this sentence the following passage has been crossed out: 'Thirdly that the strictness and Rigour of these first ingagements may not frighten and deter such person[s] as are every ways proper & desirable to be members of this Society from listing and entring themselves thereof there must be made certaine limiting Lawes that shall bound all the Laws and constitutions that shall be mad[e] for the future.'
175 'First that noe person [deleted] members shall give his vote' deleted before 'that'. Later in the sentence, 'member'[?] is deleted before 'faithfull'. The following sentence. 'This is to be a part of the oath', is emphasized by being boxed round with ink lines.
176 'before his' deleted before 'either'.
177 'member' inserted to replace the deleted 'person'.
178 Illegible deletion before '4': '40 shillings'? Previously, 'for the year insuing' deleted before 'the'.

nor be admitted into any election if he hath not before the same cleered off all arrears whatsoever that have been due for the time past.[179]

That noe person that is a forreiner shall be elected a member of this Society that shall not before such h[is][?] Election give the Society some signall testimony of his fitness and ability for it, by two or more Disc[our]ses upon such subjects as fall under the care of the Society. nor be continued longer thereof then he continueth to send quarterly some such discourse discovery experiment or Invention as shall [be ap]proved worthy to be accepted by the Society & soe declared.

That every person that is a forreiner and [180]elected fellow of this society shall before he receive any benefit or information from the said society, take an othe upon the Evangelists or take the sacrament that he will not either directly or indirectly disclose any thing that shall be communicated to him nor make any use thereof to the detriment of the Society. that he will monthly or quarterly[181] during his continuance of that S[ociety] send in an account of all such informations & Discoverys as he can be any meanes procure either by his own invention or from others information & will indeavour to procure such bookes naturall curiositys[182] observations and experiments as the Society shall desire from him being things peculiar or proper thence to be obteind and convey them speedily to the Society, by the best & safest meanes they can[,] the said Society defraying the charge of such things procurred and the charge of the conveyance of the same.

That All such books things or Letters shall be Directed to the Society residing at (   ) and shall be there first opend & publiquely Read & perused upon the usually [*sic*] dayes of meeting between the howers of 4 & 6 in the afternoon. and proper breviats thereof taken[183] for the transactions of that Day, and the whole be transcribed into bookes proper for that purpose to be perused by any member of the Society at any seasonable time.

That All bookes that shall be presented to or procured for the society shall be putt into the library with the name of the Doner or the charge of the purchase, there to be perused by the members of the Society & by none els.

That all naturall curiositys that shall be presented or procured shall be putt into the Repository of the R[oyal] S[ociety] there to be seen & perused by the members therof at convenient times & by none els.

179 'And that noe person' deleted after 'past'.

180 'admitted' deleted before 'elected'. Later in the sentence, 'other' deleted before 'othe'.

181 'monthly or quarterly' inserted, replacing the deleted 'twice[?] every year'.

182 'make such' deleted before 'observations'.

183 Illegible two-word deletion before 'for'. Later in the sentence, 'to' deleted before 'be', 'and' before 'to', while 'seasonable' replaces the deleted 'convenient'.

That all Instruments modules & artificiall curiositys [184]that shall be presented or procured shall be Recorded and put into the Repository of Art there to be seen by the Members at all convenient [?] times and by none els. / **fol. 92**

The busnesse of the Society being as I sayd before three fold to wit the perusal of Bookes, the consulting of men & the Examination and tryall of things, shall be distributed among the severall membe[rs] of the Society, according to this proportion[:] one tenth part of the whole number, shall be imployd to peruse and examine all bookes that are extant for the collecting examining trying &[185] Explaining whatsoever is to be found Recorded in them, approving & confirming what they find true & reall and damning all that they find fals and fictitious. and then for recording them in their proper places & methods.

a fifth part shall be imployd in consulting men and procuring from them either by friendship or Reward whatsoever is curious usefull & pertinent to the busness of the Society. all which informations shall be returnd to the meeting place of the Society. there to be first openly read at the usuall time and then Recorded in its proper place & the Information it self layd up in its proper place to be kept.

The other 7 tenths of the Society shall be imployd upon the examination and tryall of Naturall things and improving of Artificiall inventions and productions.[186] all whose acquisitions shall be brought into and Read in the Society at the usuall place & time & then recorded in their proper place there to be perused at any convenient time by the members of the Society & by none els whatsoever.

And that each of the said three busnesses of the Society may be the more effectually carryed on, according as the Revenue of the Society shall increase from the bounty of its own members. which tis to be hoped it may if this way be taken, there shall be severall persons elected and chosen by the sufferage of the whole society or the major part of them, to be curators of the severall parts thereof. to wit two for bookes. 4 for men 8.10.12.20[187] or more for things as the stock increaseth.

The busness of one of the first should be for the examination of all Hypotheses and Doctrines that hav[e] hitherto been published and for trying how Consonant they are with their own assertions and experi[ments] and in what they are good & in what Defective where they have asserted

[184] 'of the R' deleted before 'that'. Later in the sentence, 'Recorded' is written over something else.
[185] '&' is apparently adapted from something else; previously, '&' has been deleted between 'examining' and 'trying'.
[186] 'all which ref . . .' deleted before 'all'.
[187] '20' appears to have been deleted, or altered from something else.

truth and where falshoods and what part of them is worthy to be retaind and what is absolutely fals & absurd & to be rejected, and what is of a middle nature & deserves to be farther considerd.

The busness of the 2[d] should be for the examination of all mathematicall & mechanicall problems & for the solving any Difficulty that shall occurr to any of the other members in any of their inquirys into those subjects.

The busness of the 3[d] should be to converse with all manner of tradesmen & artists to learn & discover whatsoever is curious and usefull amongst then and to bring in a history & description thereof.

The busness of the 4th to hold correspondence with as many[188] correspondents as can be procured here in England. or in any of our own plantations as newfoundland new england virginia barbadoes Jamaica &c. to send to them what is ordered by the Society & to bring in whatever is Returnd back againe.

The busness of the 5th to hold correspondence with Italy france Spaine Portugall or any forrein plantation; belonging to them in the east or west indies.

The busness of the 6th to hold correspondence with[189] the lower & higher Germany Denmark Sweden Poland Hungaria &c. —

The busness of the rest, to be chimistry, anatomy, opticks, astronomy, mechanicks, botanicks, a [*sic*] and a thousand particular subjects which the Society shall think fit to have more particularly and exactly examined.

All these to be members of the Society and in all particulars bound strictly to observe all the Lawes & Rules of the Society and to Receive for their labour and paines taken In each of their severall Imployments according as It shall be judgd to Deserve[190] by the Society when the thing is brought in not Exceeding the sum of ( ). / **fol. 92**[v]

Every one of these Curators[191] once in a quarter besides the dayly accounts & experiments he produces should bring in a compleat and well digested Discourse upon somewhat of the subject upon which he is imployd which he shall read openly at the time & place of the meeting of the R[oyal] S[ociety] and make such experiments as are requisite to the further explication thereof whereof Breviats shall be taken by the Register. for the transactions of that Day and the whole Discourse verbatim shall be transcribed into the Register book [192]there to be perused by any member of

188 'persons as' deleted before 'correspondents'.
189 'Holland' deleted after 'with'.
190 'judgd to Deserve' inserted, replacing 'agreed with them'[?].
191 Illegible deletion of 1 – 2 words before 'once'.
192 'and the Discourse it self layd up in its proper' deleted before 'there'.

the society at any seasonable time & by none els whatever. and the Discourse to be putt up in the cell proper for it to be kept.

The breviate of all the transactions of the Society performed at their usuall meeting shall be carefully made perused & corrected by a select number of the members chosen pro tempore by the society, and once a month or oftner if the matter requires there shall be (   )[193] printed in the forme of a gazet which by the Register shall be sent to every member of the S[ociety] which is absent provided he take care to pay for the postage & conveyance thereof. and if upon seeing the breviat he desireth to be further satisfyd by having the whole Discourse it self to peruse[,] the Register shal take care to have it transcribed provided he defray[194] the charge thereof after the rate of 04^{d} for every thousand words that shall be soe transcribed fairly.

That Every member of the Society shall once a quarter either bring or returne into the Society some account and discourse of the particular province committed to his care & charge which shall be there openly read [195]either by himself or some curator whose busness is much the same and the person Reading it shall take care to have such experiments made ready to be shewn at the same time that will be necessary for its explication. the which shall be registred and reposed in its proper cell, abbreviated and put into the transactions of the month at the end thereof to be printed as is abovesaid & communicated to every member of the said Society as above.

That there shall be a select committee chosen out of the whole society by the major part of the persons present at the annuall day of Election, who shall have the care of collecting &[196] disposing the Revenue of the society for that year. and shall consider of all meanes for the increase & promoting the interest & Revenue of the said society and shall during the time of their being of the said committee be excused from all other duty or exercise which they are otherwise obliged unto save only that they shall pay the 4^{lb} for the year insuing as if they were not chosen of that Committee.

That this committee shall consider debate & digest such orders or Laws as they shall think convenient for the good of the society and that they shall afterwards propound the same to the S[ociety][197] at the usuall time & place of the Society meeting. where it shall remaine for one whole month & be once read every day of meeting and at the end of the month shall be debated

193 '200' deleted within the brackets which occur in the text. Earlier in the sentence '&' has been deleted between 'made' and 'perused'.
194 'defray' appears to have been altered to 'defraying', but the original has been left here as making better sense.
195 'by' deleted before 'either'.
196 'collecting &' has been inserted, replacing the deleted 'Sup' [?]. Two words later, an illegible deletion precedes 'Revenue': mon[ey]'?
197 'upon' deleted before 'at'.

throroughly [*sic*] by the whole society & put to the vote and soe shall remaine as the vote leaves it till the [?][198] day where if it be again affirmed by the major part of the persons present yt shall Remaine and bee a Law for the future which shall obleige All the members of the said society soe Long as they continue of the same./ **fol. 93**^{v}

[*Bottom half of page, left-hand column*]
every member to be desirous as well as able to be a member.
to pay admission & Contribution.
to make choice of something yearly to be serviceable by his own actings.
to keep secret what is soe desired
not to shew the Gazets till 12 months after.
To pay postage & 2^{d} per gazet

[*Right-hand column*]
Sutable Acquaintance
Delightfull Discourse
Pleasant Diversion by Experiments
Instructive Observations by tract
Considerable Intelligence by Letters
New Discoverys by Inventors
Solution of Doubts & Problemes.
An easy way to know what is already know [*sic*]
A Liberty of Perusing repository
of seeing & using the Library
of Perussing [*sic*] Modules & Instruments
of Perusing Letters & Discourses.
of being present at tryalls — mechanick, Optick chymick Anatomick Astronomick —
An opportunity to shew their [199]Abilitys with advantage
to Get a character of the Society
To have an account sent to him every fortnight of all is doing.

Note 1: The Authorship of Royal Society Miscellaneous Manuscripts 4.72

As has already been noted, this document is merely dated 19 October 1674 and signed 'A.B.', either for 'Amicus Benevolens' or merely as a mark of

198 'next' deleted before 'day' and replaced by an insertion which has been rendered illegible by an ink smudge. Six words later, 'confirmed' has been deleted before 'affirmed'.
199 'with advant[age]' [?] deleted before 'Abilitys'.

anonymity.[200] There are a number of clues as to the identity of the author, but none, unfortunately, is conclusive. The most important — which at least means that the author can be placed within a grouping in the Society's membership — is the author's reference to 'the Chymical History of Animals . . . which hath never yet been done; & would certainly be to good purpose, as I conclude from some new experiments I have already made'. This shows that 'A.B.' was an active researcher, and the type of research involved aligns him with the tradition of physiological research which was so prominent in mid to late seventeenth-century England.[201]

Beyond that, the clues are rather slight. The author promised substantial financial contributions in the course of the letter, and, if he really meant what he said, he must have been fairly wealthy. The solicitousness about the Society and the knowledge of its affairs that the letter displays suggests that its author must have had a fair degree of involvement with the Society's activities. The fact that the letter was addressed to the Council of the Royal Society probably implies that its author was not a Council member in 1673 – 74, and this is apparently the meaning of the second sentence of the letter, though it is rather ambiguously put.

Lastly, there is the question of handwriting. This is complicated by the possibility that the document could be a copy dictated to an amanuensis or a scribal copy of a lost original — the seventeenth-century equivalent of a photocopy made so that all present at the meeting could read the letter, perhaps — as might be suggested by the manuscript's denseness, neatness and fewness of corrections, and by the way in which the deleted material has been carefully blocked out. The few corrections that there are, however, do appear to be substantive ones, reflecting reconsideration on the author's part in the course of penning the letter: this would suggest that the manuscript is a holograph one, and hence the handwriting of potential candidates has also to be taken into account.

Within the Royal Society's membership in 1674 there seem to be four likely claimants to authorship, who are as follows.[202] Walter Needham was a medical man who was active enough in the Society to give a paper on the serum of the blood in November 1675 as part of the reform effort of the time. However, he was on the Council in 1673 – 74, and in general his research does not fit the bill particularly well.[203] Dr Edmond King also gave a paper in 1675. He was not on the Council in 1673 – 74, but he also is not especially likely from the research point of view: King was primarily a

200 For the use of the initials 'A.B.' for this purpose see, e.g., Birch, *History*, iii, 140, or Bowood House Petty MSS H (8) 8.

201 See R. G. Frank, jr, *Harvey and the Oxford Physiologists: A Study of Scientific Ideas and Social Interaction* (Berkeley and Los Angeles, 1980).

202 We have considered a number of other possible candidates among the Fellows of the Society at this time, but none seems as likely as those dealt with in the text.

203 Birch, *History*, iii, 233 – 41; Hunter, *Royal Society*, catalogue entry 230. On Needham's research see Frank, *Harvey* (n. 201), esp. ch. 8.

practical surgeon, and his 1675 paper was on animal anatomy.[204] Extant examples of the handwriting of these two men are dissimilar to that of the document in question.

Likelier is Daniel Coxe, who was concerned with chemistry, though almost exclusively with the analysis of vegetable substances.[205] Coxe gave a paper on 'the analysis of vegetables' in January 1675 and he makes other appearances in the minutes in these years. Moreover it could be significant that he was elected to the Council on 30 November 1674 for the only time while he was a member of the Society, though he never attended a meeting, and this could have been a response to the initiative he had shown in the previous month if the letter is indeed his.[206] From the point of view of handwriting Coxe is the strongest contender, since examples of his script show some similarity to that of Miscellaneous Manuscripts 4.72, within the likely parameters imposed by changes of pen and the like.[207] In terms of benevolence to the Society, however, Coxe does not seem a particularly likely figure, since he made no financial contributions between 1666 and 1699, whereas the author of the letter — unless highly hypocritical — must have been a regular subscriber.[208]

An equally likely candidate is William Croone, an influential early Fellow of the Society about whose role we otherwise know surprisingly little. Croone was not on the Council in 1673 – 74 or 1674 – 75, but he gave a paper in 1675 and he had been called on by the Society to help collect arrears in 1673.[209] More important, his research seems to fit the description given in the letter better than anyone else's, since he was an active experimenter interested in living structures and in producing a chemical theory of muscular action, of which he adumbrated on important modification in lectures given to the Barber Surgeons at this time.[210] Moreover Croone's interest in the well-being of the Royal Society — and his preparedness to place substantial sums of money at the Society's disposal — is illustrated by

204 Birch, *History*, iii, 179 – 80; Hunter, *Royal Society*, catalogue entry 212. For information on King's interests, we are indebted to Robert Frank.

205 See Daniel Coxe, 'A way of extracting a Volatile Salt and Spirit out of Vegetables', 'A Discourse denying the Præ-existence of *Alcalizate* or Fixed Salt in any Subject, before it were exposed to the Action of the Fire', and 'A Continuation of *Dr Daniel Coxe's* Discourse', *Philosophical Transctions*, 9 (1674), 4 – 8, 150 – 8 and 169 – 78.

206 Birch, *History*, iii, 177, and *ibid.*, 158, 174, 192 – 3, 221; Hunter, *Royal Society*, catalogue entry 189.

207 In addition, extant letters of Coxe's suggest a similar penchant for densely filling sheets of paper; see e.g. Royal Society Boyle Letters, ii, 52 – 79.

208 Hunter, *Royal Society*, catalogue entry 189, and see below, Document B.

209 *Ibid.*, catalogue entry 20; Birch, *History*, iii, 95, 181. Note also a reference in Hooke's diary showing that he talked to Croone about the state of the Society on 10 November 1674: Hooke, *Diary*, p. 129.

210 L. M. Payne, L. G. Wilson and Sir Harold Hartley, 'William Croone, F.R.S. (1633 – 1684)', in Sir Harold Hartley, ed., *The Royal Society: Its Origins and Founders* (London, 1960), pp. 211 – 19; L. G. Wilson, 'William Croone's Theory of Muscular Contraction', *Notes and Records*, 16 (1961), 158 – 78.

his endowment in his will of the Croonian lecture, even though this did not actually materialize until 1738. Croone's handwriting, however, is quite different from that of this manuscript.

Regrettably, the question of authorship must, therefore, be left open.

Note 2: The Dating of Robert Hooke's Papers on the Organization of the Royal Society.

Of the various papers by Hooke dealt with in this article, the one whose date seems clearest is that concerning a possible printed newsheet to be found in British Library MS Sloane 1039, fols 112 – 3. Though C. R. Weld, who printed this document in his *History of the Royal Society*, dated it to 1663, this is certainly mistaken: the fact that Weld associated other documents of clearly later date with the Society's earliest years does not inspire confidence in his judgment.[211] Many of the papers in MS Sloane 1039 date from *c.* 1680, and it seems likely that this one does too, not least since, as has already been noted, the question of gazettes and the collection of magnetic phenomena referred to in the paper are also alluded to in a letter from Hooke to John Beale of 6 March 1680. Both were also the subject of discussion at meetings of the Society at this time.[212]

This being the case, it is interesting that the paper in MS Sloane 1039 echoes Hooke's sentiments in the main group of his papers in Royal Society Classified Papers xx.50 about the need for inducements for volunteers, while the letter to Beale shows that the question of a special missive for absent Fellows, which appears so often in the papers at the Royal Society, was in Hooke's mind in 1680.[213]

A further possible dating clue is provided by the fact that one of the items in Classified Papers xx.50 (fol. 97), though overlapping in content with the others, is in the hand of an amanuensis with emendations by Hooke. This

[211] Weld, *History* (n. 14), i, 146, 148 – 9. That this dating is mistaken (and the manuscript is in MS Sloane 1039, not Additional MS 4441, as stated by Weld) has already been pointed out by Harcourt Brown: *Scientific Organizations in Seventeenth-century France* (Baltimore, 1934), p. 187 n. 2. For other misdatings by Weld see above nn. 14 and 57. It is perhaps worth mentioning here a further document printed as by Hooke in Weld, *History*, i, 146 – 8 (and reprinted from there in Gunther, *Early Science* (n. 57), vi, 122 – 3, and Lyons, *Royal Society* (n. 11), pp. 41 – 42) and also dated to 1663. However, the original, which is to be found in British Library Additional Manuscript 4441, fol. 1, is not dated, and we see no reason — on grounds of handwriting or otherwise — for attributing it to Hooke as Weld does: it is therefore treated above, on p. 208, as an anonymous memorandum of uncertain date.

[212] See above, Section 1.

[213] It is perhaps worth noting that the letter to Beale in British Library MS Sloane 1039, fol. 171, makes it clear that at this point Hooke was evidently considering both a 'gazette' for all and a separate publication exclusively for absent members.

same amanuensis also wrote a paper by Hooke concerning his dispute with Sir John Cutler, dating from 1682 or 1683, which survives among the Petty manuscripts at Bowood.[214]

Beyond that, there is little firm evidence to date the papers in Classified Papers xx.50, though there are a number of insubstantial hints. The fact that at one point Hooke leaves a blank for the place of residence of the Society might be taken to suggest a date around the time of the Society's move from Arundel House to Gresham College in 1673, or around 1700 when the Society's tenure at Gresham came under threat.[215] The papers have numerous echoes of themes that came up during the reform effort of the early 1670s,[216] but since other points can be correlated with the early 1680s when, as we have seen, some of the measures suggested by Hooke *were* implemented, it would be unwise to lay too much stress on this.[217]

A date either early or late in the 1670s might seem to be favoured by the fact that Hooke's diary for 1672 – 80 contains a number of references to proposals for reforming the Royal Society that he devised at that time. Thus on 20 June 1674 Hooke records: 'to Mr Hoskins with whom much discourse about the R.S. module'; on 8 July that year he was again at Hoskins's and 'Read to him my paper of the Society'; and on 17 November Hooke 'Drew up proposalls about Secresy and Secretary'.[218] A further group of references appears in the aftermath of Oldenburg's death, for

214 Cl.P. xx.50, fol. 97; Bowood House Petty MSS H (8) 16.

215 See Adamson, 'Royal Society and Gresham College' (n. 32), pp. 2, 7 – 11.

216 For instance, the idea of requesting Fellows to provide a lecture or to pay for a substitute if unwilling to do so themselves (Cl.P xx.50, fol. 88; see Hunter, *Royal Society*, p. 38); the request that they should advance a year's subscription (Cl.P xx.50, fols 88, 96; see above, Section 1); the advocacy of an inventory of the Society's collection of rarities and instruments (*ibid.*, fol. 86; see Birch, *History*, iii, 158 – 9, 310 – 11: but see also above, n. 47); and the hope for extra staff (see above, Section 1).

217 For instance, the wish to make the Society more exclusive: see above, p. 219. Though the concern about the accuracy of the Secretary's records, about access to these and about the Secretary's relations with foreign virtuosi (Cl.P. xx.50, fols 86, 97, and see above, Document C) all smack of Hooke's dissatisfaction with Oldenburg's running of things in the 1670s, this is quite likely to have rankled sufficiently for Hooke still to have felt concern later. The justification of severity to the Secretaries which appears at one point (*ibid.*, fol 97) *might* suggest a date outside Hooke's own Secretaryship from 1677 to 1682, but the concern about the openness of committee decisions (see above, Document C) could suggest a date when Hooke was worried that decisions were being made behind his back, which *would* apply during his Secretaryship. On the interrelationship between the different papers in Cl.P xx.50 see above, pp. 201 – 2. On discrepancies between them see n. 59: it is doubtful, however, if these are sufficiently marked to suggest differences of date within them. We are inclined to dismiss as coincidental the slight similarity between the financial calculations by Hooke on fol. 88 and those by Oldenburg in D.M. 5.14 – 15. In addition, though the fact that Oldenburg's paper of 1 November 1673 is bound up among these papers (fol. 91) might be taken to suggest a similar date for those in Hooke's hand, it seems equally likely that it was brought together with them because of the similarity of their subjectmatter.

218 Hooke, *Diary*, pp. 108, 111, 131; see also pp. 57, 109, 112, 124, 126.

instance on 7 September 1677 when Hooke 'talkd with him [Robert Boyle] about Secretary and method very much', or 16 November the same year when he 'Discoursd much about regulating Society'. In December 1678 at Jonathan's coffee house Hooke 'declared my project of the Royal Society', while in 1680 he recorded how on 6 July he discoursed to Thomas Henshaw 'about my Designe for the Royal Society, he seemd much to approve it'.[219] It should be pointed out, however, that after 1680 Hooke's diary writing activity dwindled apart from a brief revival around 1690, so this chronologically-specific group of references could be misleading.

Lastly, Royal Society Domestic Manuscripts 5.12 has been identified as Hooke's owing to its similarity in appearance to the papers in Classified Papers xx.50 (this could conceivably be taken as evidence for a comparable date for both). There are two internal clues which make it possible to date this document. The reference to 'his Present Majesty' suggests a date after Charles II's death in 1685 and before 1702, when William III was succeeded by Queen Anne — and perhaps most likely either before 1689 or after the death of Queen Mary — while the reference to the Berlin Academy indicates a date after 1700.

[219] *Ibid.*, pp. 311, 328, 391, 448; see also pp. 315, 324, 335, 411, and perhaps also 416. See also above, n. 56.

7

PROMOTING THE NEW SCIENCE: HENRY OLDENBURG AND THE EARLY ROYAL SOCIETY*

In the annals of the early Royal Society, the name of its first Secretary, Henry Oldenburg, is inescapable. Moreover that name is now inseparably linked with those of A. Rupert and Marie Boas Hall. Their edition of *The Correspondence of Henry Oldenburg* saw its completion in 1986 with the publication of its twelfth and thirteenth — and final — volumes, after twenty-three years of publishing history, over 8,000 pages of published text, and two changes of publishers.[1] Undoubtedly, this is one of the landmarks of the history of the science of its period, scrupulously and lovingly edited with a full and thoughtful apparatus of notes, translations of all letters in languages other than English, learned commentary and full indices. It has already proved an unparalleled resource for the history of ideas in the period, and undoubtedly will continue to do so for many decades to come.

This is due not least to the fact that, although this is Oldenburg's correspondence, it also represents the bulk of the correspondence of the Royal Society up to the time of Oldenburg's death in 1677. These volumes provide the principal record of the Society's external, and especially its international, relations during the first seventeen years of its existence. It was appropriate that the first volume of the edition opened with a Foreword by the then Foreign Secretary of the Royal Society. In these pages, we see the formative stage of a facet of organized science that was to be intensified in the next century and which has come to be taken for granted ever since.[2]

The names of the early Royal Society and its dutiful first Secretary are thus inseparable, but they are not indistinguishable, and there is something to be learnt about both — and about wider facets of the contemporary scientific scene — by taking this opportunity to look more carefully at the exact nature of their interrelationship, to assess how much the success of the correspondence owed to Oldenburg and how much to the Royal Society,

* I am grateful to Paul Wood for his comments on a draft of this article.

[1] A. Rupert and Marie Boas Hall, eds., *The Correspondence of Henry Oldenburg*, vols. i – ix (Madison and Milwaukee, 1965 – 73); x – xi (London, 1975 – 6); xii – xiii (London, 1986). Volume xiii includes 60 pp. of addenda and corrigenda to the earlier volumes, but, unfortunately, no cumulative index. References in both text and notes below in the form 'iii, 136' denote volume and page in the edition.

[2] For a recent survey, see J. E. McLellan III, *Science Reorganized: Scientific Societies in the Eighteenth century* (New York, 1985).

and to consider how the enterprise developed, and why and how it served the functions that it did.

At the outset it is worth stressing the real effectiveness of this correspondence as an agency for the promotion of the new philosophy throughout Europe and even the rest of the world in the late seventeenth century. From this point of view, Oldenburg will stand comparison with any of the 'intelligencers' of his period — Marin Mersenne, Samuel Hartlib or Sir Hans Sloane — and in many respects, by a mixture of luck and design, he stands above them.

Not only was he closely associated with one of the greatest generations of scientists in England — the generation of Robert Boyle, Robert Hooke, Isaac Newton, John Wallis, John Flamsteed, Martin Lister and Nehemiah Grew, all of whom figure in the correspondence. He also established direct and lasting links with the most significant natural philosophers in contemporary Europe: Johann Hevelius, one of his earliest and most assiduous contacts; Adrien Auzout, Christiaan Huygens and other thinkers associated with the Académie des Sciences in Paris; the young G. B. Leibniz; Marcello Malpighi, for whom Oldenburg was to arrange the publication of his works; and Antoni van Leeuwenhoek, for whom he provided a comparable service. Arguably the core of Oldenburg's success was his correspondence with such figures of international renown, and it is this which assures his central place in understanding the scientific debates of his day. Quite apart from its function in simply exchanging news about recent publications and discoveries, it is apparent again and again in these pages how effective the correspondence was in stimulating international research, often on the very topics that Oldenburg reported.

Oldenburg was also tirelessly active in responding to approaches that he received from minor virtuosi both at home and abroad, and many otherwise little-known figures are illuminated by their appearance here. Often such men appear only fleetingly, exchanging a handful of letters over a short period and then disappearing again, as, for instance, with the Northamptonshire cleric, John Palmer, in 1667 – 8. Others recur so frequently that they become minor celebrities of the correspondence, like the East Anglian doctor and minister, Nathaniel Fairfax, or the Durham schoolmaster, Peter Nelson, and from their missives one gains an interesting glimpse of provincial intellectual life at the time.

It is perhaps worth pointing out that Oldenburg was in touch only with a minority of such scattered enthusiasts. This is revealed even by the lists that his contacts sometimes gave of other virtuosi in their own localities, while it is interesting to note from the last volume of the edition how many widely-scattered observations of a meteor the Oxford scientist, John Wallis, was able to collect himself, though interested in collating these with what Oldenburg had received.[3] Yet, haphazard as Oldenburg's contacts in the provinces may have been — as they clearly also were with lesser enthusiasts

[3] xiii, 190, 200 – 1. For lists, e.g. iv, 337 – 8; vii, 326 – 7. On provincial virtuosi at the time see Hunter, *Science and Society*, pp. 48f., 78f.

abroad — they undoubtedly helped to promote the new science. Oldenburg's importance for such men is clear from the eagerness with which they wrote to their 'much honoured friend Mr Henry Oldenburg secretary to the Royal Society' (vi, 308). Taken together with his major international contacts, such activity gave an impression of Oldenburg as an almost universal orchestrator of scientific inquiry which enhanced the Royal Society's reputation, giving it a fillip even when times were bad at its institutional base.

Oldenburg's dominant role in carrying on the Royal Society's correspondence did not emerge overnight. In fact, though it was taken for granted from the outset that the collecting of information and exchange of views would be an important part of the institution's activity, it was not self-evident how this was best carried out. The Society's first response to this task — as with many of its other functions — was to form a committee: as early as May 1661 a committee was appointed to correspond with Prince Leopold of Tuscany, the functions of which then seem to have become generalized to deal with epistolary duties as a whole. It was also a group of Fellows who were initially requested to respond to one of the first 'official' letters that the Royal Society received, from Eccard Leichner of Erfurt in 1663, though in the end Oldenburg responded to it; a further committee for correspondence was established in 1664.[4]

In addition, in the Society's early years, a good deal of correspondence was carried out by individual Fellows other than Oldenburg. Thus Dr William Croone, the 'Register' of the nascent Society, corresponded on its behalf with the Halifax natural philosopher, Henry Power, in the early 1660s; later, Croone is to be found receiving letters from the Danish intellectual, Nicolaus Steno, who never in fact entered into correspondence with Oldenburg; while it was Croone who was requested to draw up a letter to be sent to the Duchess of Newcastle on the Society's behalf in 1668.[5] Also active in this respect was the courtier and mathematician, Sir Robert Moray: not only did Moray write officially as acting president to a figure like the Parisian virtuoso, Henri Louis Habert de Montmor, in July 1661; he also conducted an extensive correspondence from 1661 onwards with the Dutch philosopher, Christiaan Huygens.[6] In addition, particularly in the 1660s, the letters to Oldenburg among the Royal Society's Early Letters can be supplemented by a substantial number written to other active Fellows, which, languishing unpublished, have arguably not received the

[4] See Birch, *History*, i, 22 – 4, 26, 296, 305, 407. The Halls are, incidentally, therefore wrong to say of Leichner (ii, xxi) that 'as always in such cases, Oldenburg was deputed to reply'.

[5] British Library MS Sloane 1326, passim; see also Royal Society Boyle Letters, vii, 19; E.L. P.1.10. Birch, *History*, ii, 100, 102 – 3, 112, 140, 250; E.L. S.1.92. See also R. Rome, 'Nicolas Sténon et la 'Royal Society of London'', *Osiris*, 12 (1956), 244-68.

[6] Birch, *History*, i, 34 – 5. See also Royal Society Miscellaneous Manuscripts 3.120 (and ibid., 119); Royal Society Copy Letter Book, i, 1 – 3; Huygens, *Oeuvres Complètes*, iii – vi (The Hague, 1890 – 5), passim.

attention they deserve. Sometimes these were so addressed because they were punctiliously directed to the Society's President; sometimes because they concerned a topic in which a specific Fellow was known to be interested, as with John Wallis and the movement of the tides; more often, however, it was because the authors concerned had close links with the Fellows to whom they wrote, and hence it seemed more appropriate to contact them than to write more formally to the Secretary.[7]

Yet increasingly as the 1660s went on, it is Oldenburg who dominates the record. By 1665, even Moray's letters to Huygens passed through Oldenburg's hands, and from this point onwards Oldenburg rapidly superceded Moray as Huygens' Royal Society contact. Similarly, provincial naturalists like Joshua Childrey or John Newburgh, who had previously corresponded with other Fellows like Croone or the writer, Edward Waterhouse, begin from 1669 to appear in the volumes of the Oldenburg *Correspondence*.[8] Hence, after starting slowly, by the late 1660s Oldenburg's activity as a correspondent reached a peak: for 1668 no fewer than 319 incoming and outgoing letters are known, and the number continued to average around 250 per annum thereafter, falling below 200 only in one year (1674) before that in which Oldenburg died.

Oldenburg's dominance — and the success of his correspondence — was cemented and extended by his *Philosophical Transactions*, begun in 1665, as is apparent at every turn in the Halls' edition. The value of the *Transactions* as a more public version of his correspondence was almost immediately apparent, and from the mid 1660s, we frequently find letters being written specifically for the *Transactions*. In addition, it was to the *Philosophical Transactions* and its content that both established correspondents and complete strangers often responded, whether it be Adrien Auzout in Paris or the Norfolk schoolmaster, John Burton, who wrote in to the publisher of the journal without knowing him by name (iii, 100, 498 – 9). To understand the correspondence fully, it has to be read alongside this second achievement on Oldenburg's part, that 'Collection, that is like to be as lasting as usefull', in Boyle's words (iii, 145).

Why did Oldenburg come to the fore in this way? Clearly, it was because it quickly became apparent that he was ideal for the job. In part, he was suited by his background, coming as he did from Germany and hence finding natural a degree of international intercourse that might have eluded his more

[7] These letters are listed in W. E. Schuckard's catalogue in J. O. Halliwell, *A Catalogue of the Miscellaneous Manuscripts preserved in the Library of the Royal Society* (London, 1840), passim. Among these see, e.g., H.1.107 – 8 for letters to Wallis, or S.1.42f. for letters to the President. In addition, scattered references illustrate that other such letters are lost: e.g. Birch, *History*, i, 345, for a further letter from Joshua Childrey to Edward Waterhouse in addition to the one that survives as E.L. C.1.1. The Early Letters also, of course, contain many letters dating from the period after Oldenburg's death.

[8] For the Huygens letters, see ii, 538, 551, 618, iii, 7, 13, 128 – 30. For Childrey and Newburgh see E.L. C.1.1, 4 (and above, n. 7); N.1.1.; cf. v, 384 – 6, vi, 225 – 6.

insular colleagues. Oldenburg's aptitude for such work had already been illustrated in the 1650s in the context of his association with that earlier 'intelligencer', Samuel Hartlib, who came from a similarly international background. Oldenburg had first come into contact with Hartlib in 1656, and his assiduity as a regular correspondent was displayed to Hartlib and others in the late 1650s, while he travelled in France as a tutor. It is symptomatic that in 1657 Oldenburg declared himself 'hugely pleased' by a scheme to support staff who would keep intelligence and distribute rewards 'for the promotion of knowledge': this epitomizes the Interregnum prehistory of the kind of activities to which he was to devote himself after the Restoration.[9]

Like Hartlib, Oldenburg was prepared to dedicate himself wholeheartedly to merchandizing in knowledge. Indeed, such was his identification with his task that he remains somewhat shadowy as an individual outside his persona as Secretary to the Royal Society, despite the assiduity of the editors of the *Correspondence* in unearthing such information as is available about Oldenburg's private life and not least his sources of income. What comes across clearly is how phenomenally hard-working and dedicated he was. The sheer amount that he was able to do was graphically illustrated by a letter to Boyle of 17 December 1667 in which he solicited more reward for his efforts, explaining 'I am sure, no man imagins, what store of papers and writings passe to and from me in a week'. Quite apart from his correspondence, which not only involved writing letters but often also doing business and making enquiries on behalf of his correspondents, he also had to orchestrate the administration of the Royal Society's meetings, as well as carrying out other tasks like translating for Boyle. No wonder he considered that even those reluctant to pay him better had to admit 'that no man in England would doe, what I doe, upon the terms, I goe away with' (iv, 58).

Also important was the fact that, not least through his close association with Boyle, Oldenburg was well-informed about many aspects of the new science. Indeed, it is possible to see Oldenburg as a competent natural philosopher in his own right, with a coherent and quite sophisticated scientific outlook evidenced by both the correspondence and the *Philosophical Transactions*: he held a strong but not uncritical Baconian viewpoint, stressing the need to collect the materials for induction and suspicious of premature hypotheses, but never wholly eschewing theoretical questions.[10] Moreover he was very broad-minded. Though his editors have shown a slightly disdainful attitude towards some of some of his correspondents' concerns — perhaps particularly John Beale's enthusiasm for the promotion

[9] i, 137. For the background, see Charles Webster, *The Great Instauration* (London, 1975).

[10] See A. R. and M. B. Hall, 'Philosophy and Natural Philosophy: Boyle and Spinoza', in *Melanges Alexandre Koyré* (Paris, 1964), ii, 241 – 56. For a further study that deserves publication, see Paul B. Wood's chapter on Oldenburg in his thesis, 'Francis Bacon and the Experimentall Philosophy' (London M.Phil. thesis, 1978), briefly summarized in Hunter, *Science and Society*, p. 54.

of cider — Oldenburg himself seems to have been as keen to discuss projects like these as the latest astronomical observations.[11] The wide range of his enthusiasm reflects that of his correspondents and acts as a warning against the temptation to impose an anachronistic degree of specialization on the seventeenth century. Thus one finds Newton and Leibniz, for instance, wanting advice on cider-making and varnish (xiii, 71, 138, 146, 319), and this illustrates a real wholeness about the intellectual enterprise of the day which *The Correspondence of Henry Oldenburg* constantly exemplifies.

Oldenburg also had skills which made him excell as an 'intelligencer'. For one thing, he was a much better linguist than most of his correspondents. Foreign scholars frequently complained about the use of English for the *Philosophical Transactions* — 'being clumsy in the English language' (ix, 197) — so it was essential that Oldenburg should translate everything that he needed to say to them into their own vernacular or into Latin. On the other hand, an English scientist like John Flamsteed not only had no French but was also dependent on Oldenburg to correct his 'exorbitant latin' (viii, 427; x, 371).

In addition, Oldenburg could show the graces of a diplomat in knowing how best to deal with the susceptibilities of his correspondents, and it may have been partly for this reason that he came to the fore in conducting the Society's correspondence. That this was the case is suggested by comparing Oldenburg's letter to the rather reactionary Eccard Leichner with a surviving draft by the committee initially deputed to reply to him. The draft is both more abrupt in telling Leichner that he had misunderstood the Society's aims and more convoluted in explaining that not all the books that Leichner had sent had arrived and that few members could read German and Dutch in any case. Oldenburg, on the other hand, was artfully tactful in putting Leichner in his place without wholly discouraging him, in a manner that was to become typical.[12]

Yet, for all his ability for tact, Oldenburg was also aware of the value of controversy, and here he displayed something of a journalistic flair, even at times egging on his correspondents in their disagreement with one another. This is perhaps most strikingly revealed by a curious survival in the form of a copy of a letter from Auzout to Hooke which Oldenburg annotated with marginalia in which he tried to persuade Hooke to respond: 'What say you to this?', he notes at one point, for instance, or 'A handsom sting again will be necessary' (ii, 474). At times, there is also almost a sense of salesmanship: his report to the Belgian mathematician, René Sluse, on the Royal Society's blood transfusion experiments in 1667, for instance, was coined in the form, 'I cannot conceal from you that our Society has so far succeeded with that experiment . . . that a few days ago they performed it upon a man' (iii, 617). It is easy to see how he was led from letter-writing to scientific journalism.

[11] See, for instance, ix, 65. The Halls' disdain, seen earlier in patronizing remarks (e.g. ii, 12 n. 8), reaches its climax at xiii, 341, where Beale's last letter to Oldenburg, though admittedly hard to decipher, has been epitomized rather than included in full.
[12] British Library Additional MS 4441, fol. 93. Cf. ii, 110 – 11. The committee's reply is, however, only a draft, and it is possibly incomplete.

Yet all of these talents worked as well as they did because they were placed at the disposal of the Royal Society, and it was very much as the Secretary of the Royal Society that Oldenburg presented himself. In part, this was because he was the Society's more or less adequately paid employee, who had to obey the orders that he was given even if he disagreed with them. This was illustrated clearly in 1668 in connection with a letter from the French savant, Pierre Carcavy, in which he suggested a liaison with Oldenburg on behalf of the Royal Society. Probably many members of the Society were rather put off by the letter's chauvinistic tone, its aspiration to French *gloire* almost overriding its statements of commitment to international co-operation, and Oldenburg was therefore instructed to reply as a private individual rather than as Secretary. But, as he complained in a letter to Boyle, 'This I was not very well satisfied with. I see, Punctilio's retard and obstruct much good, both publike and private. But I must submit'. (iv, 156 – 9, 206 – 7).

More important, however, is the fact that, because he normally wrote in the role of representative of a public body, Oldenburg acquired a status and an authority that he would have lacked as an individual. In fact, though the Royal Society was cumulatively to gain considerable prestige through Oldenburg's prolonged enterprise as its correspondence secretary, it is paradoxically the case that from the outset Oldenburg owed much of his success to his deployment of the Society's image. This was not necessarily an accurate representation of the Society as it actually existed, but more a projection of a sort of idealized view of the Society, treading a tightrope between the ideal and what actually underlay it. But this image had a power which Oldenburg knew how to use and to which his correspondents responded, and it illustrates an important function that the Royal Society served in the international scientific community of the day.

This comes across clearly from Oldenburg's formal exchanges with his correspondents, which, though rhetorical, arguably illustrate something about the reasons for Oldenburg's success. His letters to his foreign contacts echo with a constant repetition of clichéd but revealing sentiments about the Royal Society and its role, from the missives he sent to Peter van Dam or Johann Hevelius in 1663 onwards (ii, 12 – 15, 25 – 9). In them, Oldenburg gave a grandiloquent view of the foundation of the Royal Society, armed with royal patronage and supported by the eminent and the wise, as the embodiment of the enterprise of reforming knowledge. Quoting its motto, 'Nullius in Verba', almost as a slogan of its enterprise, Oldenburg set out the Society's aim 'to scrutinize the whole of Nature and to investigate its activity and powers by means of observations and experiments': he outlined its aspiration to build up a great bank of data as raw material for the induction that Bacon had advocated, so that the great enemy, scholasticism, might once and for all be routed.

In addition, Oldenburg presented an almost apocalyptic view of the Society as the figurehead of international scientific enterprise, the leading force in what ought to be a collaborative struggle against obfuscation and ignorance. 'I seriously urge all who perceive its importance to unite in aiding and perfecting it as best they can', he told Van Dam, while to

Hevelius he added: 'if such friendship could be spread through the whole world of learning, and established among those whose minds are unfettered and above partisan zeal, because of their devotion to the truth and human welfare, philosophy would be raised to its greatest heights'. Moreover Oldenburg was to repeat this again and again in subsequent years, as in a letter to J. C. van Boineburg of Mainz in 1670: 'What we are about is no task for one nation or another singly. It is needful that the resources, labours, and zeal of all regions, princes, and philosophers be united, so that this task of comprehending nature may be pressed forward by their care and industry' (vii, 109).

What is more, his correspondents responded with almost as much enthusiasm for the great design — and, significantly, for the Royal Society as its embodiment — as Oldenburg showed himself. In Oldenburg's exchange with Hevelius, as subsequently with others to whom he was to repeat overlapping sentiments, one finds a counterpoint of mutual exhortation and compliment. 'Nullius in Verba', for instance, was a stroke of genius, and Hevelius, like others later, actually echoed this motto in responding to Oldenburg, enthusing how 'the learned world especially owes humble and eternal thanks to the King of England, a prince worthy of every high praise, because he has founded a unique Assembly of those philosophers who cultivate and advance the arts and sciences by following not tradition, but observations and experiment alone' (ii, 138). Thereafter the correspondence is littered with superlatives of this kind. 'If anything is to be hoped for in extending the limits of science, it is to be expected from them', wrote René Sluse of the Royal Society (iii, 413), while the young Leibniz, introducing himself to Oldenburg in 1670, went further: 'Who has not heard of the Society, if he is in any way drawn to an interest in true learning, the true learning I mean which under your guidance particularly is withdrawing from the groves of the critics into the fastnesses of nature?' (vii, 66).

Clearly the Royal Society — particularly in the idealized form in which Oldenburg presented it — served a great psychological need for the European scientific community of the day. Oldenburg was undoubtedly preaching to the converted as far as a significant group of like-minded intellectuals were concerned, and they welcomed this because of the moral support they thus acquired. 'I greatly rejoice', averred the Dutch physician, Sylvius, 'that the British Royal Society proceeds in the investigation and illumination of truth by the same method that I have always thought should be adopted and, indeed, with all my might have attempted'. He continued, significantly: 'I am not a little strengthened in my opinion on that point by this fact' (iv, 70). Comparable sentiments came from Peter Moller of Königsberg, who rhapsodized about how God's spirit had 'excited some minds to strive in this melancholy age to explore the secrets of Nature; this I have long hoped for and despaired to see' (iii, 557).

More specifically, it is clear that, to frustrated intellectuals feeling isolated and unappreciated in their native context, the Royal Society as depicted by Oldenburg was a kind of projection of all that they aspired to. Thus René Sluse complained to Oldenburg how, learned though his countrymen were, natural philosophy was not highly valued among them (iii, 596), while P. J. Sachs enviously contrasted the Royal Society's permanent seat in a metropolis with the way in which intellectuals like himself were scattered in smaller centres: he hoped that 'a flood of learning and experimental truth can be brought through epistolary channels from your British ocean to fertilize our more sterile Germany' (ii, 342-3, 346). At a still more practical level, Johann de Muralt of Zurich desired 'a public warrant over the seal of your Society' to impress the conservative magistracy of his city (xiii, 160).

The role that the Royal Society played through the correspondence is perhaps seen most clearly on the question of scientific arbitration, since both minor and major figures from the continent seem to have been almost uncritically keen to solicit the Society's views on their work. As the Parisian intellectual, Gilles De Launay, put it, 'If I should be so happy as to deserve your approval I should thereby expect to obtain that of all the citizens of the Republic of letters, since in order to judge physical works properly one must be of your opinion and of that illustrious and celebrated Royal Society which is known to all learned men as the first of those which have been created in imitation of it' (iv, 218). This was hardly less true of a man like Hevelius, who was anxious to submit his work 'to the high judgment and due consideration of the Royal Society', requesting the arbitration of 'those Fellows who are learned in these matters, being free from prejudice and self-interest, who are, that is to say, most impartial judges': it is not surprising that Sir Robert Moray told Oldenburg that 'I am exceedingly satisfied with Hevelius taking the Society in consideration as hee does' (ii, 138 – 9, 529; iii, 171).

As far as arbitration was concerned, there was a very satisfactory interrelationship between Oldenburg's epistolary activity and the Society's institutional existence: Oldenburg was at pains to report how he had read letters at meetings, stressing how, when arbitration occurred, it was an official, corporate step, taken by Fellows jointly rather than by himself as an individual. Indeed, on the whole, the Society lived up to its image very well in this respect, due to Oldenburg's ability to rely on expert colleagues like Robert Hooke, John Wallis or John Flamsteed for the opinions that he gave. Only rarely were there problems, as over the judgement on Hevelius' *Prodromus Cometicus* in 1665, since the intermission of meetings due to the plague in that year made it difficult to give a formal Society verdict, as against the private opinion of 'outstanding astronomers' (ii, 623).

If in this the Society's image and its actuality harmonized well, however, in other respects this was not the case, as the Society failed to live up to the aspirations of its founders which Oldenburg so faithfully reproduced in the fanfares for it that he issued. In particular, the royal endowment that in the Society's earliest years Oldenburg had assured his correspondents he confidently expected never materialized, which limited the funds available to support its activities. Equally important were the associated difficulties encountered in trying to pursue science in a body that was institutionalized but voluntary, which constantly threatened to let down the image of intellectual dynamism that Oldenburg strove to promote. This is reflected especially clearly in one of the most interesting components of the correspondence in its early years, the rather frank letters which Oldenburg exchanged with Robert Boyle, in which sentiments about the Society's international mission and its institutional weakness are poignantly juxtaposed.

As Oldenburg put it in 1666, 'I hope, our Society will in time ferment all Europe, at least; I wish only, we had a litle more Zeale, and great deal more assistance, to doe our work thoroughly', while, complaining in the following year of the 'deadnes' of the Fellows of the Society, he pointedly and dramatically expressed the hope that 'so noble and usefull an institution may not fall to the ground, in an age, so full of knowledge and curiosity, and enjoying so considerable advantages to improve and increase the same' (iii, 155, 476). Boyle comparably wrote: 'I am not sorry that the reputation of our Society increases abroad; but we shall have cause to be sorry if nothing be done at home (by those whom it most concerns) to enable us to make it good' (iv, 299).

On a later occasion, Oldenburg is to be found using his foreign correspondents' acclaim for the Society to read lessons to those in positions of authority about the need for more support at home. As he explained in 1673 to Sir Joseph Williamson, Secretary of State and F.R.S.: 'the most Eminent men in most parts of the World' concurred in their esteem of the Royal Society, 'which, as it adds not a litle to the renown of England, now admired abroad above other nations for advancing Experimental knowledge as well as Academical Learning, so I hope it will at length induce at home all that know what value to put upon vertu and knowledg, to contribute to the support and encouragement of so excellent a Foundation' (x, 175 – 6).

In fact, the Society's domestic problems were to continue, but it soon became apparent that, so long as there was something to report in the way of research and publication, it was paradoxically irrelevant to the vitality of the correspondence how healthy the Royal Society as an institution actually was. What was more important was Oldenburg's ability to cite discoveries and inventions made by Fellows of the Society, whether or not they attended meetings, while at times he even supplemented this by reporting the work of non-Fellows such as Ralph Bohun and Sir Samuel Morland and implying that this, too, was part of the Society's achievement (viii, 529 – 30). The only thing that would have been disastrous would have been complete collapse, as was rumoured in 1668 — when Henri Justel reported from Paris how he had heard 'that your Society no longer works seriously . . . that most of the members attend no longer, and that soon it will be quite dispersed' — or in 1673, when one correspondent was worried because he did 'not so much as heare whether the Royall Society bee living or no' (v, 39, x, 129). Oldenburg's answers to these particular letters are unfortunately not extant, but one can imagine how anxious he would have been to scotch any such rumours, even if aware of the element of truth that underlay them, especially in the early 1670s, when 'the failing or Continuance of that Society' really was felt 'to be now brought to a Crisis'.[13]

[13] Hunter, *Royal Society*, p. 51.

In the event, the Society survived such difficulties, but, if the Society was permanent, Oldenburg was not, and the fragility of the arrangement that had long worked so well was illustrated by its collapse in the aftermath of his death in September 1677. Up to a point, the tradition of correspondence that Oldenburg had pioneered had by this time acquired a kind of momentum of its own, and it is clear from the Royal Society's minutes and letter-book that there was an expectation that things should continue much as they had under Oldenburg, with links being retained with his regular correspondents and a substantial amount of time at meetings being given over to the reading of letters. As a result, pressure was put on those active in the Society to keep things going: a jointly-drafted circular letter was sent to the chief correspondents early in 1678, and over the following four years Nehemiah Grew, Robert Hooke, Thomas Gale and Francis Aston acted as correspondence secretaries, trying more or less adequately to ensure that the expectations which Oldenburg had created in the Society's correspondents were satisfied.[14]

In fact, however, none of these found it possible to match Oldenburg's record. As John Flamsteed reported to Richard Towneley in February 1680, 'Our Meetings at the Royall Society want Mr Oldenburgs correspondencys and on that account are not so well furnisht nor frequented as formerly'.[15] It is symptomatic that, in casting around to try to deal with the problem, the Society reverted at one point to the pattern of its earliest years, appointing a committee to supervise the Society's correspondence, while soon after Thomas Gale was given the job of correspondence secretary, he had to be instructed on what kind of information to send correspondents.[16] Indeed, things really only reverted to a state of affairs like that under Oldenburg when a similar individual came onto the scene, in this case Sir Hans Sloane in the 1690s. The position with *Philosophical Transactions* was comparable. After limping on from Oldenburg's death till the beginning of 1679, bulked out by extensive space-fillers, it then collapsed and was only partially replaced by Hooke's *Philosophical Collections* (1679 – 82). Thereafter, it was temporarily revived in the 1680s (assisted by extra funds that the Royal Society had at that point), but it was only in the 1690s that it was finally and permanently re-established.[17]

Not the least of the problems at this stage was that of conflicting commitments on the part of the Society's officers. In 1678 Grew was given the task of cataloguing the Royal Society's 'repository' or museum, and this

[14] British Library MS Sloane 1942, fols. 1 – 3 (for 'The Circular Letter sent to divers of the forreine Correspondents': for a published version, see Huygens, *Ouevres Complètes*, viii (The Hague, 1899), 66 – 7); cf. xiii, xxviii-ix; Royal Society Copy Letter Book, viii, passim; Birch, *History*, iii, 369, 515, 518.

[15] Flamsteed to Towneley, 13 Feb. 1680, Royal Society MS 243, no. 45.

[16] Birch, *History*, iii, 518, iv, 9.

[17] *Phil. Trans.*, xii – xvii (1677 – 93); on the finance, see Hunter, *Royal Society*, p. 42. A proper history of the *Philosophical Transactions* is a desideratum.

distracted him from his epistolary duties, while he was probably also beginning the move towards medical practice which, within a few years, was to divert him from scientific work altogether.[18] As for Hooke, not only was he expected to continue to display experiments to the Society as he had throughout the period when Oldenburg had managed the correspondence: he also acquired extra responsibilities outside the Society just at this time, quite apart from being distracted by legal wrangling over Sir John Cutler's erratic benefaction to him.[19] The result was that by December 1679 complaints from foreign correspondents had reached the point that Hooke was relieved of responsibility for this facet of the Society's activities, and Gale was instead requested to write 'to the several persons in foreign parts, who had formerly corresponded with the Society . . . and to assure them of a constant correspondence for the future'.[20]

Such problems constantly plagued the early Royal Society, and they add to one's respect for Oldenburg in successfully transcending them. Of course, he had his lapses, apologizing to one correspondent in 1676 for an oversight, 'the multiplicity of letters, I am obliged to write, making me sometimes forget, whether I have written such and such letters, or not' (xii, 364). In addition, a degree of conflict between his public and private roles is in evidence in at least one of the mutually beneficial arrangements that he made with his foreign correspondents, that concerning the patent for Huygens' spring-balance watch: for this went sour, and Oldenburg became involved in a vitriolic dispute with his colleague, Hooke, which is fully documented in the pages of this edition and which was undoubtedly distracting to both. What is significant, however, is Oldenburg's success in carrying on a truly epic correspondence in spite of such setbacks, and to this the published edition is a proper tribute.

What one can learn from both Oldenburg's success and his successors' failures is just how precarious and dependent on individuals the Society's attainments were in its formative years. For if the debacle after Oldenburg's death shows the negative side of the coin, the positive side is displayed by the successful symbiosis that he established between the institutional and the personal during his lifetime, from which both the Royal Society and the advancement of learning gained immeasurably. As we have seen, while Oldenburg contributed skill and sheer hard work, his efforts were rewarded the more abundantly because of the enhanced authority that he derived from his successful deployment of the ideal of organized science that the Royal Society represented. The result was an achievement for which Oldenburg deserves the highest praise, but which was also typical of the way in which the early Royal Society evolved, with institutional ideals and personal initiatives forming a fruitful, if often fragile, combination.

18 See below, 'Early Problems in Professionalizing Scientific Research', pp. 275f.

19 See below, 'Science, Technology and Patronage' and 'Reconstructing Restoration Science', pp. 350 – 1.

20 Birch, *History*, iii, 515.

NOTE

On Oldenburg and Millenarianism

The intention of this note is to draw attention to some material relating to Oldenburg to be found among the papers formerly owned by the eighteenth-century scholar, Thomas Birch, which are now in the British Library. This usefully supplements the *Correspondence* concerning Oldenburg's religious views. In particular, it adds further evidence about his role as a purveyor of and commentator on millenarian ideas which is worth summarizing, even if it is not entirely clear what conclusions should be drawn about Oldenburg's attitude to such material.

In general, Oldenburg's religious views are revealed clearly if briefly by his correspondence: they seem to have been of a conformist, middle-of-the-road kind which harmonized well with his overriding commitment to the pursuit of natural knowledge. The clichés in his letters about how God's glory might be furthered through study of His works fit naturally with the moralistic asides that appear elsewhere, for instance when he complained to Boyle in 1667 about how the Christmas festivities were 'so much generally abused to idlenes and irreligion' (iv, 78); Oldenburg's religious proclivities are also well illustrated by comments on books like Sir George Mackenzie's *Religio Stoici* (1663), which, he thought, 'hath some strains of wit, and some very honest and healing maxims for broken Christianity' (ii, 291).

But it is interesting to note that such hints may be supplemented by a short religious treatise by Oldenburg now in the Birch collection, 'Admonitions and Directions of a good Parent to his Child, especially a Son': this was evidently addressed to Oldenburg's son, Rupert, and it was apparently at one time accompanied by comparable documents intended for his wife and daughter which no longer survive.[21] The 'Admonitions' comprise a set of revealing, if commonplace, injunctions to piety and sobriety which are almost worthy of publication for the light they shed on Oldenburg's prudentialist but sincere religious outlook.

More important, however, is the evidence available from the same archive concerning Oldenburg's role in purveying millenarian ideas. The bulk of this dates from 1666, when the supposed messiah, Sabbatai Sevi, was active in the Near East and it was widely believed that this might signal the conversion of the Jews and the end of the world.[22] Oldenburg's main

21 British Library Additional MS 4458, fols. 108 – 116. See also ibid., fol. 144, for theological notes. On the lost manuscripts, see ix, xxviii – ix, and Birch, *History*, iii, 355 – 6: unlike the Halls, I see no reason to doubt that these once existed although they are now lost.

22 On the European reception of the news about Sevi, see particularly the now definitive study of Peter Serrarius by E. G. E. Van der Wall, *Der Mystieke Chiliast Petrus Serrarius (1600 – 1669) en Zijn Wereld* (Leiden, 1987), ch. 10. See also Michael McKeon, *Poetry and Politics in Restoration England* (Cambridge, Mass., 1975), pp. 206-8.

informant on this topic was the Dutch chiliast, Peter Serrarius, and a letter from Serrarius to Oldenburg of 5 July 1667 was included at the appropriate point in the *Correspondence* (iii, 446 – 7). A further relevant item, a letter from Oldenburg to Sir Joseph Williamson of 10 November 1666 in which he reported news about Sevi received via Amsterdam, has been included in the supplement to the earlier volumes in the recently published final volume of the edition (xiii, 407 – 8). But additional related material survives in the British Library, and, though some of it is referred to in the commentary on this letter, its significance is not there fully brought out.

In their note to Oldenburg's letter to Williamson the editors suggest that the news it gives is likely to have derived from a letter from Serrarius, and they add: 'another letter of about the same date and with similar news perhaps also from Serrarius (or perhaps even the same letter) is in B.M. MS 4476, f. 141, a much torn sheet addressed to 'Monsr Grubendol A Londres' in another hand, which has also written directions for disseminating the news'. In fact it is worth stating here that this is certainly not the same letter, since the missive in Additional MS 4476 contains different news. Though it is somewhat damaged and its date is unclear, almost all of it can be deciphered, and it comprises a report of Sevi's reception on his arrival at Constantinople, which took place in March 1666, quoting a letter from a merchant there to his brother in Amsterdam that he was 'beyond all suspicion of doubt persuaded of the approaching restitution of the Kingdome of Israel'; it also noted related events in Venice. Though the text of the letter is in an unknown hand, it is interesting to find that Oldenburg has added short passages in his own, as if he was collating more than one copy and was anxious to have as full an account as possible.

On the reverse, in addition to the address and an instruction concerning the delivery of an enclosure, there is a long, though damaged, passage in Oldenburg's hand. This is described in the note to vol. xiii as 'a summary of milleniarist doctrine', but in fact it seems to comprise a report on the French chiliast, Jean de Labadie, who is known to have arrived in Amsterdam in July 1666, which is perhaps derived from another, lost letter from Serrarius.[23] This details Labadie's preaching, remarking on the passion with which he urged listeners to repentance and assured them that the millenium was imminent. His message on the latter topic is reported in particularly striking terms: 'He also shews, that the time is at hand, and that we are as a woman with Child, that is past her reckoning . . . I believe, that those, who shall never dye, but remain till Christ with his Saints come downe to creat his Kingdome, are [now] living'.

[23] On Labadie see Van der Wall, *Serrarius*, chs. 10 – 11. The endorsement is as follows: 'This Inclosed be pleased to direct in Wales, or to send it to Mr Jackson or Mr Isaac Gray and to impart them also these News, and by them others may enjoy it likewise', while under the address are the words: 'in Ladlane[?] at the Swan with 2 necks for Brereton'. The address and the main endorsement, unlike the text of the letter, are in the same hand as the letter published in iii, 446 – 7, from PRO SP 84/ 183, no. 4.

In addition, another volume of miscellaneous letters in the Birch collection contains copies in Oldenburg's hand of two further letters from Amsterdam, perhaps also from Serrarius, dated 18 and 29 October 1666. These again report news about Sabbatai Sevi and his activities from letters from Constantinople, adding information on the impact that the affair was having on the Jewish community in Amsterdam, and the author again showed a real commitment to the possibility of an imminent apocalypse: 'We have reason to lift up our heads, and to believe, that our redemtion is at hand, and the Iudgement of Babylon already begun'.[24]

It has seemed worth itemizing all this material here because, taken together, it shows that Oldenburg was much more actively involved in transmitting information of this kind in 1666 – 7 than would have been apparent from the relevant portion of the main text of *The Correspondence of Henry Oldenburg.* This seems worthy of note, despite the fact that it is not entirely clear what construction one should place on Oldenburg's role. Obviously, it is possible that he was disinterested or even sceptical, and that he was simply recording information to pass on to Williamson, as in the letter of 10 November 1666. But it is equally likely that he was genuinely concerned that the end was nigh, as his informant, Serrarius, certainly was. By way of background, it is worth noting how, in his correspondence in the 1650s, Oldenburg had juxtaposed somewhat non-commital reports on specific millenarian theories — as in a letter to Menassah ben Israel in 1657 (i, 124) — with occasional comments that make clear his expectation of an imminent apocalypse: thus he told Samuel Hartlib in a letter of 12 September 1658 how 'It is not unlike, that a sore persecution hangs yet over the Protestant Churches, before the destruction of AntiChrist' (i, 180).

Lastly, it is worth noting another remarkable but perplexing document which survives among the Birch papers. This comprises three pages of excerpts made by Oldenburg from a book of apocalyptic prophecies by the Dutch chiliast, Johannes Rothe, in which Rothe predicted an imminent apocalypse in the context of the international politics of the 1670s: this is of special interest because the source of the material is said to be the third edition of the work in question, published in Amsterdam in 1673, and it thus takes us into the very last years of Oldenburg's life.[25] In part, Oldenburg may have been interested in this for its commentary on the hostilities between England and Holland during the Third Dutch War, and it is even conceivable that the notes were intended for the use of Sir Joseph Williamson and the government, either with a view to censoring Rothe's book, or for propagandist purposes.[26] But Oldenburg's notes also deal with

24 British Library Additional MS 4299, fols. 50 – 1.

25 British Library Additional MS 4458, fols. 146 – 7. The work in question is *Eenige Prophetien en Revelatien Godts, Aengaende de Christen Werelt in dese Eeuw* by 'Johannes Den Dienstknecht Godts', of which an English translation had been published in 1672. On this work and its author, see K. H. D. Haley, 'Sir Johannes Rothe: English Knight and Dutch Fifth Monarchist', in Donald Pennington and Keith Thomas, eds., *Puritans and Revolutionaries* (Oxford, 1978), pp. 310 – 32, esp. pp. 316 – 7.

26 See ix, xxix – xxx. No item on this subject appeared in the *London Gazette* between the start of 1673 and the end of the war.

Rothe's general predictions of awful things in store — 1672 was, in Rothe's view, to be 'the beginning of sorrows' — and it does not seem unlikely that Oldenburg was at least open-minded about the possible validity of Rothe's utterances.

Hence Oldenburg may well have retained a commitment to the idea of an imminent apocalypse into the later years of his life. Certainly, there would be nothing surprising about this, and neither would there be any incompatibility between this and the rest of what we know about Oldenburg's religious outlook, for recent scholarship has stressed how important millenarian ideas were even for figures in whom such beliefs might at one time have seemed surprising.[27]

[27] For a useful synthesis, see Charles Webster, *From Paracelsus to Newton* (Cambridge, 1982), ch. 2.

8

EARLY PROBLEMS IN PROFESSIONALIZING SCIENTIFIC RESEARCH: NEHEMIAH GREW (1641 – 1712) AND THE ROYAL SOCIETY, WITH AN UNPUBLISHED LETTER TO HENRY OLDENBURG*

Nehemiah Grew's *The Anatomy of Plants* (1682) was one of the most important botanical books published in the seventeenth century. In this compendium of his research, Grew combined an unprecedentedly detailed study of the structure of plants with an attempt to understand their functions. Working independently of, but in parallel with, his Italian contemporary, Marcello Malpighi, Grew justifiably saw himself as a pioneer in a new field, and between them these two scientists revolutionized biological knowledge.

Grew's outstanding work has attracted various appraisals, the most detailed of them appearing in the pages of this journal,[1] but no full scrutiny has ever been made of the role of the Royal Society in assisting these investigations. Not only did the Society arrange for the publication of Malpighi's *Anatome Plantarum* in two parts in London in 1675 and 1679, in itself a tribute to the Society's role in European science. In the case of Grew, its influence was even greater, for a decisive stimulus to the completion and publication of his work was provided by the Society's offer of a research post to him in 1672, and by its continued support thereafter: Grew himself claimed that without the Society's encouragement in the project, he 'would scarcely have ventured upon it'.[2] Indeed the 1672 episode has a wider significance, as one of the earliest cases of a scientific institution sponsoring a specific research project. The Royal Society had by then employed Robert Hooke for nearly a decade, during which he had published research of great importance: but the capacity in which he had been retained by the Society had been more that of a demonstrator, and other employees had been clerical or more menial in function. In France, of course, the state support

* I am indebted to Sir William Paton, Mrs Marie Boas Hall and Dr Moti Feingold for their helpful comments on this article.

1 I.e., *Notes and Records of the Royal Society*, where this paper originally appeared: see Jeanne Bolam, 'The Botanical Works of Nehemiah Grew, F.R.S. (1641 – 1712)', *Notes and Records*, 27 (1973), 219 – 231. See also Conway Zirtle, 'Introduction', in N. Grew, *The Anatomy of Plants* (Sources of Science No 2. New York: Johnson Reprint Corporation, 1965), and C. R. Metcalfe, 'Nehemiah Grew', in C. C. Gillispie, ed., *Dictionary of Scientific Biography* (New York, 1970 – 80), v, 534 – 6. References to the older studies of Agnes Arber and others will be found in these articles.

2 N. Grew, *An Idea of a Phytological History Propounded* (London, 1673), sig. A3v.

provided when the Académie des Sciences was founded in 1666 assured a number of salaries for scientists: but the voluntary nature of the Royal Society imposed financial constraints which made such patronage far harder to organize.[3]

In illustrating hesitant steps towards what has since become an almost universal practice, this early piece of sponsorship is therefore of considerable interest, and new light can now be thrown on various aspects of the episode by a hitherto unknown letter from Grew to Henry Oldenburg, Secretary of the Royal Society, which has recently come to light, having been overlooked by the editors of the definitive edition of Oldenburg's correspondence.[4] This letter is above all revealing of the difficulties involved in patronage of this kind which have not hitherto been apparent: but its discovery provides a good opportunity to collect together all available evidence about its context, outlining Grew's career before, during and after his employment by the Royal Society, and the way in which it was affected by that episode.

Nehemiah Grew was born in 1641 and brought up in Coventry, where he evidently received his schooling. His father, Obadiah Grew, was a prominent figure in local dissent: he became Vicar of St Michael's, Coventry, in 1644, and, after resigning to avoid having to comply with the Act of Uniformity in 1662, he continued to preach in the city. Coventry was a centre of nonconformity — the Bishop of Lichfield and Coventry complained in 1669 of the 'numerous and dangerous conventicles' there — and Obadiah was a popular figure, who risked defiance of the Restoration penal laws to pursue his ministry. Indeed in 1682, ironically the year in which his son's *magnum opus* was published with a dedication to King Charles II, Grew was imprisoned for six months.[5] Nehemiah inherited his father's dissenting convictions, retaining them throughout his life.[6]

In 1658, Nehemiah went up to Pembroke College, Cambridge, acquiring a B.A. in 1661 – 1662. He then presumably returned to Coventry, but little is known of his career over the next decade except for what he later recorded about the development of his scientific interests. Though he claimed many years afterwards to have 'addicted my self, to the Contemplation of Nature, from my Youth upward', there is no evidence of any association between him and others with related interests while he was at Cambridge, and it was apparently only in 1664 that he began his

[3] Hunter, *Science and Society*, p. 41 and ch. 2, *passim*; Roger Hahn, *The Anatomy of a Scientific Institution* (Berkeley and Los Angeles, 1971), ch. 1.

[4] Since this paper was originally published, this letter has been included among the 'Additions and Corrections to Earlier Volumes' in *Oldenburg*, xiii, 431 – 3.

[5] *Calendar of State Papers Domestic, 1668 – 9*, p. 665; *Victoria County History of Warwickshire*, viii (1969), 373 – 6; A. G. Matthews, *Calamy Revised* (Oxford, 1934), p. 236. On Nehemiah's schooling, almost the only clue is a Greek-Latin glossary compiled by him in 1656, now British Library MS Sloane 1945.

[6] Hunter, *Science and Society*, p. 116; John Shower, *Enoch's Translation. A Funeral Sermon Upon the Sudden Death of Dr Nehemiah Grew* (London, 1712).

botanical investigations.[7] In 1668 he showed what he had done so far to Henry Sampson, his half-brother and another nonconformist, who had been ejected from his living in 1662 and had travelled abroad, studying medicine at Padua and Leiden, where he had acquired a doctorate earlier in 1668.[8] Sampson's up-to-date knowledge of medical and scientific developments is shown by two surviving letters from him to Edmond (later Sir Edmond) King, F.R.S., and, when Grew showed Sampson his work shortly after Sampson's return to England, he not only encouraged Grew but was also able to draw his attention to a seminal passage in a book by Francis Glisson which further stimulated Grew.[9]

It was also Sampson who, two years later, acted as a mediator between Grew and the Royal Society. Whether or not Grew himself suggested this, perhaps stimulated by relevant research recently published in *Philosophical Transactions*, is unknown.[10] But, in any case, he showed Sampson work that he had by now completed and Sampson showed it to the Society's Secretary, Henry Oldenburg, that indispensable 'intelligencer' of scientific projects. Oldenburg showed it to John Wilkins, Bishop of Chester, who had long been one of the leading figures of the Society, and it was Wilkins who produced it at a meeting of the Society and desired to see more. It was also read by the Society's President, William, Viscount Brouncker, and at a Council meeting on 11 May 1671, Grew's *The Anatomy of Vegetables Begun* was licensed for publication and ordered to be printed by the Society's printer.[11]

How Grew had been making a living during these years is unknown. It is probable that he had already entered medical practice in the Coventry area, though unfortunately no evidence of this appears to survive. If so, however, he must have learnt his skills by apprenticeship to another practitioner, as he himself had no academic qualification in the field, and in 1671 he resolved to obtain one. Excluded from Cambridge by his nonconformist convictions (and perhaps also by the time and expense involved in getting a doctorate there), he followed in Sampson's footsteps by going to Leiden, where doctorates were much more readily available — where, indeed, candidates who could prove medical knowledge and present a thesis were given a degree with very little formality. Possibly the stimulus to Grew's decision came from Sampson, who is known to have been on the continent in May 1671.[12] Grew was registered as a candidate at Leiden on 6 July 1671, and on 14 July he submitted his doctoral disputation, *De Liquore Nervoso*: this was

7 N. Grew, *Cosmologia Sacra* (London, 1701), Sig. A1v; *The Anatomy of Plants* (London, 1682), sig. a1. On Grew and Cambridge science, see C. E. Raven, *John Ray, Naturalist* (2nd ed., Cambridge, 1950), pp. 54 – 5, 200 – 1, and compare Charles Webster, *The Great Instauration* (London, 1975), pp. 144 – 153.

8 Matthews, *Calamy Revised* (n. 5), p. 425.

9 Sampson to King, Montpellier, 9/19 Dec. 1665, Leiden, 3/13 Jan. 1668, E.L. S. 1. 39 – 40; N. Grew, *Anatomy of Plants* (n. 7), sig. a1v.

10 See Raven, *John Ray* (n. 7), p. 188.

11 Grew, *Anatomy of Plants* (n. 7), sigs. a1v – a2; Birch, *History*, ii, 480 (there are no earlier references to Grew).

12 Kisner to Oldenburg, 12 May 1671, *Oldenburg*, viii, 38.

issued in printed form and there is a copy of it in the British Library.[13]

Proudly bearing the title of Doctor, Grew returned to England. Later in 1671 he was proposed Fellow of the Royal Society by Robert Hooke, whom he must already have met, since he refers to him in the preface to *The Anatomy of Vegetables Begun* — which must by this time have been in the press — as his 'much Honoured Friend'. Grew was elected on 16 November and admitted in person on 30 November.[14] Shortly afterwards, on 7 December, Spencer Hickman, the printer, presented copies of *The Anatomy of Vegetables Begun* to the Society, to the President, Brouncker, to the Secretary, Oldenburg and also to Wilkins: the latter sent Grew a letter of acknowledgement on 12 December 1671, praising the book and urging Grew to continue his researches.[15] Meanwhile Grew had returned to Coventry for the winter, whence he wrote a letter to Henry Oldenburg on 12 March 1672 with 'some brief Observations touching the nature of Snow'.[16]

During the early months of 1672 Wilkins must have given further thought to Grew's project and the ways in which the Royal Society could help him with it while also doing itself a good turn. Since Grew's research had been brought to the Society's attention, similar findings had also been submitted to the Society by Malpighi from Italy, thereby driving home the potential of research in the field.[17] Sometime at this stage Wilkins hit on the idea of trying to raise a salary for Grew by soliciting special subscriptions from Fellows, so that he could come to London and complete his work under the Royal Society's auspices: he may have recalled that extra contributions had earlier been invited from members to help support Hooke in the 1660s, which did for a short while materialize.[18]

Grew later cited a letter from Wilkins of 18 February 1672 telling him that 'the *Society* were pleased to engage me to proceed' with his work. Whether the idea of subscriptions was mooted at that point is unknown, however (the letter itself does not survive), since it was not until 10 April 1672 that a proposal along these lines was formally put to the Society's Council.[19] But it was then approved, and on 18 April the Council's decision was relayed to the Society as a whole: the appointment was to be for a year; the salary was to be £50; and a form was 'now ready for the subscriptions of such, as would freely contribute to so good a work' (Wilkins, the proposer of the scheme, was 'desired to take care of the subscriptions').[20] The

[13] R. W. Innes Smith, *English-speaking Students of Medicine at the University of Leyden* (Edinburgh, 1932), p. 102; British Library shelf-mark 1185.g.11 (19).

[14] Birch, *History*, ii, 485, 490, 496; Grew, *The Anatomy of Vegetables Begun* (London, 1671), sig. a1.

[15] Birch, *History*, ii, 498; Grew, *Anatomy of Plants* (n. 7), sig. a2.

[16] Grew to Oldenburg, 12 Mar. 1672, in *Oldenburg*, viii, 586 – 9; see also Birch, *History*, iii, 41.

[17] Birch, *History*, ii, 499, 501; Grew, *Anatomy of Plants* (n. 7), sig. a2v.

[18] Birch, *History*, i, 453, 472 – 3; see below, ch. 9, pp. 289 – 90.

[19] Grew, *Anatomy of Plants* (n. 7), sigs. a2v – a3; Birch, *History*, iii, 42.

[20] Birch, *History*, iii, 47.

response to this invitation is evident from a letter from Grew to Oldenburg of 13 September 1673, which shows that the following Fellows volunteered, presumably in addition to Wilkins himself, who had died by 1673 and was therefore irrelevant to the question of renewed subscriptions which was then being canvassed: Lord Brouncker; Robert Boyle; Samuel Pepys; George, Baron Berkeley, a statesman slightly active in the Society; Daniel Colwall, the Society's Treasurer; Charles Howard, brother of the Duke of Norfolk; Seth Ward, Bishop of Salisbury; and Thomas Barrington and Sir Kingsmill Lucy, two Fellows of gentry status who otherwise contributed little to the Society. The exact sum that each promised is uncertain, but if each paid an equal contribution it would have been £5.[21]

It is easy to see why the scheme appealed to Wilkins. Quite apart from his undoubted disinterested concern to promote the advancement of scientific knowledge, there were also potential benefits for the Royal Society. The Society had long been attacked by sceptics questioning 'What have they done?'[22] and a research programme like Grew's, directly sponsored by the Society, would help to answer such objections. Certainly Grew's findings were subsequently proudly relayed by Oldenburg to his correspondents, while Grew's activities also provided Oldenburg with an opportunity to stimulate the kind of exchange of views by letter in which he specialized, in this case involving Grew and the York naturalist, Martin Lister.[23] An additional incentive related to the Society's meetings, at which activity was flagging by 1672, and Wilkins, as one of the Society's principal organizers, must have seen botanical investigations like Grew's as particularly suitable for report and exhibition stage by stage as an 'entertainment' to Fellows: this was an attraction lacking in more austere scientific research. Indeed, Grew's subject-matter was of special fascination: though he had so far made his observations with the naked eye — only while sponsored by the Royal Society did he start using a microscope — his findings were similar to the revelations about the structure of everyday things in Robert Hooke's famous *Micrographia*, which had aroused much curiosity since its publication in 1665.[24] One can only speculate whether Wilkins, an indefatigable advocate of a more comprehensive Church,[25] was particularly keen to patronize a young nonconformist like Grew.

By the time this offer was made, Grew was definitely engaged in medical

21 *Oldenburg*, x, 200. On these Fellows, see Hunter, *Royal Society*, p. 15 and catalogue entries 1, 2, 187, 34, 48, 123, 29, 279 and 268. On the sums promised see also n. 43.

22 Hunter, *Science and Society*, p. 90.

23 *Oldenburg*, vols. viii *et seq.*, Index s.v. 'Grew, Nehemiah'; ibid, x, xxv, 93 – 7, 200 – 10, 301 – 9, 385 – 97.

24 Hunter, *Royal Society*, pp. 37 – 8; Lindsay Sharp, 'Sir William Petty and Some Aspects of Seventeenth-century Natural Philosophy' (Oxford D.Phil. thesis, 1977), pp. 255 – 260; M. H. Nicolson, 'The Microscope and English Imagination' in *Science and Imagination* (Ithaca, 1956), pp. 155 – 234. On the differing suitability of subjects for presentation at meetings see R. G. Frank, jr., 'Institutional Structure and Scientific Activity in the Early Royal Society', *Proceedings of the 14th Congress of the History of Science, 1974* (Tokyo, 1975), iv, 91 – 4.

25 See Hunter, *Science and Society*, pp. 115 – 6.

practice in Coventry. But now he was persuaded by the Society's offer of a stipend to abandon his career there and come to London to pursue his researches, and at a series of meetings in the summer and autumn of 1672 he displayed his ongoing work to the Society.[26] As Henry Oldenburg reported in a letter of 6 July 1672 to the virtuoso Francis Willughby, 'Dr Grew is now amongst us at Arundel-House [the residence of the Duke of Norfolk, where the Royal Society met from 1667 to 1673], making now and then very good Observations upon Plants', and he reported in similar terms to Wilkins on 10 July and to Lister on 31 October.[27]

November 1672 saw the death of Wilkins, the main figure behind Grew's employment and the person responsible for collecting the subscriptions for it. Not only was this a great loss to the Royal Society as a whole, which had owed much to the Bishop's indefatigable efforts on its behalf:[28] it was a particular blow to Grew, who was now left high and dry in London, only partly into an immense research project and with the certainty only of a single year's subscriptions, not all of which had been paid. In fact, as is underlined by this failure of all the contributions to materialize, 1672 had been a particularly unfortunate year for the launching of the project, since the Stop of the Exchequer by the Treasury Commissioners that January had led to a great scarcity of ready money.[29]

This is the context of the newly discovered letter from Grew to Oldenburg, filed among the Royal Society's Miscellaneous Manuscripts, volume 4, item 71, and evidently delivered by hand since Grew has merely endorsed its cover 'For my honoured Friend Henry Oldenburge Esq.'. Though it is undated, this must be the missive that stimulated a consideration of the future of Grew's employment at a Council meeting on 18 December 1672.[30] The text of the letter is as follows, and it illustrates clearly the difficulties that Grew now felt about what he had undertaken, and particularly the problems about patronage of a long-term project by an institution like the Royal Society.

> Sir, That you may the better comprehend the business I spake to you of the other day, I have here sent you my mind more fully; which I desire may be reported at a Meeting of the Councel of the Royal Society; vizt
>
> That when I was sat down & setled to the practise of Physick in Coventry; I was sent for hither by the Bishop [of] Chester in the name of the R.S. by whome I have here beene imployd for the space of a year & upward. In whose service, except I be continud at least for 5 years more; as I have beene here already, so I shall be still to my further loss: & that for these reasons,

26 Birch, *History*, iii, 49, 50, 51, 55, 56, 58.

27 *Oldenburg*, ix, 147, 153, 311.

28 Hunter, *Royal Society*, p. 36.

29 See D. C. Coleman, *Sir John Banks* (Oxford, 1963), p. 63. The shortfall in Grew's subscriptions is mentioned in the letter quoted below.

30 Birch, *History*, iii, 69 – 70.

1 The subscriptions made, if they had been all paid would not (as I have tryd) have amounted to the profit which would have already arose from my practise in the Country.

2 And my expences here have beene much greater, then there they would have beene; where I should have had my Fathers House & Table free.

3 By Leaving the City & Country, where I was educated, & where the number of my friends & acquaintance grew up with me for the space of 30 years, I have given them occasion to choose them another Physitian in my stead, & one at least hath made my absence his opportunity to present himself to their choyce, by comeing amongst them. Nor is it to be doubted, but that he will with the chiefe Physitian of the City & Country about it (with whome I had good correspondence) enter friendship, for the common good both of themselves & their Patients. Whence my Practise will certainly be much diminished, if not in a manner wholly lost.

4 To my loss of profit, I shall accumulate the hazard of my reputation also in some respects; some may think me to be changable & unsetled whilst they see me rowle from place to place; & many others; that haveing not answered the expectations of the R.S. they have turned me downe; many I say, whome I shall neither have an opportunity, nor list to satisfy to the contrary. And I should be loath to be accounted giddy, unless I had made myself so: or unserviceable unto those, whose service I sought not; verily because I durst not, think my my [sic] self unworthy [sic] of it; not did in the least designe or think of, before I had the honour of haveing it offerd. For which reasons, had I foreseene the uncertainty of my longer continuance here, I should not have stired from the Country, where I knew what to trust too. And if I must goe away a looser, & hazard all; I will still choose to doe it amongst my old friends. Where notwithstanding, I shall have no higher ambition then to my utmost ability to subserve the Interest & Designes of the R.Society.

But if they think fit to honour me with their continued trust (what the Bishop of Chester, speaking I presume the mind of the Society, as well as his own, told me at my first comeing up, was intended to be); I crave leave to make these two propositions;

First, that since what I have undertaken will be a work of time, two hundred & fifty pounds, in consideration for five years, may be raised, as the Councel shall see best, & payd me now all together.

Or if this cannot be done, then that fifty pounds per annum, be continud for five years. This my Lord Brouncker hath already in his very great favour engaged for, some time since when I proposed it to his Lordship.

But because my Lord through important business continually upon his hands may easily forget; I desire therefore, that his Lordship, in pursuance of his promise, will be pleasd so soone as may be, to call a Councel, & that my propositions may be reported, & that some way may be thought of for the security of the full, & quarterly payment of the monys.

And although I account my worke, & desire[31] my salary, but for five years; yet I fore see it will take up more time, which however I shall take my self obliged to dedicate to the Societys service.

I remaine, Sir,
Your affectionate obliged friend
& humble servant
Neh. Grew

Perhaps most important is the clarity with which the letter reveals that, but for Wilkins' intervention, Grew would have happily buried himself in provincial medical practice in the city of his birth. Though this might have been intended at least partially as a device to force the Council's hand, his comments have an air of plausibility. The letter elaborates on the financial sacrifices that he had been obliged to make in leaving Coventry and the threat that his absence presented to his clientele there, though my enquiries in the Coventry area have failed to throw further light on the other local physicians to whom he refers.[32] The longevity of his medical practice at Coventry is unfortunately not clarified: his worries about accusations that he was 'changable & unsetled' might seem to suggest that he had left his practice soon after beginning in it, but his comments about the threat from a newcomer imply that he himself had a well-established clientele, presumably built up over a period of years. The reference to the likely profit 'which would have already arose' from his practice is puzzling, since later parallels suggest that even a starting annual income for a country doctor might have been expected to be substantially higher than the £50 offered by the Royal Society.[33] It should be noted in passing, incidentally, that his claim that he had been in London 'a year & upward' must be an exaggeration.

Equally important is the uncertainty that Grew now saw in the future of his London employment, which he anticipated would damage his reputation in Coventry even if he could get his patients back. The letter emphasizes Wilkins's personal role in the matter, leaving Grew unsure whether, now that Wilkins was no longer on the scene, the Society would continue its support. Though Wilkins had promised continuance — 'speaking I presume the mind of the Society, as well as his own' — this had evidently never been

[31] This replaces 'except', which has been deleted. After 'monys.' in the line above, the words 'So as' have been deleted, and in the line above that, 'such' has been deleted between 'some' and 'way'.

[32] For help in this connection I am indebted to Dr Joan Lane and to the staff of the Warwickshire County Record Office, the City of Coventry Libraries Department and the Lichfield Joint Record Office.

[33] See Edmund Hobhouse, ed., *The Diary of a West Country Physician* (London, 1934), p. 26.

formally confirmed. We thus see clearly the personal structure of the early Royal Society, its dependence on individuals for major initiatives of this kind, and the problems that could sometimes result.

Hence Grew's proposal that in order to complete the project, 'a work of time', he be guaranteed five years' income — either by the immediate payment of a single lump-sum, 'raised, as the Councel shall see best', or by a guarantee that the subscriptions of his voluntary supporters would continue for five successive years. His tactful reference to Brouncker perhaps implies a certain lack of trust in the Society's President, while his stress on the need for the security of quarterly payments of the money must have been based on the disappointing outcome of the promises of the 1672 subscribers. The five-year covenant proposal must in itself have been problematic, since it is doubtful if even quite wealthy Fellows would have wanted to saddle themselves with this continuing commitment, and it is clear from the upshot that Grew was disappointed in this and at best was only promised renewals on an annual basis.[34] The £250 lump-sum must have been still less feasible, unless Grew had already had wind of Wilkins's £400 legacy to the Society and wondered if that might be devoted to this purpose.[35] Yet the letter shows how difficult it was, without such assurances, for a young man to commit himself to a long-term research project.

As this illustrates, in this as in so much else, the Royal Society suffered in the Restoration period from a gap between reputation and reality. Grew had been lured from Coventry by the Society's prestige, as is evident not only from his claim that he had 'no higher ambition then to my utmost ability to subserve the Interest & Designes of the R.Society' but also from his presumption that in the event of the project folding, people would assume that it was the Royal Society that had turned him down rather than vice versa. Yet in actual fact, the Society's dependence on voluntary subscriptions made it vulnerable to the whims of members even when it came to the support of a research project as apparently popular as Grew's. The outcome of Grew's plea for financial security shows how little members could be relied on in such circumstances, and it is not surprising to find the organizers of the Royal Society in these years casting repeated wistful glances at the more effective way in which science was managed across the channel.[36]

When Grew's letter was discussed at the meeting on 18 December, the Council expressed its satisfaction with his achievement so far, 'and that the council would farther recommend him to the Society, to continue him another year, if the subscribers would please to continue their contributions'.[37] It was perhaps also suggested, either at the meeting or thereafter,

[34] See Oldenburg to Grew, 6 Nov. 1673, *Oldenburg*, x, 336, which refers to promises of 'a new contribution'.

[35] See Birch, *History*, iii, 70, 93, 110 – 11, and Sir Henry Lyons, 'The Society's First Bequest', *Notes and Records*, 2 (1939), 43 – 6.

[36] Hunter, *Science and Society*, p. 40.

[37] Birch, *History*, iii, 69 – 70.

that it would be helpful if Grew could outline more precisely what his entire research plan would involve. So, at two successive meetings after Christmas, Grew presented a work-scheme presenting in intelligible form a complete programme for the study of plants, incorporating both the research he had already done and that which he yet hoped to carry out.[38] This was published in 1673 as *An Idea of a Phytological History Propounded*, and it was subsequently placed at the start of *The Anatomy of Plants* (1682) as an introduction to Grew's work as a whole.

In it, Grew argued passionately for the importance of what he was doing, avowing his hopes 'of being somewhat instrumental to the improvement of Medicinal and other wholesom Knowledge: if peradventure, as we increase herein, we may become better and more happy', as he put it in his fulsome Epistle Dedicatory 'To the Right Honourable William, Lord Viscount Brouncker, The President, And to the rest of the Fellows of the Royal Society'.[39] But the point of such an exposition at this stage was presumably to show Fellows why it was important to go on supporting the project, and at the second meeting at which the *Idea* was read out, on 15 January, 'the Society considering . . . that this prosecution of the anatomy of plants was very suitable to their design, and that Dr Grew was very fit to be farther employed therein, ordered, that Mr Collins should be desired to attend the several subscribers for a contribution to this work, and from the Society recommend to them the continuance thereof' (John Collins was a government clerk and mathematician who was exempted from normal membership dues and hence could be called on to do administrative chores for the Society).[40]

What happened next is a little uncertain. In May and June 1673 Grew made two further displays of his work in progress at Royal Society meetings, but by August he had left London for Coventry.[41] The first set of subscriptions, promised as a result of the meeting of April 1672, must have come to an end in the summer of 1673, and Grew's departure from London suggests that Collins's efforts had failed to lead to definite promises of renewal.

Writing to Oldenburg from Coventry on 13 September 1673, Grew declared: 'I shall be loath to stirr from Coventry, till I have further ground to believe the subscriptions will be made up, then yet I have had'. In this published letter, Grew enumerates the former contributors, of whom only two — Brouncker and Boyle — had so far agreed to continue for a further year. It is interesting that he was keen for Oldenburg to present copies of his *Idea* to other former subscribers, thus confirming that the book was at least

38 Ibid., p. 72. The *Dictionary of National Biography*, and following it, Bolam, 'Botanical Works' (n. 1), p. 221, mistakenly date these readings to 1672 rather than 1673, thereby giving a misleading picture of the chronology of Grew's work.
39 Grew, *An Idea of a Phytological History* (n. 2), sigs. A3 – 4.
40 Birch, *History*, iii, 72. On Collins see Hunter, *Royal Society*, catalogue entry 235.
41 Birch, *History*, iii, 87, 91 – 2; *Oldenburg*, x, 170 (reference to a lost letter from Oldenburg to Grew of 30 Aug. 1673).

partly intended to stimulate support for the continuation of the project.[42] By 6 November, as revealed by a letter of that date from Oldenburg to Grew, two more had subscribed (Howard and Barrington) and another two were almost certain to (Colwall and Ward), thus nearly making up the whole sum. But, as Oldenburg added in as pessimistic a vein as Grew's, 'However, I must not be looked upon as promising any certainty, or engaging for any body, upon the long experience I have of the lubricity of all humane affaires'.[43] Indeed Grew delayed his return to London longer than Oldenburg expected, as Oldenburg explained to Martin Lister when accounting for the delay in dealing with a letter that he had held in London rather than forwarding to Coventry in expectation of Grew's imminent arrival.[44] Throughout the autumn of 1673, the whole project seemed in jeopardy.

It is clear that — particularly without the powerful personal partisanship of Wilkins — the idea of sponsoring research by special voluntary subscription was impractical; one should recall that by the early 1670s mounting arrears even of Fellows' ordinary dues were becoming a major worry for the Society's organizers.[45] Indeed, it is significant that the Royal Society seems never again to have attempted an enterprise like this sponsorship of Grew. Once again, the gap between good intention and actual outcome in the Society's programme was revealed and the novel idea of a voluntary institution directly sponsoring research proved over-ambitious until aided by major benefactions in modern times.

Instead, Grew's research was saved by intervention from a different quarter. This was announced by Oldenburg in his letter of 6 November, when he told Grew: 'Mr Hook desired me to give you notice, that he hath some thing in view worthy of your acceptance, which when he shall come to speake with you about, you will, he saith, acknowledge to deserve the making a step to London . . . Mr Hook seems to speak so promisingly of the busines . . ., and I am apt to think it so consistant with your care of Botanic Experiments for our Society, that I find more cause for animating you to come and see, than for the contrary'.[46] Just what was involved is revealed by an entry in Hooke's diary for 10 October 1673. Evidently Jonathan Goddard, Professor of Physic at Gresham College, was seeking a deputy to lecture for him for an annual emolument of £40. Hooke's laconic note

[42] *Oldenburg*, x, 200.
[43] ibid., x, 336 – 7. Oldenburg here implies that a total of eight subscribers would be sufficient to make up the total: if so, and if each gave an equal sum, it must have been £6 5*s* 0*d*. See also above p. 265.
[44] See Oldenburg to Lister, 18 Nov. 1673, *Oldenburg*, x, 364, and ibid, x, 397n., Oldenburg's endorsement to a letter from Grew to Oldenburg of 11 Dec. 1673. Oldenburg's letter to Grew of 6 Nov. (ibid, x, 337) reveals that at that point Oldenburg was expecting Grew to return to London in connection with a translation of his book (*An Idea?*): but no such translation materialized.
[45] Hunter, *Royal Society*, p. 37.
[46] *Oldenburg*, x, 336 – 7. This allusion is not elucidated by the editors, who wrongly presume from Grew's return to London in December 1673 that the scheme for renewing subscriptions was successful (p. 201, n. 1).

implies that he spoke to Goddard on behalf of another young physician, Richard Blackburne, but in the event Goddard chose Grew for the job.[47] The offer was evidently relayed to Grew, and by 11 December 1673 he was back in London, by 18 December once again displaying his botanical findings to the Royal Society.[48]

The nature of this employment deserves elucidation, for it illustrates how — failing grants specially raised for important research — a substitute could be found through the manipulation of endowments already available. Gresham College had been founded in the Elizabethan period by Sir Thomas Gresham to supply educational facilities in the City of London: his bequest provided both for the erection of a building and for the establishment of professorships in various subjects. Over the years, the professorships had tended to become attractive as a source of income and lodgings for academics and others, who frequently displayed a rather casual attitude towards the teaching they had been appointed to do, instead devoting themselves to their own affairs: London could not offer much preferment of an academic nature at this time, which made the Gresham posts all the more inviting. By the Restoration, many professors were absentees, and it was evidently commonplace for proxies to be hired to give lectures — the only retrospective justification perhaps being that this provided stipends for men like Grew and John Flamsteed (who, as Astronomer Royal, was another inadequately-endowed researcher) and hence spread the college's revenues more widely.[49]

Grew's employment as Goddard's deputy came to an end when Goddard died suddenly on 24 March 1675. An election was held on 27 March to replace him, in which Grew was a candidate. But the long list of aspirants' names given by Hooke in his diary reveals just how popular these virtual sinecures were, and the chances of a young outsider like Grew were small. In the event, the doctor and religious writer John Mapletoft was chosen, and he held the Professorship of Physic till 1679.[50] Whether Mapletoft followed Goddard in using Grew's services at all is unknown. For at least part of his tenure of the post Mapletoft was in residence in the college, and extracts from the lectures that he gave survive: but in 1676 – 7 he went abroad and he perhaps then required a deputy.[51]

By 1679, when Mapletoft resigned, Grew had married and hence was no longer eligible to stand for a Gresham professorship; moreover Henry

[47] Hooke, *Diary*, p. 64.

[48] Cf. Grew to Oldenburg, 11 Dec. 1673, *Oldenburg*, x, 385 – 97; Birch, *History*, iii, 115, 116.

[49] Ian Adamson, 'The Foundation and Early History of Gresham College, London, 1596 – 1704' (Cambridge Ph.D. thesis, 1975), esp. ch. 7. See also his 'The Royal Society and Gresham College, 1660 – 1711', *Notes and Records*, 33 (1978), 1 – 21. On Flamsteed see Hunter, *Science and Society*, p. 133, and below p. 273.

[50] Hooke, *Diary*, pp. 154, 155.

[51] Adamson, 'The Royal Society' (n. 49), p. 5; John Ward, *The Lives of the Professors of Gresham College* (London, 1740), p. 275 and Appendix, pp. 120 – 32.

Paman, Mapletoft's successor, is known to have lectured in person.[52] But at least by 1680, Grew had found other employment through Gresham College. A note from Walter Pope, Gresham Professor of Astronomy, to Grew of 21 January 1680 happens to survive among Grew's papers because he made notes on the back, and this reveals that Grew was then deputizing for Pope: he may have done so earlier, since Pope is known to have been infirm and from 1681 onward employed John Flamstead as his deputy.[53] The fact that this arrangement is known only from a chance survival makes it likely that Grew also found other sources of employment about which nothing is known. Some earnings may have come through Sir William Petty, to whom Grew wrote on 20 February 1678: 'You may wonder, that one, who is already so much obliged to you; should presume to sollicite your further trouble'.[54] It is also possible that Grew made some money by practicing his medical skills in London in the mid-1670s, though no evidence of this survives.

Grew's income from such sources was supplemented by haphazard payments from the Royal Society. Whether any further subscriptions under the Wilkins scheme materialized after 1672 is unknown. The only payment to Grew from an individual Fellow definitely known after that date came from Lord Brouncker under a different rubric. At a meeting of 3 December 1674, Council members were requested to present an experimental discourse to the Society themselves, or, if they preferred, to pay for someone else to do so, and the Society paid Grew £4 on Brouncker's behalf for a lecture of 28 October 1675 on 'the nervous liquor, its origin, nature, motion, and uses in the body', perhaps an English version of his 1671 Leiden disputation.[55] Grew had earlier been paid £4 for his lecture of 25 February 1675 on 'the structure of the cortical, ligneous and medular part of trees', and the cost of this may have been at least partly covered by contributions from other Council members.[56]

These sponsored lectures formed part of a deliverate campaign to improve the quality of business at the Royal Society's meetings in 1674–5 by arranging for a series of papers to be presented: Grew also participated in this by giving a paper on 10 December 1674 which was published in 1675

[52] Op. cit., pp. 279–80. Grew's marriage certificate, dated 20 April 1678, will be found in British Library MS Sloane 1941, fol. 17.

[53] British Library MS Sloane 1963, fol. 86v. On Flamsteed's employment, see Adamson, 'Foundation and Early History' (n. 49), p. 227; E. G. Forbes, ed., *The Gresham Lectures of John Flamsteed* (London 1975), p. 1 and passim.

[54] British Library MS Sloane 1942, fol. 9.

[55] Birch, *History*, iii, 158, 228–9, 243. See also Oldenburg to Lister, 20 June 1674, *Oldenburg*, xi, 36.

[56] Birch, *History*, iii, 192. The accounts reveal that the Earl of Ailesbury and Sir John Lowther each paid £2 in lieu of giving a lecture, which would exactly cover the fee paid to Grew, but there are complications: Lowther did not pay his £2 till 29 April 1675, and though Ailesbury's reached the Society on 20 January it was earmarked for a lecture on 11 February, a date on which William Croone performed — but there is no reference to the payment of any fee to Croone in the accounts. See Birch, *History*, iii, 175, 176, 181, 243.

as *A Discourse . . . Concerning the Nature, Causes, and Power of Mixture.*[57] On the whole — like others of its kind in the Society's early history — this attempted revitalization flagged after a few months,[58] but, for Grew alone, it had more lasting effects. The precedent of paying for research papers set during the reformist episode of 1674 – 5 was taken up by the Society afterwards, and between April 1676 and December 1677 the Society's accounts show that Grew was paid a total of £48 from the Society's general funds for twelve lectures in which he adumbrated his research, a practice that was as unique in the Society's early history as had been Grew's support by voluntary subscriptions in 1672.[59] In other words, although on a piecemeal basis and strictly in return for 'entertainment' at meetings, the Society effectively provided a small honorarium for Grew, supplementing his earnings from Gresham teaching and possible other sources, and thereby encouraging him to finish the work that he had begun.

By more piecemeal and haphazard means, Grew was thus provided with a livelihood, and there is no evidence that, after 1673, he was ever tempted to return to Coventry to settle there. Hence his botanical programme went forward. In August 1674 Oldenburg reported to a correspondent that 'Dr Grew poursues his inquiries concerning the Texture of Trees and other Plants; which he does with great exactnes and curiosity', and the results were reported to the Royal Society in successive lectures in 1674 – 5, while the major part of his findings of this period reached print in 1675 as *The Comparative Anatomy of Trunks.*[60] In 1675 – 6 there is something of a gap — perhaps while Grew was preoccupied with earning his living — but in the next two years he returned to his task, aided by the lecture fees already referred to, moving from roots and stems onto what was to be published as the fourth part of his *Anatomy of Plants* in 1682, dealing with flowers, leaves and seeds. He also reported on related chemical investigations, which ranged from vegetable to animal and mineral substances: all these lectures

57 Ibid., iii, 161.

58 Hunter, *Science and Society*, pp. 45 – 46; Sharp, 'Sir William Petty' (n. 24), pp. 261 – 273, 401 – 7.

59 The accounts record payments of £4 for lectures on the following dates: 13 Apr. 1676 (Birch, *History*, iii, 313), 1 June 1676 (iii, 318), 26 Oct. 1676 (iii, 321), 9 Nov. 1676 (iii, 323), 21 Dec. 1676 (iii, 328), 18 Jan. 1677 (iii, 329 – 30), 8 Feb. 1677 (iii, 333: interestingly, the accounts record that only eight were present, a fact not noted in the minutes; might this be taken to imply that only good attendance justified expenditure on a lecture?), 8 Mar. 1677 (iii, 335 – 6), 3 May 1677 (iii, 338 – 9), 14 June 1677 (see below, n. 63), 21 June 1677 (iii, 342) and 6 Dec. 1677 (iii, 359). Some of these bills also included expenses for Grew's demonstrations, which were also separately paid for on 24 May 1673, 11 Feb. 1675, 16 April 1676, 2 Dec. 1676 and 13 Oct. 1678 (cf. Birch *History*, iii, 502). Between February and October 1675 Grew gave two lectures for which he was not paid (see ibid, iii, 195, 223). The accounts for 1674 are unfortunately missing, so any payments made to Grew in that year are not recorded.

60 Oldenburg to Herbert, 19 Aug. 1674, *Oldenburg*, xi, 70; Birch, *History*, iii, 122, 124, 130, 132, 192, 195 – 6, 223.

were included in *The Anatomy of Plants*, while two reached print in 1678 as *Experiments of Luctation*, 'a Specimen of a Natural History of Materia Medica' (to quote its preface) which reveals Grew's continuing medical concerns.[61]

By this time, Grew may have felt that he was approaching the end of his programme: the five years that he projected in 1672 were, after all, almost up, although one modern commentator has pointed out that he had hardly completed the entire scheme outlined in *An Idea of a Phytological History*, 'which would occupy centuries, not a mere handful of years'.[62] Hence in 1677 Grew is to be found embarking on a quite different line of research, 'the Comparative Anatomy of Stomachs and Guts'. This complete shift from plants to animals was welcomed by the Royal Society, who declared themselves 'very well pleased with this design, and this beginning of the execution thereof'.[63] But in the event Grew only ever gave three lectures on the subject, and the failure of this second major project to get off the ground indicates how fortunate had been the stimulus provided by his initial research grant in 1672, though new developments now added further complications.

In September 1677, matters were suddenly transformed by the death of Henry Oldenburg, Secretary of the Society since 1663. On 30 November Grew was appointed as joint Secretary with Hooke, and over the succeeding months Grew spent much time in carrying on the correspondence that Oldenburg had for so long transacted with scientists and virtuosi at home and abroad.[64] This duty seems to have prevented Grew from presenting any research of his own at meetings of the Society in 1678. His efforts in this capacity were rewarded with a £10 gratuity from the Society, though he failed to get the full 'Establishment' which he requested on 18 July 1678 according to Hooke, one of numerous laconic notes revealing some strain between the two men at this time which appear in Hooke's diary.[65] In 1678 (perhaps partly in order to dispel this tension), Grew was also given the job of cataloguing the 'repository' of natural objects and mechanical devices that the Society had accumulated since 1666: this also seems to have taken up a great deal of time, so that he increasingly left general secretarial duties to Hooke.[66]

61 Ibid., iii, 313, 318, 321, 323, 328, 329 – 30, 332, 335 – 6, 338 – 9, 359. Grew, *Experiments of Luctation*, ed. M. A. Hoskin (Heffer Scientific Reprint, Cambridge 1962), sig. A4v.

62 Bolam, 'Botanical Works' (n. 1), p. 226.

63 Birch, *History*, iii, 333. See also ibid, iii, 342, for some minutes which Birch surmised must refer to a presentation by Grew: the accounts refer to a payment of £4 to Grew for an anatomy lecture on that date and also for another on 14 June, and the garbled minutes in Birch, *History*, iii, 341, must refer to this.

64 Ibid., iii, 353. For Grew's letters see British Library MS Sloane 1942 and Royal Society Original Letter Book, viii, 34 – 94, *passim*; see also Grew to Lister, 20 Feb. 1678, Bodleian Library MS Lister 34, fol. 202.

65 Birch, *History*, iii, 450; Hooke, *Diary*, p. 367 and *passim*.

66 See Birch, *History*, iii, 450. On the repository see also ibid, iii, 310 – 11, and

Grew continued these epistolary and taxonomic activities till the middle of 1679, and from April 1679 his contributions return to a regular place in the Society's minutes, reaching a climax and then suddenly ceasing in the summer of that year.[67] His offerings at this time fall almost entirely into two categories: he presented a great deal of the material that he was dealing with for his catalogue of the Society's museum; in addition, he read papers on mineral waters in the London area and elsewhere, and this suggests that by now, if not earlier, Grew had begun to practice medicine in London (a passage in a letter to his wife of 26 August 1678 might be construed to indicate the same).[68] Since by this time the Royal Society had ceased to pay for his lectures, he must almost certainly have had some source of revenue in addition to that coming from his teaching at Gresham.

Why Grew's contribution to the business of the Royal Society virtually ceased in the summer of 1679 is not known: the fact that he continued to attend meetings but not to present papers thereafter suggests that it was not due to personal tensions of the kind that were so debilitating to the early Royal Society,[69] and it seems likelier that his withdrawal was due to an increasing preoccupation with other business. This was probably his medical practice, and from this point of view a significant development occurred on 30 September 1680, when Grew became an Honorary Fellow of the College of Physicians, thus greatly enhancing his status as a London doctor. Honorary Fellows had been introduced to the College in the Restoration period as a way of allowing physicians without Oxford or Cambridge doctorates to become part-members without taking the normal oaths: it was thus the only route into the College for a nonconformist like Grew. In 1675, further such elections were vetoed by the Crown, evidently in response to a petition for membership of this kind by a whole group of doctors with scruples about the oaths: the names of those involved unfortunately do not survive, but it is quite possible that Grew's was among them. This decision was overturned when the College sought fresh legal opinion in 1680, and it was now that the nonconformist Grew was able to become a Fellow; though he was suspended during the reign of James II, he was reinstated in 1689.[70]

With his election to the College and presumably a growing medical practice, it must have been clear to Grew that his time as an active scientist was over, and his two folios of 1681 – 2 — *Musæum Regalis Societatis* and *The*

Hunter, *Science and Society*, pp. 66 – 67. Hooke, *Diary*, p. 381 (24 Oct. 1678) notes: 'Grew promised to resign his Secretarys place'. Royal Society Original Letter Book, viii, shows Hooke replacing Grew as the main recipient of letters to the Society in 1679.

67 Birch, *History*, iii, 475 – 6, 480, 481, 482 – 5, 486, 489, 490 – 1, 492, 494 (p. 496 may also be relevant, but the minute is incomplete).

68 British Library MS Sloane 4062, fol. 192.

69 Hunter, *Science and Society*, pp. 44 – 45.

70 Sir George Clark, *A History of the Royal College of Physicians of London* (Oxford, 1964 – 66), i, 312 – 17, 339 – 40, 347 – 48, 357, 372.

Anatomy of Plants — have a certain finality about them: as Grew himself put it in a letter to Martin Lister before *The Anatomy* came out in 1682, 'I am now about to print one more Book; & then it will be time to have done'.[71] The *Musæum Regalis Societatis* included as an appendix such as he had completed of 'The Comparative Anatomy of Stomachs and Guts', while *The Anatomy of Plants* brought together all his research on plant anatomy and related topics, whether published or not. Both, it may be noted, were lavish illustrated books published by subscription, and they were almost the first scientific books to be brought out in this country by such means. It is interesting to find Grew thus associated with this extension of subscription from the limited group who had supported him in 1672 to a much wider circle of scientific enthusiasts, each of whom made a smaller contribution to get expensive works of limited appeal into print. Grew himself organized and collected the contributions, and his successful employment of this publishing technique looks forward to its commonplace use in the eighteenth century: indeed the contrast between the difficulties of the 1672 scheme and the successful promotion of these books could itself be seen as symptomatic of a shift away from private to more public patronage in the Restoration period.[72]

Though Grew attended the Royal Society frequently in the 1680s and occasionally in the 1690s and gave a couple of papers, he was no longer doing much research.[73] Already in 1682, when arranging the publication of *The Anatomy of Plants*, he was complaining of lack of time, and this may explain why, although various appendages to the *Musæum Regalis Societatis* are mentioned, none ever appeared in print.[74] Such writing on scientific topics as Grew did thereafter was fitted into the spare time from his medical practice, as he pointed out in the preface to his last published book, *Cosmologia Sacra*, an elaborate exposition of God's goodness in his creation published in 1701.[75] He prepared a few manuscript tracts on religious and

[71] Grew to Lister, 8 April 1682, Bodleian Library MS Lister 35, fol. 67.

[72] See Hunter, *Science and Society*, p. 69; F. J. G. Robinson and P. J. Wallis, *Book Subscription Lists: A Revised Guide* (Newcastle upon Tyne, 1975). Copies of the prospectuses for Grew's books will be found in British Library MS Harleian 5946, fols. 162 – 3, 164 – 5. Various extant letters deal with the collection of these subscriptions: see particularly Grew to Lister, 18 May 1681, Bodleian MS Lister 35, fol. 49, which gives a list of Yorkshire subscribers to his *Musaeum* and shows that the list of subscribers included in the published prospectus is far from complete.

[73] For the papers, see Birch, *History*, iv, 225; Royal Society Copy Journal Book, viii, 57, 299; *Phil. Trans.* 14 (1684), 566 – 7, 17 (1691), 543 – 4. See also ibid, 17 (1693), 760 – 1, 815, and Grew, *Tractatus de Salis Cathartici Amari in Aquis Ebeshamensibus . . . Natura et Usu* (London 1695).

[74] Grew to John Evelyn, 5 May 1682, Christ Church, Oxford, Evelyn Collection MS 3.2, fol. 16. On the appendages to the *Musaeum* see Grew to Lister, 18 May 1681, Bodleian MS Lister 35, fol. 49; Aston to Lister, 29 Mar. 1683, ibid, fol. 92; Birch, *History*, iv, 171, 250, 455, 557; Royal Society Copy Council Minutes, ii, 99.

[75] Grew, *Cosmologia Sacra* (n. 7), sig. A1v.

economic topics and collected botanical specimens and engraved gems,[76] but medicine was his main concern: though only fragmentary information on his practice survives, it was as an active and conscientious doctor that he was celebrated in the funeral sermon preached after his death in 1712.[77]

In the way that medicine gradually ousted scientific research, Grew can be readily paralleled among physician-scientists of his day: this was no secret to contemporaries, as is shown by Oldenburg's aside when reporting to Martin Lister on the research of Dr Daniel Coxe, that more results were to be expected 'if his practice doe not divert him from it'.[78] What makes Grew's case unusually interesting is the means by which he was deflected from the provincial medical practice to which — as it is clear from the letter published in this article — he would otherwise have committed himself, and the crucial role played by the Royal Society's patronage in ensuring that his research matured and that Grew made his mark as one of the leading botanists of the day before his move into London medical practice in the 1680s. Hardly less revealing is the contrast — here as in so many matters — between the Society's ambitions and its actual performance. The scheme launched in 1672 almost proved abortive, and only the growing public audience for science made possible the publication of Grew's lavish books of 1681 – 2: but the Society nevertheless achieved a certain success by muddling through, combining the benefits of contact with Gresham College with sporadic payments to Grew for entertainments at meetings. The difficulties are revealing, but they only increase one's respect for what the Society achieved with the limited resources at its disposal — with regard not only to Grew's research but to the promotion of the new science in late seventeenth-century England as a whole.

76 Hunter, *Science and Society*, pp. 116, 122 – 23; on botanical collecting see esp. British Library MS Sloane 2145, fols. 61, 90, 91 and *passim*. His collection of gems was sold with his library after his death: no copy of the catalogue of this auction now appears to survive, but it is referred to by the Coventry antiquary, William Reader, in a note on Grew in Bodleian MS. Top. Warwicks. c. 5, fol. 60 (I am indebted to Drs Lane and Alcock for this reference, and to David McKitterick for help in attempting to trace the catalogue).

77 Grew appears in the lists of London doctors published *passim* in John Houghton's *Collection for Improvement of Husbandry and Trade*, from vol. 5, no. 97 (8 June 1694) to vol. 8, no. 171 (8 Nov. 1695). British Library MS Sloane 1949 is a medical notebook of Grew's, while MS Sloane 2145 contains letters, etc., concerning his medical practice in 1709 – 10; there are recipes attributed to Grew in MSS of Richard Mead and Sir John Pringle at the Royal College of Physicians, MSS 173 and 496. See also papers in British Library MSS Sloane 1941 and 1963 and John Shower, *Enoch's Translation* (n. 6), esp. pp. 12, 13.

78 Frank, 'Institutional Structure' (n. 24), p. 97; R. G. Frank, jr., *Harvey and the Oxford Physiologists* (Berkeley and Los Angeles, 1980), pp. 208, 283; Oldenburg to Lister, 20 June 1674, *Oldenburg*, xi, 36.

9

SCIENCE, TECHNOLOGY AND PATRONAGE: ROBERT HOOKE AND THE CUTLERIAN LECTURESHIP*

The lectureship endowed by Sir John Cutler and held by Robert Hooke in conjunction with his post as Curator of Experiments to the Royal Society from 1664 onwards is one of the most famous examples of scientific patronage in the seventeenth century. Moreover, the benefaction was undoubtedly important. It was important to the Royal Society because it seemed to exemplify the kind of support which it was hoped that the new science would enjoy in contemporary society, not least among 'men of *Trafick*' like Cutler, though in the event no other such benefaction followed.[1] As it was, Cutler's gift enabled the Society to enjoy Hooke's services at less cost to their own tight budget than would otherwise have been the case, while its regular activities were valuably supplemented by the presentations provided by Cutler's gift. For Hooke, the lectureship was a worthwhile source of income, while it also inspired his major publication programme during his lifetime, apart from his celebrated *Micrographia* (1665) — his various tracts published in the 1670s and collected in 1679 under the title *Lecturae Cutlerianae*. In addition, it will be argued in the course of this paper that various other lectures by Hooke, some of them published since his death while some remain unpublished, were stimulated by the demands of the Cutlerian Lectureship: many of these are characterized by their attempt to give a programmatic view of the desiderata of science, and they provide important information on Hooke's methodological views which he might otherwise not have recorded.

If Cutler's initial endowment is well-known, it is equally familiar that it proved problematic. Hooke's grievance about the non-payment of the annual salary that Cutler was supposed to provide for him is familiar from his diary for the 1670s, while we are told by his biographer, Richard Waller, that the law-suit which ensued caused Hooke much distress in his later

* I am especially grateful to Christopher Whittick for his careful attention to this paper from a legal point of view. Dr J. H. Baker also kindly answered queries relating to it, while John Harwood, David Oldroyd and Paul B. Wood helpfully commented on it in draft. Lastly, I am indebted to the library staff at the Royal Society, especially Alan Clark and Sally Grover, and to the staff at the Public Record Office for their help with the manuscript sources utilized here.

[1] Sprat, *History*, pp. 129 – 30, cited, for instance, in Margaret 'Espinasse, 'The Decline and Fall of Restoration Science', *Past and Present*, 14 (1958), 86 – 7. For a caveat, see Hunter, *Royal Society*, p. 127, n. 10.

years.[2] In the past, it has been presumed that these problems were caused entirely by Cutler, and that Hooke and the Royal Society were innocent, injured parties. Certainly, Cutler's curious mixture of philanthropy and miserliness was notorious, as we shall see. But in fact, new manuscript evidence from the records of the Courts of King's Bench and Chancery and from the papers of Sir William Petty at Bowood House suggests that the picture was somewhat more complicated, and that the right was not as entirely on the side of Hooke and the Royal Society as they implied at the time and has generally been presumed since. Moreover, if the documentation familiar hitherto is re-examined in the light of this new data, this too contains clues that it was not only Cutler's miserliness that caused the problems associated with the lectureship. Rather, some of these were caused by the parties with whom he had to negotiate, and they throw interesting light on Hooke and on the early history of the Royal Society.

In part, Cutler had reason to suspect that Hooke simply failed to maintain his side of the bargain by not lecturing as often as he was required to under the terms of the endowment. Though this is something which it is now almost impossible to assess fairly, it was connected with a certain confusion as to how the lectureship was supposed to relate to Hooke's job as Curator of Experiments to the Society, which stems from what might be described as a deliberate fudge on the part of the Society's organizers. Anxious not to miss the opportunity of being the beneficiaries of Cutler's generosity, they seem to have tried to conflate his original intention for the post with Hooke's responsibilities as Curator in a manner which meant that Cutler could legitimately claim that the two had become indistinguishable and hence that it was not clear what specific activity his endowment was paying for.

Linked to this was the further complication that, although Hooke was the beneficiary of Cutler's foundation, it was with the Royal Society that its details were agreed: evidently it was considered more appropriate for an initiative for promoting the advancement of learning to be ratified by a public institution rather than a private individual. This, however, made the agreement rather lopsided, for while Cutler was contractually bound by a guarantee that he agreed to sign in 1665, Hooke for his part was able to argue that there was no privity of contract between Cutler and him, as against the Royal Society. Moreover, since the legal relationship between him and the Society was never very clearly defined, it was far from obvious how far Cutler could insist on specific obligations on Hooke's part, other than those decreed by the Society. The result is that the dispute that arose over arrears of payment of the stipend in the 1680s throws interesting light on the law of contract and the way in which this was interpreted in the late

[2] Hooke, *Posthumous Works*, pp. xxiv – v. See also below, pp. 306f.

Plate 12. Ink drawing, heightened in grey wash, showing details of 'the manner of making Salt at a Saltern in Hampshire'. Royal Society Classified Papers xx.40, reproduced by permission of the President and Council of the Royal Society. The text which this drawing illustrates was evidently delivered by Robert Hooke as a Cutlerian Lecture in 1666: see below, pp. 299 – 301.

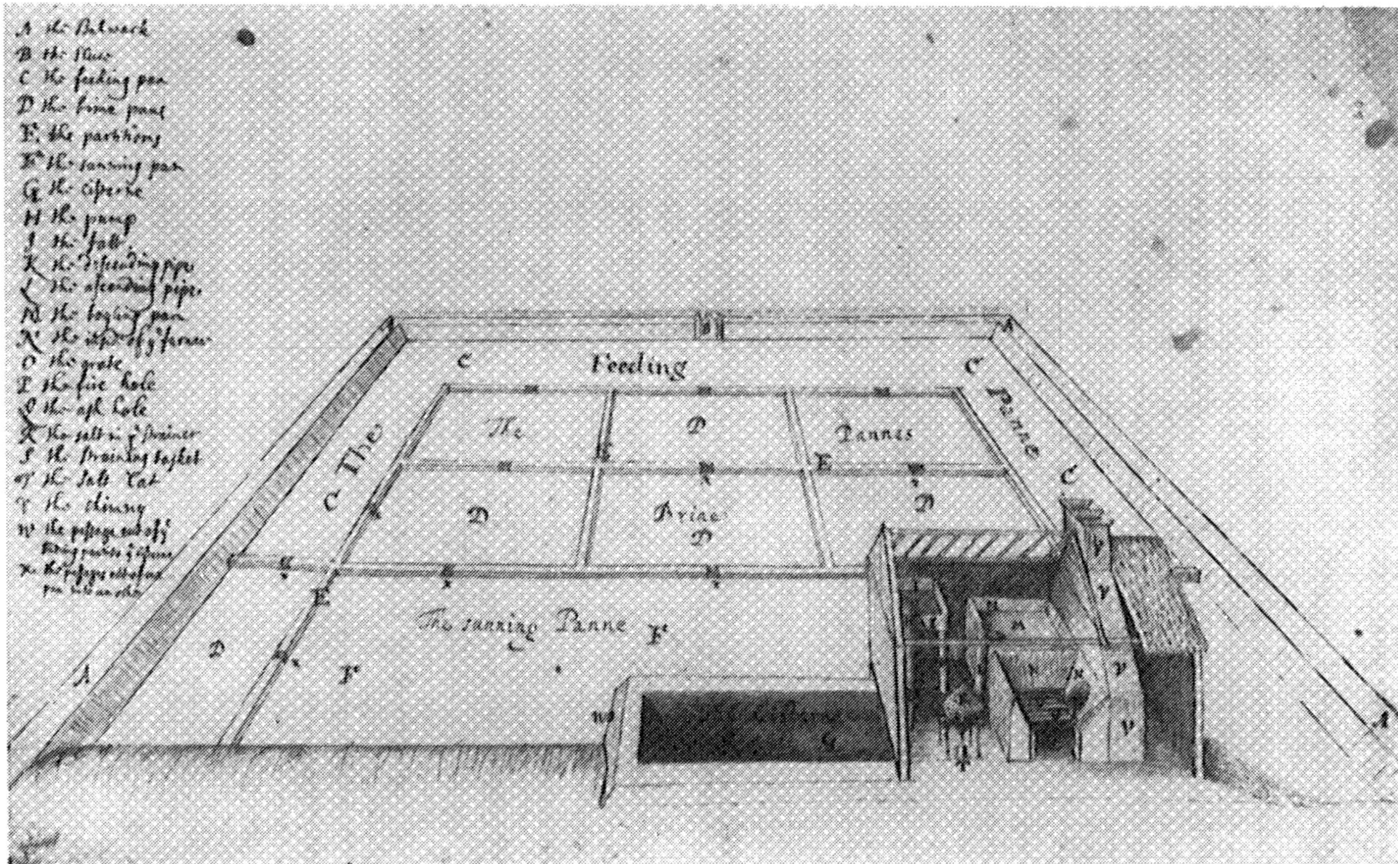

A the Bulwark
C the feeding pan
E the partitions
H the pump
O the grate
P the fire hole
S the straining basket
T the chimney
The Feeding Panne
The Brine Pannes
The sanning Panne

seventeenth century: for while Cutler argued that Hooke *had* a contractual obligation which should be more clearly ratified, Hooke argued strenuously that specific and legally binding commitments had never been the original intention of the agreement.[3]

The negotiations also reveal a further area of disagreement, as to what topics the lectureship should deal with. Cutler seems to have seen his foundation as being for the propagation of improved technical practices through the so-called 'History of Trades'; the Royal Society, though eager to encourage such technological study, sought to redefine the subjectmatter of the lectureship so that it covered science as well, and this was something which Hooke — after initially wavering — was soundly to endorse. Indeed, during the negotiations in the 1680s, he was stimulated to one of the clearest statements as to what he considered the proper relationship between science and technology that he ever made, which will be quoted in full below. This document is crucial not least for clarifying the issues involved, enabling one to be a little more discriminating as to what is meant by 'technology' than has often been the case in the past. For some aspects of this were arguably more germane to scientific enquiry and more susceptible to assistance by scientific intervention than others, and, while Hooke was to exemplify again and again how easily inventions made in the course of scientific research could be given practical applications, it may be argued that the value of the History of Trades programme was much more dubious.[4] Moreover, this was something which the Society in general and Hooke in particular arguably learnt in the course of the dispute, having previously rather uncritically seen the whole of technology as susceptible to immediate improvement by their efforts.

Lastly, a narrative of Hooke's tenure of the Cutlerian Lectureship entails a broader consideration of his employment by the Royal Society, a facet of Hooke's career which is important both in its own right and in connection with the history of the institution with which he was associated for over forty years. Arguably, the leading figures of the Society compounded Hooke's problems in his later negotiations with Cutler by the rather cavalier way in which they treated their loyal employee in their initial negotiations over the benefaction, and it is revealing that a tone of grievance against the Society was well as against Cutler creeps into some of Hooke's comments on the subject in the 1680s. On the other hand, the episode also sheds interesting light on the formative years of the Royal Society itself. For one thing, it illustrates the ingenuity that the Society deployed in its efforts to secure funding for scientific research and exposition. In addition, the case of Cutler is interesting for showing the Society utilizing its benefactor's offer to extend its activities in a direction — of providing lectures — which it had not up to that point seen as falling within its remit, and this illustrates the extent to which the Society's early development was the result of grasping

[3] On the law of contract in this period, see J. H. Baker, *An Introduction to English Legal History* (London, 1971), ch. 16, esp. pp. 194f.

[4] See J. A. Bennett, 'Robert Hooke as Mechanic and Natural Philosopher', *Notes and Records*, 35 (1980), 33 – 48; Hunter, *Science and Society*, ch. 4.

unforseen opportunities as they presented themselves. For all these reasons, a detailed account of this episode may be justified, using existing sources as well as the new documents that have come to light to piece together the negotiations and disagreements entailed in the complex history of Sir John Cutler's benefaction.

The Founding of the Lectureship

The first reference to Cutler's proposed endowment in the Royal Society's minutes occurs in a report of a Council meeting on 22 June 1664, when the Fellows Abraham Hill and John (later Sir John) Hoskins were ordered to confer with Sir William Petty and John Graunt 'concerning the manner and form, in which it might be most proper for Sir John Cutler to put in execution his promise of giving fifty pounds a year to Mr Hooke during his life for the reading of the histories of trades in Gresham-College'.[5]

What do we know by way of background to Cutler's offer? Sir John Cutler was a prominent member of the Grocers' Company of the City of London: born in 1609, son of another grocer, Thomas Cutler, he had risen to be Master of the Company in 1652 – 3, a post he was to hold again in the late 1680s. But it seems that he abandoned trade for finance at an early stage in his career, specializing in business in large-scale conveyancing and brokerage of the kind also associated with Sir Robert Clayton, whose activities in this field have recently been the subject of definitive treatment by Frank T. Melton.[6] The profits that he made from such sources enabled Cutler to invest in estates in Yorkshire in the 1650s, where he was said to have lived for a while in miserly seclusion; later, he moved to Westminster, where he was reputed by his neighbours an 'old curmudgeon'. Indeed, Cutler's parsimony was to become notorious: it was elaborated in such stories as that, by having a maid do his darning in silk, he transformed a pair of worsted stockings into silk ones, while it was immortalized — if a little unfairly — by Pope in his *Epistle to Allen, Lord Bathurst* 'Of the Use of Riches' of 1733.[7]

Cutler's public role was first in evidence at the approach of the Restoration, when he took an active part in promoting the subscription

[5] Birch, *History*, i, 442.

[6] F. T. Melton, *Sir Robert Clayton and the Origins of English Deposit Banking, 1658 – 1685* (Cambridge, 1986). Biographical information on Cutler is provided by *DNB*; J. R. Woodhead, *The Rulers of London, 1660 – 89* (London, 1965), pp. 55 – 6; G. E. Cockayne, *Complete Baronetage*, iii (London, 1903), 128; and B. D. Henning, *The House of Commons, 1660 – 1690* (London, 1983), ii, 183 – 4.

[7] Henning, loc.cit.; *DNB*; Alexander Pope, *Epistles to Several Persons*, ed. F. W. Bateson (2nd ed., London, 1961), pp. 119 – 20.

raised by the City of London for the use of Charles II, a service for which he was knighted on 17 June 1660, being created a baronet on the following 12 November. Then, in 1663, he gave £1500 towards repairs to St Paul's Cathedral and was made receiver of voluntary contributions for the purpose: according to contemporary gossip, it was in order to secure this post that he made his own donation, the proceeds of the receivership being worth three times what he himself contributed.[8] Subsequently, Cutler was to lavish his generosity on his own company, whose premises he rebuilt at his own expense in the aftermath of the Fire of London; on the parish church of St Margaret's, Westminster, for which he erected a gallery; and on the College of Physicians. For the latter he paid for an anatomical theatre which was named the 'Theatrum Cutlerianum' in his honour when it was opened in 1679, though the 'gift' turned rather sour after Cutler's death, when his executors decided that the £7000 involved had in fact been a loan and the College was forced to repay £2000 of it: the College retaliated by erasing the inscription that they had erected commemorating Cutler's generosity.[9]

This benefaction to another learned institution provides a clue as to why Cutler may have seen the endowment of a lectureship at Gresham College as a suitable use for his funds. Also relevant is his acquaintance, for Cutler had contacts with various Fellows of the Royal Society, most notably the draper and statistician, John Graunt, who had been elected to the Society in 1662 after presenting it with his *Natural and Political Observations . . . upon the Bills of Mortality*.[10] The references to Cutler in the diary of Samuel Pepys almost invariably note his presence at coffee houses and the like in the company of Graunt; on one of these occasions, Pepys records an astute comment on the state of the nation's trade which is syntactically ambiguous, being presumed — probably rightly — by the diary's editors to have emanated from Graunt, though it could conceivably have come from Cutler.[11] It was clearly because of this friendship with Graunt that a leading role in the negotiations between the Royal Society and Cutler was taken by Graunt and by Graunt's close friend, Sir William Petty; a subsidiary role seems to have been played by Dr Daniel Whistler, a prominent Fellow of the College of Physicians who evidently belonged to the same circle, having presented Graunt's book to the Society in 1662.[12]

It may also have been these men who were responsible for the suggestion that the lectureship should be for the 'histories of trades', as was stated in the original proposal to the Society and reiterated at a further Council

[8] Pepys, *Diary*, iv, 430.

[9] See *DNB* and Sir George Clark, *A History of the Royal College of Physicians of London* (Oxford, 1964 – 6), i, 330f.

[10] Birch, *History*, i, 75, 77.

[11] Pepys, *Diary*, iv, 22, 64 – 5, 100, 256. Cf. ibid., 365 (this records Cutler's experience with putting a piece of iron on barrels of beer during thunderstorms).

[12] Birch, *History*, i, 75. See also ibid., i, 442 (suggesting that Whistler first reported Cutler's intentions), 484. John Wilkins was presumably involved because of his prominence in the Society, though he may also have had links with Cutler.

meeting on 27 July, when it was stated that Cutler intended to establish Hooke as 'professor' of this subject.[13] The History of Trades was the programme for studying industrial processes and craft practices advocated by Francis Bacon as a crucial part of the enterprise of reforming knowledge.[14] This project had proved of considerable interest to various early members of the Royal Society, and Petty, in particular, was one of its most effective protagonists: he had waxed lyrical about the importance of 'the History of Arts or Manufactures' in his *Advice of W.P. to Mr Samuel Hartlib* (1648), and his 'Apparatus to the History of the Common Practices of Dying', read before the Society on 7 May 1662 and published in Sprat's *History of the Royal Society* in 1667, was one of the best pieces of writing of this kind that the Royal Society was to produce; in addition, both Graunt and he were on the Committee for the History of Trades established by the Society in 1664.[15]

But, though we know frustratingly little about Cutler's own views, it is quite likely that, as a man of affairs himself, it was his own idea that the History of Trades should form the subject-matter of the lectureship attached to his name, on the grounds that this was an area of study of greater utility than the more rarified studies in which the Royal Society also indulged. That this *was* Cutler's view is suggested by what can be deduced about his later views on the actual use to which the lectureship was put by Hooke and the Royal Society, about which more will be said in due course. It is also interesting that the lectureship was to be at Gresham College: it was almost as if Cutler was appointing Hooke as a kind of extra Gresham professor to supplement the foundation of Sir Thomas Gresham in the previous century and to bring it up to date in dealing with topics of contemporary relevance: it is unlikely to be coincidental that the stipend mentioned — of £50 — was the salary that the other Gresham professors enjoyed.[16]

Turning to the other parties in the negotiation, Hooke and the Royal Society, here the background is provided by the essential role that Hooke had by this time come to play as the Society's Curator of Experiments. As we saw in the Introduction, when Hooke forsook his earlier employment as assistant to Robert Boyle to take up this appointment in November 1662, it had had an immediate effect on the Society's proceedings, giving them a new degree of single-mindedness through his research on pneumatics; also

[13] Birch, *History*, i, 453.

[14] See W. E. Houghton, 'The History of Trades: its Relation to Seventeenth-century Thought', *Journal of the History of Ideas*, 2 (1941), 33 – 60; Hunter, *Science and Society*, ch. 4; K. H. Ochs, 'The Royal Society of London's History of Trades Programme: an Early Episode in Applied Science', *Notes and Records*, 39 (1985), 129 – 58.

[15] William Petty, *The Advice of W.P. to Mr Samuel Hartlib* (London, 1648), p. 18 and passim; Sprat, *History*, pp. 284 – 306. Cf. Cl.P. iii(1)6. Birch, *History*, i, 83, 407. See also above, 'An Experiment in Corporate Enterprise'. Whistler's contributions to the Society's proceedings as recorded in Birch, *History*, passim, do not show any particular penchant of this kind, though he was on two committees on technological topics (ibid., i, 172, 207).

[16] See E. G. Forbes, ed., *The Gresham Lectures of John Flamsteed* (London, 1975), p. 2.

crucial were the microscopical observations that Hooke made over these years, which were published in 1665 as *Micrographia*.[17]

In addition, Hooke carried out more miscellaneous work for the Society, making astronomical and meteorological observations, and sometimes testing matters suggested by other Fellows. Moreover, increasingly from the middle of 1663 onwards, he brought forward a series of mechanical inventions that he had made — sounding instruments, for example, or a 'whale-shooting engine', quite apart from a machine for grinding optic glasses and others intended to assist with scientific observation.[18] It is thus hardly surprising that when the Society was casting around for ways of making money in February 1664, it was Hooke who was asked to help with 'some experiments, that might be acceptable and useful to the public', while he was also central to the 'Mechanical Committee' set up later that year.[19] On the other hand, my scrutiny of the pages of Birch's *History of the Royal Society* does not suggest that at this point Hooke had devoted any time to the History of Trades. While other Fellows showed considerable interest in that project, Hooke's role lay rather in his profuse performance of experiments and his fecundity of invention, and it was clearly for this that the Society valued him. From the outset, therefore, one might feel that there was an element of tension between the Society's expectations of Hooke and the subjectmatter of Cutler's proposed foundation.

A further complication from the Royal Society's point of view was that, crucial as Hooke was to the Society, he was not actually being paid by the Society at this stage. His appointment had initially been made on the understanding that he took the job, 'expecting no recompence till the society should get a stock enabling them to give it' — which by 1664 had still not materialized.[20] No figure for a salary was mentioned when he took the post, and a subsequent statement by Hooke of his account as Curator begins only at Midsummer 1664, a date the significance of which will be considered shortly.[21] For the previous year and a half, his support must have come from other sources. Possibly he was deputizing for Walter Pope, Professor of Astronomy at Gresham College, throughout this period, as he certainly was later in 1664 and as Nehemiah Grew and John Flamsteed were to do at a later date.[22] In addition, he may have continued to be retained by Robert Boyle, who had received 'the thanks of the society for dispensing with him for their use' when Hooke was first appointed.[23]

[17] See above, Introduction, pp. 22 – 3.

[18] Birch, *History*, i, 287, 332, 365, 477 and passim. For many of these devices, see Bennett, 'Robert Hooke' (n. 4).

[19] Birch, *History*, i, 379; see above, 'An Experiment in Corporate Enterprise', pp. 87f.

[20] Birch, *History*, i, 123, 124. For hopes of endowment at this time, see Hunter, *Science and Society*, pp. 38 – 9.

[21] Birch, *History*, i, 123, ii, 148.

[22] See below, p. 290; on Grew, see above, 'Early Problems', pp. 272 – 3; on Flamsteed, see Forbes, *Gresham Lectures* (n. 16).

[23] Birch, *History*, i, 124.

It is thus hardly surprising that when, in May 1664, the resignation of the Cambridge mathematician, Isaac Barrow, as Professor of Geometry at Gresham College necessitated the filling of the vacancy, both Hooke and the Royal Society must have seen this as an ideal opportunity for Hooke to obtain a position which would have made it easy for him to carry on his work as Curator to the Royal Society. In fact, however, Hooke failed to achieve the position when the election took place: instead, it went to Arthur Dacres.[24] Moreover a connection between this event and Cutler's proposal is illustrated by a retrospective account of the affair which Hooke submitted as evidence to the Court of Chancery in 1683, the basic gist of which seems plausible enough, even if Hooke may have been a little evasive on certain points, including the lectureship's subjectmatter. Hooke recalled how in or about May that year, 'being in Company with one Captaine Graunt at a Publicke house neere Burchin lane', he accidentally met Cutler, 'who as it seemes was verie intimately acquainted with the said Captaine Graunt[;] amongst other discourse this Defendant [Hooke] happening to mention some disappointment he had mett with at an Election of a Geometry Reader in Gresham Colledge in the said Month of May by the deportment of some of the Electors to whome or one of them the Complainant [Cutler] had some relation[,] the said Complainant was so concerned thereat that he bid this defendant not trouble himselfe at the same disappointment for that he himselfe as he was pleased then to say would give this defendant as good an allowance as the vallue of that place for his encouragement to proceed with those matters this defendant had allready so successefully begunne as he had bynn enformed or much to the same effect'.[25]

It will be noted that in this account, Hooke presented the matter as a personal approach to him by Cutler, though the 'matters . . . allready so successefully begunne' refer to his work as Curator to the Society. On the other hand, it is worth pointing out that the version of events which Cutler presented in 1683 made more of his intention of doing good for the Royal Society. He explained how, 'designeing to give Incoragement to Learning and Ingenuity and more particulerly for the Benefitt and Advantage of the Royall Society Incorporated by his Majestye and Consisting of many persons of great honour Quallity and Ingenuity Inhabiting in & about this City of London and who frequently mett att Gresham Colledge in the sayd Citty for the Advancement of such Learning as was by them from tyme to tyme thought fitt to bee Insisted upon', he treated with Hooke through Petty and Whistler 'or one of them', 'thinking him to be A Person fitt for the

[24] See John Ward, 'Memoirs relating to Gresham College', Book I, British Library Additional Manuscript 6195, fol. 70v. For an account of this episode which postulates the connection for which evidence is here provided, see Patri Pugliese, 'The Scientific Achievement of Robert Hooke: Mechanism and Mechanics' (Harvard Ph.D. thesis, 1982), p. 2. See also below, p. 297.

[25] PRO C7/564/29, item 2. See further below, pp. 317f. For the names of the electors, see Ward, 'Memoirs' (n. 24), fol. 70v: it is unclear to which Cutler was related.

sayd Designe'.[26] Moreover the fact that the matter was brought to the Society in June for a decision as to how best to implement the offer suggests that it was from the outset taken for granted that the Society should be involved in the matter.

Clearly, however, there was some lack of clarity from the outset as to whether or not the benefaction was a personal one. At one point during the negotiations which were to ensure, it was specifically stated that Cutler had suggested the benefaction, 'intending a particular kindness to Mr Hooke':[27] but the Society, true to its ambition to try to establish an institutional infrastructure for the new science with a permanence beyond the life of any individual, seems to have wished to establish itself as the primary beneficiary. Thus its organizers are constantly to be found attempting to impose themelves between Cutler and Hooke in a supervisory capacity, while a slightly obscure passage in a later report of the negotiations suggests that at one point the Society persuaded Cutler to promise that, on the basis of their success in managing the lectureship during Hooke's lifetime, he might decide to endow it in perpetuity.[28]

Evidently, this was linked to a wish to display that it was not so much an individual as the new science in general — as symbolized by the Society — which had the confidence of a public figure like Cutler. For not only was this the first such benefaction of its kind: it was also crucial in terms of the Society's aspirations concerning its role in contemporary society, in that it came from one of the '*men of Converse and Traffick*' to whom it was hoped that the new science would especially appeal. This was spelt out most clearly by Hooke himself in the passage towards the end of the preface to *Micrographia* — probably written late in 1664 — in which he noted 'the *munificence* of *Sir John Cutler*, in endowing a Lecture for the promotion of *Mechanick Arts*, to be governed and directed by This *Society*', continuing: 'it cannot now be objected to them, that their Designs will be esteemed *frivolous* and *vain*, when they have such a *real Testimony* of the *Approbation* of a *Man* that is such an *eminent Ornament* of this renowned City, and one, who, by the *Variety*, and the *happy Success*, of his negotiations, has given evident proofs, that he is not easie to be deceiv'd'.[29]

Quite apart from the fact that Hooke may have come to regret the latter comment in the light of developments to be surveyed in this paper, such grandiloquence belies the quandary in which Cutler's offer initially placed the Society. For it suddenly made it glaringly obvious that though Hooke, the individual named in it, had for the past two years been playing a crucial role as Curator in giving substance to the Society's institutional ambitions,

26 PRO C7/564/29, item 1.

27 Birch, *History*, i, 503. Cf. ibid., 484, where he was thanked for 'his particular favour to a worthy member, and for that respect and confidence he hath hereby expressed towards the whole body'.

28 Birch, *History*, i, 503, quoted below, p. 292.

29 Hooke, *Micrographia* (London, 1665), sig. g1v. The imprimatur is dated 23 Nov. 1664; cf. also Hooke to Boyle, 24 Nov. 1664, Boyle, *Works*, vi, 499.

the basis on which he had been doing so remained highly provisional. Not only had no stipend been either arranged or paid: in addition, he had not even been formally elected to the post along the lines which had been elaborately laid down in the statutes that the Society had drawn up in 1663.[30] This explains the slightly curious way in which the Society initially reacted to Cutler's offer, for it suddenly decided that its relations with Hooke needed to be formalized as a matter of urgency. At a Council meeting on 27 July 1664 it was ordered 'That at the first opportunity Mr Hooke be put to the scrutiny for the place of curator: That he should receive eighty pounds *per annum*. as curator to the society by subscriptions of particular members, or otherwise: That he forthwith provide himself of a lodging in or near Gresham-College: And That these orders and votes be kept secret, till Sir John Cutler shall have established Mr Hooke as professor of the histories of trades'.[31]

The fact that these decisions were not to be disclosed to Cutler suggests that this was a kind of preemptive strike, intended to cement Hooke's relations with the Society more firmly than hitherto and to establish his primary commitment as to his Curatorship, while at the same time leaving open the possibility that Cutler should endow him. The secrecy may have reflected the Society's embarrassment at being rather caught on the hop: but the fact that, at the outset, the Society indulged in some cloak and dagger business to maneouvre itself into a favourable position for the negotiations that were to ensue was itself a poor augury for the future. Whether any significance should be attached to the requirement that Hooke should live at Gresham College is doubtful: this may simply have reflected the feeling that, if on the premises, he could do his job as Curator better (he had in fact earlier been asked to live on the premises temporarily in connection with preparations for the abortive visit of the King, being paid twenty shillings a week for his trouble).[32]

After this, no more is then heard of the matter until the autumn, but on 5 October there were fresh moves in connection both with Cutler's offer and with Hooke's appointment as Curator. First, Abraham Hill, the Society's Treasurer, was ordered to collect 'the arrears of the Society . . . as also those of the subscribers for Mr Hooke; and that in the mean time he pay Mr Hooke both what he had received already from the said subscribers, and, as soon as conveniently he could, what should yet remain due to him; the whole being £80 *per annum*'. This may be elucidated by Hooke's later statement of his account as Curator, which — significantly — begins with the quarter ending on Midsummer Day 1664, in other words 24 June, just two days after the Society had heard of Cutler's offer. By Michaelmas (29 September), there were thus two quarters' salary, or £40, owing, and in his

30 See *The Record*, pp. 296 – 8.

31 Birch, *History*, i, 453.

32 Ibid., i, 315, 340. On the King's visit, see also above, p. 75.

account Hooke records his receipt of £39 from Hill, at least part of which evidently came from special subscriptions.[33]

Arrangements were also made for the Cutlerian Lectureship, in which the Society made clear the managerial role it intended to take. Hooke was told to 'prepare an oration upon the account of Sir John Cutler's founding of a mechanical lecture; and that, between that and the next meeting, he think upon a method to proceed by in his lecture, which, upon the Council's approbation, he might give a hint of in the said oration'. Further information is provided by a letter from Hooke to Boyle written the following day, in which he reported how the lectureship was being established, 'the person I formerly told you of, namely, Sir *John Cutler*, having very nobly and freely, without any compulsion or excitement, not only kept his word, but been better than it, sending me yesterday a half year's salary beforehand, as an earnest of his intention'.[34] He continued: 'The most I think I shall be able to do in this business this term (being engaged to read for doctor *Pope*[35]) will be only to make a short speech, both in praise of Sir *John*, my noble patron, and of the excellency and usefulness of the design itself, and of what method and course I shall take in it'.

At another Council meeting on 2 November, John Wilkins, Dean of Ripon and one of the leading lights in the Society, 'related, that Sir John Cutler had declared to him, that he was firm in his resolution to settle upon Mr Hooke £50 *per annum*. for such employment, as the Royal Society should put him upon', and Wilkins was ordered to draw up a form of thanks to Cutler for his donation, 'declaring him withal an honorary member of the society, to see whether it be according to his sense'.[36] The last phrase is lightly ambiguous: it is not clear whether it refers to Cutler's election as honorary Fellow, or to the proposed nature and supervision of Hooke's employment. As will be seen, the latter represented quite a significant shift from a commitment to the history of trades to a much vaguer license to the Royal Society to put Hooke onto any tasks they liked, which would presumably therefore have been perfectly compatible with his continued employment as Curator using Cutler's funds.

A week later, Cutler was elected honorary Fellow, and a deputation comprising Petty, Wilkins, Whistler and Graunt was 'appointed to attend him in the name of the society', with an 'order' which noted Cutler's resolution to give Hooke £50 per annum for life, and the fact that he had already advanced £25 for the current half year, 'in consideration of which

33 Birch, *History*, i, 472 – 3, ii, 148. The Royal Society's accounts have no reference to a payment of £39 at all. A payment of £20 to Hooke is, however, recorded without date in the account covering the period Nov. 1664 to April 1666, which could be for a quarter's salary under the £80 p.a. arrangement, thus leaving only £19 unaccounted for.

34 Birch, *History*, i, 473; Hooke to Boyle, 6 Oct. 1664, Boyle, *Works*, vi, 493. On the payment, cf. Birch, *History*, i, 484.

35 I.e., the Astronomy Professor at Gresham College, for whom Hooke was evidently deputizing at this point: see above, p. 286.

36 Birch, *History*, i, 479.

yearly stipend, the said Sir John Cutler is willing to refer it to the president, council, and fellows of the Royal Society of London for improving natural knowledge, to direct and appoint the said Mr Hooke how many lectures he shall read, and when, and upon what subjects; or what kind of inquiries, by way of experiment, he shall be engaged to prosecute'. It went on to thank Cutler 'for his particular favour to a worthy member' and his confidence in the Society and to offer him honorary membership, stressing 'what a sense they have of his favour, which they have the more reason to value, as being the first donation they have been intrusted with of this kind, and which they hope will prove a leading example to others'.[37]

By 16 November, Cutler had evidently accepted the offer of honorary membership, and Wilkins was instructed to draw up a diploma in English ratifying this. At the same meeting the suggestion concerning the Curatorship made in July was taken up again, it being 'resolved, that there shall be a curator by office; and that Mr Hooke be proposed to the society as such'.[38] Moreover at this point the interrelationship between the Curatorship and the Cutlerian endowment was explicitly laid out, when, on 23 November, the President, Lord Brouncker, told the Society 'that the Council had found it necessary to have a curator' and Wilkins formally nominated Hooke to the post. For Brouncker added 'that they would allow him, *pro tempore*, thirty pounds a year, the present revenue of the Society not being well able to bear a greater allowance; and that Mr Hooke was contented with that sum for the present, Sir John Cutler having settled upon him, during life, fifty pounds a year, upon the account of reading lectures of experimental philosophy: which addition had made him willing to accept, *pro tempore*, of thirty pounds a year from the Society for performing the office of a curator'.[39]

Even this did not represent the end of the procedure, however, and just why it was so long drawn out is unclear. On 7 December Wilkins again proposed Hooke for the Curator's place; the Council approved him on 28 December; and he was then recommended to the Society on 11 January, being 'elected curator, by office, to the society, and that for perpetuity, with a salary of £30 per year pro tempore'.[40] From the financial point of view, however, the arrangement thus ratified had evidently been deemed to start from 23 November, for — as Hooke's later account shows — it was from that date that the Society terminated its payment to him at the rate of £80 a year which had been in force since the start of the second quarter of the year: from 23 November onwards the amount due to him from the Society was

[37] Birch, *History*, i, 484 – 5. Birch's text exactly follows that in the original journal book, except that 'his particular favour' should be 'this'. Cf. Oldenburg to Boyle, 10 Nov. 1664, *Oldenburg*, ii, 297, which reports on the benefaction and clearly states how Cutler had resolved 'to referre to the Society the direction of the kind of Imployment the Stipendiat shall be putt upon'.

[38] Birch, *History*, i, 488.

[39] Birch, *History*, i, 496. Cf. ibid., i, 490.

[40] Birch, *History*, i, 499, 507, 510; ii, 3, 4. The statutes had allowed for curators to be re-elected after a year's probation for perpetuity: *The Record*, p. 297.

calculated at the rate of £30 a year.[41] Moreover, it is worth pausing here to comment on this change. When Hooke's salary had first been fixed that summer, the implication had clearly been that the £80 that he was to be paid was for his work as Curator of Experiments. The Society may have set the figure artificially high in the expectation of its being subsidized by the stipend that Hooke was likely to acquire from Cutler, perhaps rather optimistically presuming that, if that endowment failed to materialize, Hooke's support could come in part — as it had over that summer — from members' extra subscriptions. But the change now made nevertheless created a problematic anomaly: for Hooke was later to have some justification for claiming that it was for his curatorial work that he was entitled to £80, and, this being the case, he might well have felt that the lectures that he was expected to give according to the terms of the Cutlerian benefaction represented an unwelcome extra burden.

Meanwhile, at the meeting on 7 December, Petty, Wilkins and Graunt had been ordered to 'draw up a form concerning the name and subject of Mr Hooke's lecture, and shew it to Sir John Cutler, the founder thereof, to receive his thoughts upon it'.[42] A week later, Petty came back with a report, using the term 'reader' to describe the office for which Hooke would receive his £50 salary; noting Cutler's desire 'that he might not be more burthened than the other readers of Gresham-College are, but read only as many lectures in the vacations, as they do in term-time: and farther, that Sir John Cutler had intimated, that the management of this lecture by the Society, during Mr Hooke's life, should be a measure to him, to make it perpetual, or not'.[43]

The minutes continue: 'Hereupon the subject and number of these lectures were much debated; and it was agreed upon, that the subject of them should be the *History of Nature and Art*: but with regard to the number of the lectures, it was not resolved upon, some urging, that Mr Hooke should be ordered to read once upon all the ordinary weekly meeting-days of the vacations, except those of the three months of August, September, and October, the greater part of half an hour, beginning about two of the clock: others pressing, that Mr Hooke might read but as many lectures as other professors do; and that the rest of the Wednesdays might be left to be endowed by some other benefactor for another philosophical professor.'

The reference to 'meeting-days' makes it plain that it was intended that the lectures should be juxtaposed with meetings of the Royal Society, which took place at 3 p.m. Also interesting is the Society's stated hope for a further

[41] Birch, *History*, ii, 148; Royal Society accounts, passim, which include (in the accounts respectively for April to Nov. 1666 and for 1667) the payments of £30 and £45.3.4 referred to in Hooke's account, together with various smaller ones for expenses. They also show him being paid £30 a year regularly thereafter. On the earlier payment of £20, which may have formed part of the £39 noted in Hooke's account, see above, n. 33.

[42] Birch, *History*, i, 499.

[43] Birch, *History*, i, 503. 'Measure' is obviously intended in the sense of 'criterion': see *OED*.

benefactor to endow 'another philosophical professor'. Hooke himself had also noted, in connection with Cutler's generosity, 'the expectation which I have of the *efficacy* of the *Example*', a sentiment echoed both by the 'order' drawn up by the Society on 9 November and by Oldenburg in a letter to Boyle written the next day.[44] Though none was to materialize, this reference is significant, for it implies that at least some in the Society seem to have seen Cutler's benefaction, not simply as a cheap way of paying Hooke for what was effectively an extended version of his existing work as Curator, but as the first step towards a new venture on the Society's part, namely of being associated with one or more professorships in natural philosophy. The very title suggests a similarity with what went on at the universities and at Gresham College, and it implies an expository or even educational role for the Society in propagating knowledge of the new philosophy which it had hitherto lacked.[45] Just who these lectures were meant for is unclear. The fact that they were timed to precede meetings of the Royal Society suggests that the main clientele was expected to be the Fellows of that body, while in his published *Lecturae Cutlerianae* of 1679, Hooke specifically stated that these were 'made before the *Royal Society* on several Occasions at Gresham Colledge'.[46] Elsewhere in the same volume, however, Hooke records how at some of his Cutlerian Lectures 'were present, besides a great number of the *Royal Society*, many Strangers unknown to me': they must, therefore, also have been open to the public, and this is borne out by evidence of attendance from Hooke's diary and other sources.[47]

What of the lectures' subjectmatter? Here again the references in the minutes reveal a significant process of negotiation, which is worth recapitulating, since the question was to recur in later disputes about the lectureship: again, it shows the Royal Society engaging in prevarication which was to cause complications at a later date. In the first instance, as we have seen, it had been quite specifically stated that Cutler intended to endow 'the reading of the histories of trades' by a 'professor of the histories of trades': so it was reported in the Society's minutes on 22 June and 27 July, while in those of 5 October it was described as 'a mechanical lecture', which — in the light of seventeenth-century usage — may have been intended as a synonym.[48] Indeed, it was thus that Hooke categorically described it to Boyle the day after he had been told at that meeting to 'think upon a method to proceed by in his lecture', in terms which are quite at odds with what —

[44] Hooke, *Micrographia* (n. 16), sig. g1v; Birch, *History*, i, 485; *Oldenburg*, ii, 297. See also Birch, *History*, ii, 9, a reference of the following 25 January: 'It was resolved, that the business of procuring benefactors, and the manner of well managing the same, should now be begun to be seriously considered of'.

[45] For a more directly pedagogical parallel, dating from the 1680s, see Petty's proposal discussed in 'Towards Solomon's House', above, p. 202.

[46] Hooke, *Lecturae Cutlerianae* (London, 1679), title-page.

[47] Hooke, *Lecturae Cutlerianae*, reprinted in R. T. Gunther, ed., *Early Science in Oxford*, viii (Oxford, 1931), 149; Hooke, *Diary*, e.g. pp. 75 (11.12.73), 299 (5.7.77), 331 (29.11.77). Non-members also appear in some of the attendance lists noted below, n. 153.

[48] Birch, *History*, i, 442, 453, 473.

as we shall see — he was to report about Cutler's intentions twenty years later: 'I am now engaged in a very great design, which I fear I shall find a very hard, difficult, and tedious task, and that is, the compiling a history of trades and manufactures'. He added, after reporting on his lack of time for the project at that particular juncture: 'by God's assistance, I shall endeavour to the utmost of my power to go as far in it as I am able, being resolved wholly to apply my mind and endeavours to it'.[49]

Hooke repeated this more publicly in a passage at the end of his preface to *Micrographia*, evidently written about the same time, in which he alluded to Cutler's endowment of 'a Lecture for the promotion of *Mechanick Arts*'. For he went on: 'This Gentleman has well observ'd, that the *Arts* of life have been too long *imprison'd* in the dark shops of Mechanicks themselves, & there *hindred from growth*, either by ignorance, or self-interest: and he has bravely *freed* them from these *inconveniences*'; continuing with a paean of praise to Cutler for the way in which 'He hath not only obliged *Tradesmen*, but *Trade* it self: He has done a work that is worthy of *London*, and has taught the chief City of Commerce in the world the right way how Commerce is to be improv'd'.[50] The reference to 'the *Arts* of life' and the 'shops of Mechanicks' is an unmistakeable allusion to the History of Trades programme.

However, as has already been noted, the History of Trades had not hitherto been central to Hooke's work for the Royal Society, which had mainly related to general natural philosophy or to the devising of new machines and gadgets. It was evidently in recognition of the fact that it was here that Hooke's true metier lay — and that it was for this that he was most valuable to the Society — that during the negotiations with Cutler, the subject on which Hooke was to lecture was altered. By 2 November the £50 was said to be 'for such employment, as the Royal Society should put him upon', while a week later the door was opened wider still, as shown by the passage from the minutes already quoted: not only was the Society to appoint 'how many lectures he shall read, and when, and upon what subjects', but also 'what kind of inquiries, by way of experiment, he shall be engaged to prosecute' — almost a license to conflate the lectureship with Hooke's Curatorship.[51]

Either Cutler or some of the Fellows or both must, however, have stuck out for 'lectures'. By 23 November, Hooke's stipend was said to be for 'reading lectures of experimental philosophy', though on 14 December, after Petty, Wilkins and Graunt had shown Cutler the 'form concerning the name and subject of Mr Hooke's lecture' and 'the subject and number of these lectures were much debated', 'it was agreed upon, that the subject of them should be the *History of Nature and Art*'.[52] How much one should read into these phrases is not entirely clear: but undoubtedly the rubric had shifted from the History of Trades narrowly defined to a broader

49 Hooke to Boyle, 6 Oct. 1664, Boyle, *Works*, vi, 493.

50 Hooke, *Micrographia* (n. 16), sig. g1v.

51 Birch, *History*, i, 479, 484.

52 Birch, *History*, i, 496, 499, 503.

conception of the Society's activities, expressed most forcibly in 'experimental philosophy', and it may be that the formula finally decided on 14 December is to be seen as a compromise.

It is likely, however, that some confusion remained — in Cutler's mind if no-one else's — as to just what Hooke was supposed to be doing. After all, Hooke's own statement on the subject in *Micrographia* appeared only in 1665, though penned the previous year, while a further public statement to similar effect was to appear two years later in a passage of Sprat's *History of the Royal Society* (1667) about the support which the new science might expect '*From our Citizens*'. For this expressed a view of the aim of the lectureship which was much closer to what had been originally mentioned than to what was ultimately agreed. Sprat wrote of 'the *establishment* made by Sir *John Cutler*, for the reading on *Mechanicks*, in the place where the *Royal Society* shall meet', and he commented: 'This is the first *Lecture* that has been founded of this kind, amidst all the vast *munificence* of so many *Benefactors to Learning*, in this later Age. And yet this was the most necessary of all others. For this has chiefly caus'd the slow progress of *manual Arts*; that the *Trades* themselves have never serv'd *apprentiships*, as well as the *Tradesmen*: that they have never had any *Masters* set over them, to direct and guide their works, or to vary, and enlarge their operations'.[53] It would be hard to devise a clearer manifesto for a lectureship devoted specifically to the History of Trades.

After 7 December, no further discussions of Cutler's donation appear in the Royal Society's minutes at this point, but this was clearly not the end of such deliberations, whether formal or informal: writing to Boyle on 13 December 1664 Hooke apologized for his 'hasty scribble, for it was near eleven a clock this night, before I could get from some company, met about the business of Sir *J.Cutler*'.[54] What was decided on this occasion is unclear, but further discussion must have ensued thereafter, of matters including that left pending the previous Wednesday — the frequency of the lectures. Then, on 29 June 1665, a formal agreement between Cutler and the Royal Society was drawn up in which it was specified what had been agreed on this and other aspects of the lectureship. This agreement was secured by a bond, which does not survive, but to which repeated reference was made during the litigation between Hooke and Cutler that occurred from 1682 onwards, and from this its content can be reconstructed.

The bond evidently recorded that it had been arranged that, 'upon & for the allowance or yearely payment hereby secured to be paid him', Hooke 'hath promised and (upon & for the allowance or yearely payment hereby secured to be paid him) undertaken dureing his life to reade in the vacation time in Gresham Colledge or in such other place as the Royall Society shall meet in Sixteen Lectures per Annum in order to the advancement of the knowledge of Arte and Nature (the said Society haveing byn desired to direct the Particular matter of the said lectures) by reading one each weeke dureing so many weeks successively next after each of the foure severall termes in the yeare as were weekes in the then last proceeding terme upon

53 Sprat, *History*, pp. 129 – 30.

54 Boyle, *Works*, vi, 501.

such day of each weeke as the said Royall Society shall usually meet upon'.[55] The rather convoluted rubric is due to the fact that the four annual law terms of Hilary, Easter, Trinity and Michaelmas varied in their starting-dates and length according to the ecclesiastical calendar and other traditional constraints: a rider recorded at another point during the litigation reveals that matters were still further complicated by the fact that the vacation was sometimes too short to accomodate the requisite number of lectures.[56] The bond apparently also elaborated on how the stipend was to be paid in two equal instalments at Midsummer and Christmas, and how Hooke was to collect it from Cutler's house. Most crucial of all, as the later lawsuit reveals, was that in the bond Cutler gave a guarantee that he would forfeit £1000 if he or his executors failed at any stage to honour his promise to pay Hooke £50 per annum for life.[57]

Though the bond concerned Hooke, and although by the 1680s it was in his hands and formed the basis of his attempt to sue Cutler, the bond was not in fact between him and Cutler but between Cutler and the Royal Society. We know that the agreement of 29 June was drawn up by Petty, Wilkins and others on the Society's behalf, and they were evidently also the signatories to the bond; Hooke, on the other hand, was later to claim that he was not even consulted about its final wording.[58] This no doubt stemmed from the ambivalence already referred to as to whether the benefaction should be to an individual or an institution, but the result was to complicate matters, since it left the agreement rather lopsided, with Cutler legally obligated while Hooke was not, except insofar as he was contractually bound to the Royal Society as its Curator.

Eighteen months later, however, after the lectureship had been in operation for a while, this was to be at least partially rectified. The Council minutes for 21 December 1666 record 'That Mr Hooke be desired to promise by his hand-writing to observe the ends, for which the report from Sir John Cutler entered in the journal-book of the society, Nov. 9, 1664, affirms the fifty pounds a year to be given him by Sir John'.[59] The President, Brouncker, was desired to draw up a form for this, which was brought to the meeting of 17 January 1667 and was included in the minutes on that occasion. In fact, it was evidently based more on the bond of 29 June 1665 than on what was entered in the minutes on 9 November 1664: it recorded Hooke's obligation to give lectures in words virtually identical to those already quoted, and went on: 'I do hereby renew the said promise, and undertake to read the said lectures, upon such particular matters, as the said

[55] PRO C7/564/29, item 2.

[56] PRO C7/564/29, item 1. On the law terms, see C. R. Cheney, *Handbook of Dates for Students of English History* (London, 1945; corrected reprint, 1981), pp. 65f.

[57] PRO C7/564/29, item 2.

[58] PRO C7/564/29, item 2.

[59] Birch, *History*, ii, 135.

society shall direct. In testimony whereof I have hereunto set my hand and seal'.[60]

Hence Hooke had now also signed an obligation, though whether the initiative for this had come from Cutler or from the Royal Society is unclear. It was to the Council that Hooke delivered the signed and sealed paper on 25 January, and there is no specific reference to its being forwarded to Cutler.[61] Moreover, since Cutler did not refer to the paper later, it may be that he did not know of its existence. Clearly, however, this was not just a matter between Hooke and the Society, for at the meeting at which Hooke was asked to sign this document, it was ordered that 'a copy of the said report' — in other words, presumably that of 9 November 1664 — 'as also of the thanks, that were to be returned to Sir John Cutler, be forthwith made and delivered to Sir Robert Moray or Sir Paul Neile, to give it the lord bishop of Exeter [Seth Ward], to shew it to the said Sir John Cutler, that he might declare, whether it was really his intention to intrust the society with the management of the fifty pounds given by him to Mr Hooke'.[62] If we did not know from the later legal case that Cutler apparently *was* paying Hooke at this stage, this might suggest that the Society was beginning to wonder whether Cutler was really going to honour his promise of paying Hooke on a regular basis at all. Instead, the crucial passage must be that dealing with the question of whether Cutler intended 'to intrust the society with the management' of the endowment. Was the Society suggesting that it should be through them that the money was paid, rather than directly to Hooke? And was the document which Hooke was asked to sign therefore for the benefit of the Society, so that *it* could enforce his obligations on him?

The sources are unfortunately unclear on this point, but what is perfectly possible is that either Cutler or the Society or both may have felt that, even at this early stage, Hooke was flagging in his obligation to give sixteen lectures a year. Moreover in this connection it is worth noting the sequel to the disputed election for the Geometry Professorship at Gresham College in 1664 which, as we have seen, may have helped to inspire Cutler's benefaction in the first place. In March 1665 Hooke had petitioned against the preferment of Dacres on the grounds of irregularities committed at his election; as a result, the election was declared invalid and Hooke got the job, thus adding to his commitments and arguably reducing his incentive to carry out his obligation to give lectures in the vacation on Cutler's account as well as those he had to give as Geometry Professor during the term.[63] A further call on Hooke's time arose in 1666, when, in the aftermath of the Fire of London, he became City Surveyor, yet another task to add to his

[60] Birch, *History*, ii, 141: the wording quoted is identical with that cited in the Chancery case of 1683 as the agreement of 29 June 1665, with minor verbal differences; it does not tally at all with the document drawn up on 9 November 1664 (see ibid., i, 484 – 5). In the additional clause, Birch omits a comma after 'lectures' which appears in the original journal book and has been inserted here.

[61] Ibid., ii, 144.

[62] Ibid., ii, 141. On Ward's later role in connection with Cutler, see below, pp. 307 – 8.

[63] Ward, 'Memoirs' (n. 24), fols. 70v – 1.

work as Curator and as Gresham Professor, which must have made him even more reluctant to find time for his Cutlerian Lectures. Indeed, this could explain why it was in 1666 that the matter arose, though, since he was appointed Surveyor only on 4 October 1666, one would have thought that there had hardly been time by 21 December for matters to become serious.[64]

The Early Cutlerian Lectures and their Content

What do we know, then, about Hooke's delivery of Cutlerian Lectures in the 1660s? The answer is, frustratingly little. We cannot even be sure exactly when he started giving them: since they were not part of the Royal Society's proceedings, they were not recorded in the minutes at this stage, and neither do we have any record of the Royal Society supervising what Hooke lectured on once the lectures' general subject-area had been decided. Indeed, apart from the negotiations already summarized, virtually the only reference in the Society's minutes that may be relevant occurred in February 1665, when it was suggested that a 'lecture' by Hooke — evidently one dealing with the comet that had recently appeared — 'might be perfected and printed'.[65] But no such publication occurred, and it is not even clear whether or not the lecture in question was delivered under the Cutlerian foundation.

One thing that is clear is that, even if Hooke had started giving lectures before the agreement between Cutler and the Royal Society was signed on 29 June 1665, there must then have been an intermission in them, since on 28 June, the very day before the bond was signed, the Society's proceedings had been suspended because of the outbreak of the Great Plague, not to be recommenced till February 1666. Since it had been specifically stated that the Cutlerian Lectures were supposed to be juxtaposed with meetings of the Society, therefore, none can have taken place in the interim. Moreover, since Hooke is known to have left London and stayed for several months in Surrey during this period, the Lectureship might have been regarded as being temporarily suspended, presumably being revived with the Society's activities in the spring of 1666.[66]

64 T. F. Reddaway, *The Rebuilding of London after the Great Fire* (London, 1940), p. 55.
65 Birch, *History*, ii, 16, 19. Cf. Pepys, *Diary*, vi, 48, and Gunther, *Early Science* (n. 47), p. 223. Note also Moray to Oldenburg, 10 Oct. 1665, *Oldenburg*, ii, 560, in which refers to Hooke and 'his lectures for the press'. For a reference to Hooke being ordered to read what was evidently a Cutlerian Lecture in February 1667, see Birch, *History*, ii, 145.
66 Birch, *History*, ii, 60, 63.

What of the lectures' subjectmatter? Apart from the lecture on comets already referred to (assuming it *was* a Cutlerian Lecture), another clue is provided by Hooke's retrospective claim that 'I did, in the year 1664, read several of my first *Cutlerian Lectures* upon that subject [i.e., the controversial spring-balance watch], in the open Hall at Gresham Colledge', which could be true, though no such lectures now survive.[67] On the other hand, a few items *are* extant which appear to be Cutlerian Lectures of the 1660s. The only dated one is a paper on saltmaking read on 25 July 1666, which is almost certainly connected with the Cutlerian foundation, since it specifically refers to the fact that it belongs to a vacation lecture course.[68] In addition, some other surviving lectures are certainly to be connected with the earliest years of the Cutlerian benefaction, as is proved beyond all doubt by the fact that one of them opens with a lengthy passage praising Cutler for funding the lecture series to which it belonged, which is quoted in full in the appendix to this essay.[69] These certainly date from the mid 1660s, though it is unclear whether they were delivered before or after the Plague, since they are clearly antecedent to a further text on related topics published in Hooke's *Posthumous Works* which may be confidently dated to 1668.[70] In addition, it may not be coincidental that one of them survives in juxtaposition with a lecture by Hooke that can be dated to June 1665.[71]

These extant lectures apparently illustrate an interesting sequence of development on Hooke's part. As we have seen, it had originally been categorically suggested that the subjectmatter of the lectureship should be the History of Trades, and Hooke's initial acceptance of this is shown by the passages already quoted from his letter to Boyle of 6 October 1664 and

67 Gunther, *Early Science* (n. 47), p. 149. Cf. ibid., p. 207, though the date of publication of Sprat's *History of the Royal Society* is there incorrectly given, and the exact date mentioned by Hooke in the quotation in the text is possibly also to be treated with caution. Related material of about this date survives in Trinity College, Cambridge, MS O.11a.1[15], but it does not appear to be a lecture: see the text published in Michael Wright, 'Robert Hooke's Longitude Timekeeper' in Michael Hunter and Simon Schaffer, eds., *Robert Hooke: New Studies* (Woodbridge, 1989).

68 Cl.P. xx.40 and 42. Cf. Birch, *History*, ii, 105, which shows that the first item on the agenda at the meeting of the Royal Society on 25 July comprised experiments on a related topic. See further below, pp. 300, 301.

69 For details of these lectures, preserved in Cl.P. xx.50 and City of London Guildhall Library MS 1757, see below, n. 79, and, for the best extant account of them and their mutual relationship, Pugliese, 'Scientific Achievement' (n. 24), pp. 10f. Other extant lectures of the 1660s are Cl.P. xx.41 (cf. Birch, *History*, ii, 91 – 2) and Cl.P. xx.50b, fols. 111 – 2, on which see below, p. 337.

70 See Pugliese, 'Scientific Achievement' (n. 24), pp. 9f. Pugliese's dating of the lectures published in Hooke, *Posthumous Works*, pp. 1f., corrects that of M. B. Hesse ('Hooke's Philosophical Algebra', *Isis*, 57 (1966), 68) and is to be preferred to the suggestion of Paul Wood in 'Methodology and Apologetics: Thomas Sprat's *History of the Royal Society*', *British Journal for the History of Science*, 13 (1980), 24 n. 45.

71 Pugliese, 'Scientific Achievement' (n. 24), p. 11.

from the preface to *Micrographia*.[72] It is therefore revealing that the Cutlerian Lecture of 25 July 1666 is based on a paper which is almost unique among Hooke's extant manuscripts in being a genuine history of trades. This was 'a Description of the manner of making Salt at a Saltern in Hampshire', and it gave exact details of where and how salterns were laid out and how brine was evaporated at them, precisely detailing each stage of the process and even including an illustration.[73] Hooke's papers contain only one parallel to this distinctive item in its presentation and its subjectmatter, and this is another account of a craft practice, a pair of leaves describing 'The way of making felts', which is almost identical in format and has a matching illustration, on which there is a further page of commentary. Probably, this is to be identified with the 'good lecture of Mr Hookes about the trade of Felt-making' which Samuel Pepys records in his diary that he heard on 21 February 1666 and which he considered 'very pretty'.[74] Indeed, it is even possible that this and the lecture on salt-making formed part of a systematic survey of trades which Hooke might have begun in response to Cutler's foundation, as suggested by his affirmation to Boyle that he was 'compiling a history of trades and manufactures', 'a very hard, difficult, and tedious task'.[75]

Quite apart from the fact that these are the only two items of their kind among Hooke's extant remains, however, there is evidence even within them that Hooke was not particularly comfortable with the kind of work that a proper history of trades necessitated. Thus he opened his account of the feltmaker's trade by promising to draw conclusions both as to 'the operations of Nature' and to the improvement of technical practices, but he never reached this point, abandoning the project after carrying out only the narrative part.[76] The account of saltmaking in Hampshire also ends rather abruptly, while it is arguably significant that the Cutlerian Lecture that Hooke based on this pasted it together in a rather unsatisfactory manner with general introductory remarks with which it is rather ill at ease. In these, he explained about the different kinds of salt and particularly the

[72] See above, pp. 293 – 4.

[73] Cl.P. xx.40. The illustration is reproduced as plate 12, above. For a reference to a later presentation to the Society of similar material by Hooke, see Royal Society Copy Journal Book, viii, 250.

[74] Cl.P. xx.96; Pepys, *Diary*, vii, 51. The illustration is reproduced in E. N. da C. Andrade, 'Robert Hooke', *Proceedings of the Royal Society*, A, 201 (1950), facing p. 470. It is perhaps worth noting that Andrade is wrong to presume (ibid., p. 470) that the failure of this lecture to appear in the minutes was due to negligence or hostility on Oldenburg's part: it was simply that at this point Cutlerian Lectures were not regarded as part of the Royal Society's regular proceedings.

[75] Hooke to Boyle, 6 October 1664, Boyle, *Works*, vi, 493. A reference in a letter from Moray to Oldenburg of 16 Nov. 1665 to the fact that Hooke should finish his existing tasks 'rather then to learn a dozen trades' should possibly be read as referring to this: *Oldenburg*, ii, 610. Note also that Cl.P. xx.96 starts by describing tools unique to that particular trade, as if this was part of a set pattern.

[76] Cl.P. xx.96. See also Hunter, *Science and Society*, p. 103.

difference between the bay salt that was produced naturally in the sun in hot climates, the white salt produced artificially in salterns in temperate countries like England, and salt mined underground. He then included an excursus on the chemical reaction that produced salt, before finally turning to the description of the craft practice.[77]

In an opening passage to the lecture as a whole, Hooke indicated a broader context, showing how — even when complying with Cutler's wish that he should deal with the History of Trades — he placed it in the framework with which he had been provided by the formula devised by the Royal Society in its negotiations with Cutler, namely 'the History of Nature and Art'. He explained how 'The Disigne [sic] of this Lecture as I have severall times hertofore Declard, is for the promoting and collecting of Naturall & Artificiall History', and how his method 'has been to begin the vacation with some Instances of Naturall history And to finish them with some specimina or instances in Artificiall'. In that particular vacation he had already considered how nature produced figures in petrified bodies, and how sand and other substances, even including water, could be turned to stone (a topic which overlapped, incidentally, with matters that he had dealt with in *Micrographia*). He went on: 'Now that the History of the productions of Art may not much differ from the historys I have Deliverd already of the productions of Nature I have made choice of the History of making Salt out of Sea water by boyling as It is practised at the Salturnes upon the south parts of England & particularly upon the south shores of Hamshire, for the subject of this present Exercise'.[78]

Hooke was thus attempting to achieve a balance in his subjectmatter between the natural and the artificial. But, even within the lecture from which these extracts have been taken, one may perceive more enthusiasm on Hooke's part for understanding nature than for mastering technical practices. Moreover Hooke seems to have come to see that he could legitimately use the rubric that he had been given for the lectures to talk more broadly about the project of a natural history, in other words with that vast descriptive enterprise which was so important to the Royal Society in its early years and into which the description of technological practices could be easily fitted. It is to this that the other surviving group of items that can definitely be identified as Cutlerian Lectures of c. 1665 – 6 are devoted. One, preserved in the Classified Papers of the Royal Society, has recently been published in full by David Oldroyd; a second, from the same source, is printed for the first time as an appendix to this paper; while a third, which

[77] Cl.P. xx.42, which clearly postdates and was written as an introduction to ibid., 40.

[78] Cl.P. xx.42: after 'vacation' 'Lectures' has been deleted, as has 'instances or' before 'specimina', while a small deletion between 'history' and 'And' is now illegible. After 'historys', 'of' has been deleted. Cf. Hooke, *Micrographia* (n. 29), pp. 107f.

survives in a volume of Hooke's papers in the Guildhall Library, has been discussed at length by Patri Pugliese in his unpublished thesis on Hooke.[79]

The first of these emphasizes the importance of trades for the study of natural history: indeed David Oldroyd, in his commentary on it, has suggested that a link with the Cutlerian benefaction is suggested by the disproportionate emphasis that this point receives. In the main, however, they deal with the proper aims and techniques of a 'Philosophicall History', advising the historian 'in what manner he ought to proceed, what he ought to observe, how to examine it, how to preserve & register it, How to range and order it & lastly how to apply & use it'. Topics dealt with include the proper qualifications of a natural historian, the broad classification of matters to be dealt with and the equally important question of just what it was necessary to learn about each phenomenon: the latter was something which Hooke illustrated by examples, in the lecture now in the Guildhall Library doing so by giving a series of queries about the comet of 1664 – 5. Lastly, in the fragment published here, he also devoted thought to how the piecemeal observations of individual naturalists were to be collated and utilized in an effective manner: he suggested a strategy which is not dissimilar to the *Philosophical Transactions* that Oldenburg was just beginning at this time, and to Hooke's own publication programme in the late 1670s, including, perhaps significantly, the *Lectures and Collections* which he issued in 1678 as part of the Cutlerian Lectures.[80]

These items, whose link to the Cutlerian foundation is clear, bear a close relationship to a much more famous work by Hooke on the same subject, the '*General Scheme, or Idea* of the Present State of Natural Philosophy, And How its Defects may be Remedied', of c. 1668 which Richard Waller was to publish in Hooke's *Posthumous Works*. In its current form, this sustained treatise does not appear as if it was intended for oral presentation, though it is possible that it originated as lectures. But there is an unmistakeable overlap between its content and the lectures itemized here: as Patri Pugliese has put it, 'the essay itself presents many of the same ideas as Hooke's earlier lectures, but they have been reorganized and developed'.[81] One may therefore suggest a close link between Hooke's methodological writings and the demands of the Cutlerian Lectureship. This is not to say that Hooke had had no previous interest in such matters, having already declared an

[79] Cl.P. xx.50a (fols. 99 – 109), printed in D. R. Oldroyd, 'Some Writings of Robert Hooke on Procedures for the Prosecution of Scientific Inquiry, Including his "Lectures of Things Requisite to a Natural History" ', *Notes and Records*, 41 (1987), 151 – 9; Cl.P. xx.50b, part 1 (fol. 110), printed below, pp. 337 – 8; City of London Guildhall Library MS 1757, item 11; Pugliese, 'Scientific Achievement' (n. 24), pp. 16f. See also above, n. 69.

[80] Oldroyd, 'Some Writings' (n. 79), pp. 151, 156 – 8, 159 and passim; Guildhall MS 1757, item 11, passim; below, p. 338; Gunther, *Early Science* (n. 47), pp. 209f. For Hooke's lectures on the comet in 1665, see above, p. 298. On the beginning of *Philosophical Transactions*, see Hunter, *Science and Society*, pp. 51 – 3.

[81] Hooke, *Posthumous Works*, pp. 1f.; Pugliese, 'Scientific Achievement' (n. 24), p. 22.

intention to write a discourse on 'the manner of compiling a Natural and Artificial History' in *Micrographia*.[82] But it seems likely that the need to give a sequence of presentations on related topics acted as a significant stimulus to Hooke to put his ideas on such subjects in order. It also seems highly likely that Hooke's choice of this theme reflected what he knew of his listeners, and that he saw programmatic statements of this kind as particularly appropriate to the audience made up principally of Fellows of the Royal Society that he was addressing.

Indeed, one could argue that the forum which the Cutlerian foundation imposed on Hooke may help to answer the question posed by David Oldroyd, 'why it was that one such as Hooke thought it appropriate or needful to write about "method" '.[83] It may also help to explain something that has puzzled many commentators on Hooke's methodological writings, namely why he spent such a disproportionate amount of time laying down the ground-rules of philosophical enquiry and so little outlining the superstructure of 'philosophical algebra' to which he claimed ultimately to aspire: for what he was doing was not simply expounding his own method, but trying to devise a methodology appropriate to a corporate body devoted to the pursuit of natural knowledge. Moreover, he knew that he would be presenting this to an audience containing many whose approach to nature was much less sophisticated than his own.[84]

Turning from the 1660s to the 1670s, in the latter decade we have more direct evidence on the content of Hooke's lectures, due largely to the fact that a series of his 'Lectures: Physical, Mechanical, Geographical & Astronomical' were published in these years, to be collected together in 1679 under the title *Lecturae Cutlerianae*, which is undoubtedly what they were. Moreover the first of these, Hooke's *Attempt to prove the Motion of the Earth from Observations* of 1674, was opened by a fulsome dedication to Sir John Cutler, followed by an epistle 'To the Reader' which recapitulated how Cutler had founded a '*Physico-Mechanical Lecture*' 'for the promoting the History of Nature and of Art', in connection with which Hooke had collected 'many Observations both of the one and the other kind'.[85] Apart from the *Attempt*, the other tracts in the series were Hooke's *Animadversions on the first part of the Machina Coelestis Of . . . Hevelius* (1674), his *Description of Helioscopes, And some other Instruments* (1676), his *Lampas: or, Descriptions of Some Mechanical Improvements of Lamps & Waterpoises* (1677), his *Lectures and Collections*, including *Cometa* (1678), and his *Lectures De Potentia Restitutiva*

[82] Hooke, *Micrographia* (n. 29), sig. b1v. Cf. Oldenburg to Boyle, 27 January 1666, in which he reports that Hooke had shown him 'a Method for writing a Naturall History': *Oldenburg*, iii, 32 – 3.

[83] Oldroyd, 'Some Writings' (n. 79), p. 162.

[84] Hesse, 'Hooke's Philosophical Algebra' (n. 70), esp. pp. 68 – 9; Oldroyd, 'Some Writings' (n. 79), pp. 145 – 7; R. S. Westfall, introduction to 1969 Johnson Reprint Corporation edition of Hooke's *Posthumous Works*, pp. xxiv – vi. For alternatives in sophistication see, for example, Michael Hunter, *John Aubrey and the Realm of Learning* (London, 1975), ch. 2.

[85] Hooke, *Lecturae Cutlerianae* (n. 46), sigs. A2 – 4.

(1678).[86] In addition, the unpublished manuscript of a Cutlerian Lecture which evidently dates from 1673 is to be found in a volume of Hooke's papers in the British Library.[87]

By now, Hooke's view as to what it was proper for him to lecture on appears to have altered somewhat (except that, if he *did* lecture on his spring-balance in 1664, as he claimed, there may have been a degree of continuity with at least some of his earlier presentations).[88] The lectures of the 1670s could be seen as a different — though equally valid — variation on the theme of 'the History of Nature and Art', comprising an account of Hooke's continuing work on the classic problems of natural philosophy, with particular reference to the use of instruments to supplement the senses in obtaining exact information and in testing phenomena. His unpublished lecture of 1673 dealt with the work of Huygens and of Hooke and his English contemporaries on the use of a pendulum for time-keeping at land and sea; the *Attempt* sought to show how exact measurement might make possible a new account of the motions of the heavenly bodies; the *Animadversions on Hevelius* argued for the use of telescopic rather than open sights as a means of maximizing observational accuracy; *Lampas* dealt with mechanisms to provide fuel at a regular rate so that a source of heat could be accurately controlled for experimental purposes.[89] In addition, in many of these works Hooke described other devices that he had come up with, or alternative — often practical — ways of applying the principles that they represented. For instance, Hooke's universal joint appeared in connection both with the drive for his equatorial quadrant and with a lathe, while there was also the notorious section on clocks, including the controversial spring-balance.[90]

To this was added more miscellaneous material — for instance, on microscopy in *Lampas* — while the two final volumes of the series ranged more widely still, from astronomical observations by Hooke and others to

86 All are printed in Gunther, *Early Science* (n. 47).

87 British Library MS Sloane 1039, fols. 129 – 30. That this dates from 1673 is shown by Hooke's reference to the fact that he had only recently received the book dealt with in it, Huygens' *Horologium Oscillatorium*, published in that year. In addition, this item may be linked to Hooke's reference to writing and delivering a lecture or lectures on 'Zulichems book' on 25 June and 3 July 1673 (*Diary*, pp. 48 – 9). In an introductory section which has been crossed through, Hooke refers to reading the Cutlerian Lecture 'againe in this Place where it was first begun' — i.e., presumably Gresham College, although the Royal Society did not remove its weekly meetings from Arundel House to Gresham until 1 December that year. Why the lecture was not published is unclear, though it was perhaps subsumed by the controversy with Huygens over the spring-balance, which broke out shortly afterwards.

88 See above, p. 299.

89 On this, see R. S. Westfall, 'Robert Hooke, Mechanical Technology, and Scientific Investigation', in J. G. Burke, ed., *The Uses of Science in the Age of Newton* (Berkeley and Los Angeles, 1983), esp. pp. 94f.

90 Gunther, *Early Science* (n. 47), passim.

disquisitions on a variety of subjects by authors at home and abroad which look forward to the *Philosophical Collections* that Hooke was to produce from 1679 to 1682. Indeed, if these were indeed public lectures that Hooke gave, some of them must simply have comprised the recapitulation of data sent in by others, and, however justifiable this might have been in terms of the collation of useful data that Hooke had alluded to in his lectures of the 1660s, it might have stimulated Cutler to wonder whether he was getting value for money for his endowment.

Most of Hooke's lectures of the 1670s, however, were not only original but in many ways highly topical to the interface between science and practice that could be deemed appropriate as the subject of a 'mechanical lecture'. Hooke could presumably argue that, though the devices which he divulged in these lectures were predominantly conceived for their potential in assisting with natural philosophical tasks, their spin-offs often had direct practical application, and hence could be argued to exemplify the spirit, if not the letter, of Cutler's foundation. Indeed, on the basis of these lectures, more than one author has argued for an intimate relationship between Hooke's inventive skills and his approach to the natural world, while the most recent and cogent of these, J. A. Bennett, has even seen the way in which Hooke conceptualized nature as being a fruit of his mechanical mode of thought.[91]

On the other hand, in the light of the discussion that we have already surveyed about the proper subjectmatter of the lectures, it is interesting to find that Hooke remained defensive about the topics that he dealt with. At one point in his exposition he actually inserted a broadside on this issue specifically referring to his patron, Cutler, and arguing how: 'We see therefore the necessity of the conjunction of Physical and Philosophical with Mechanical and Experimental Knowledge, how lame and imperfect the study of Art doth often prove without the conjunction of the study of Nature, and upon what rational grounds it was that Sir *John Cutler*, the Patron and Founder of this Lecture, proceeded in joyning the contemplation of them both together'.[92] He also explained 'To the Reader' in *An Attempt* how 'This first *Discourse* is upon an Observation of Nature, and may therefore be properly referred to that Head, though it contein somewhat of the Improvement of Art: The second speedily to follow, will more properly be referrable to *Artificial Improvements*, though it will contein also many Observations of Nature; and I design alwayes to make them follow each other by turns, and as 'twere to interweave them, being apart but like the Warp or Woof before contexture, unfit either to Cloth, or adorn the *Body of Philosophy*'.[93]

91 See Bennett, 'Robert Hooke' (n. 4), p. 43 and passim. See also M. 'Espinasse, *Robert Hooke* (London, 1956), ch. 4.
92 Gunther, *Early Science* (n. 47), p. 16.
93 Hooke, *Lecturae Cutlerianae* (n. 46), sig. A4v.

The Hooke-Cutler Dispute, Round 1: Arbitration

It is ironic that during the period when these latter lectures were delivered, Hooke was not in fact receiving any remuneration for them. In the lawsuit that was to ensue, Hooke alleged that, after paying his annuity regularly throughout the 1660s, Cutler ceased to pay it as from Midsummer 1670, and this is confirmed by the figures for Cutler's owings that Hooke mentions in the diary that he kept from 1672 onwards.[94] Indeed, the publication of the lectures itself evidently formed part of Hooke's campaign to try to persuade Cutler to pay his mounting arrears, particularly their collection under the general title of *Lecturae Cutlerianae* in 1679, a point when, as we shall see, Hooke's relations with Cutler were beginning to reach crisis-point. Doubtless the intention was to provide a bargaining counter in the dispute and to prove that he *had* indeed been giving lectures according to his obligation.

Just how conscientious he was in his duties should be easier to assess for the 1670s than for the 1660s, since for most of this decade Hooke kept a detailed diary; in addition, in contrast to the state of affairs earlier, his lectures were now occasionally minuted in the Royal Society's journal book as if they formed part of the Society's proceedings. In his diary, Hooke sometimes records that he 'Read Sir J.Cutlers Lecture', in a handful of cases mentioning subjectmatter that overlaps with the published series of the 1670s.[95] Matters are complicated, however, by the fact that the diary juxtaposes this with a similar record relating to Hooke's obligations as Geometry Professor at Gresham College (these lectures also took place on Thursdays, but during term). Moreover neither Hooke's record nor the Royal Society's can be presumed to provide a complete account of the presentations that he gave, so one cannot attempt a systematic assessment of his lecturing activity, especially towards the end of the decade, when the diary entries become sparser.

What the diary does suggest is that, although Hooke indeed gave numerous lectures in these years — as the published series itself testifies — he was not entirely conscientious in observing the letter of his agreement with Cutler. Thus on 18 December 1673 he notes how he 'Read not Sir J. Cutlers Lecture but Discoursd it to the Society' — in other words, he presumably gave an informal presentation rather than a public lecture.[96] In

94 See PRO C7/564/29, item 2, and Hooke, *Diary*, e.g. pp. 49 (30.6.73: '3 years'), 80 (15.1.74: 'between 3 and four year').

95 Hooke, *Diary*, p. 331. The diary entries for the following dates certainly relate to Cutlerian Lectures: 26.6, 3.7, 11.12.73; 22, 29.1.74; 28.1.75 (cf. Birch, *History*, iii, 179 and perhaps also 121 concerning 22 Jan. 1674); 3, 10.2, 20.4 and perhaps 22.6.76; 5.7, 29.11.77; 1.8.78. The following references are ambiguous, but they could well concern Cutlerian Lectures: 12, 19, 26.3, 7.5.73 (cf. Birch, *History*, iii, 78, 85 – 7); 12.11, 3.12.74; 4, 11, 25.11.75 (cf. Birch, *History*, iii, 229 – 31); 5.6.79 (cf. ibid., iii, 489), 30.10, 6, 13, 20.11.79; 17.5.82 (in City of London Guildhall Library MS 1758, fol. 76).

96 Hooke, *Diary*, p. 75.

addition, it appears that — contrary to the terms of the endowment — Hooke sometimes read lectures on days on which there was no meeting of the Society, while on other occasions he seems to have delivered what were clearly Cutlerian Lectures during term rather than in the first four weeks of the vacation.[97] Such laxity on Hooke's part is understandable enough, particularly in the light of Cutler's failure to pay him, but it was potentially a dangerous hostage to fortune when it came to a legal contest. Indeed, Cutler was apparently aware of this and may have been quietly collecting evidence to prove a dereliction of duty on Hooke's part, as on 5 July 1677, when Hooke recorded in his diary: 'Attend lecture of Sir J.Cutler in the Hall with Mr Haak. Drew scheme of comet in the hall. Capn. Panton came for Lecture but noe more', going on to note 'Axe here to spy' — in other words, Thomas Axe, steward to Cutler's son-in-law, Sir William Portman, who appears to have acted on Cutler's behalf in this and other connections at this time.[98]

The diary also provides profuse information about Hooke's attempts from 1673 onwards to persuade Cutler to pay him his mounting arrears. On 30 June that year, Hooke noted: 'Sir J. Cutler promised me payment for 3 years, Dr Whistler witnesse at Kings armes by the pump in thredneedle street'. Comparable references follow, while shadowy hints in the diary and elsewhere suggest that later in 1673 the Royal Society may have raised with Cutler the issue of the bond that he had signed in 1665 promising £1000 in the event of discontinuing his payments to Hooke.[99] This would explain why just at this time a memorandum by Henry Oldenburg noted the possibility that the Society might acquire £1000 from Cutler: perhaps the Society offered to end the arrangement and cancel Cutler's bond if he provided this much-needed injection of capital, and it may well be that an attempt at an amicable compromise was attempted, with the virtues of helping the Society out being urged on Cutler as a positive gesture in its own right by prominent ecclesiastics and others, including Seth Ward, Bishop of Salisbury, whose involvement in the matter in the 1660s has

[97] Matters are complicated by the fact that the Society's minutes are incomplete for parts of the 1670s, but the following diary entries appear to refer to Cutlerian Lectures on days when the Society did not meet: 26.6, 3.7.73; 29.1.74; 20.4 and perhaps 22.6.76; 5.7, 29.11.77. For lectures given during term, see 28.1 and perhaps 4, 11.11.75, 3 and perhaps 10.2.76, 30.10 and 6, 13, 20.11.79 (some of the latter may also have failed to coincide with Royal Society meetings, though this may be due to lacunae in the minutes), and perhaps 17.5.82 (see n. 95).

[98] Hooke, *Diary*, p. 299. Mr Haak is Theodore Haak, F.R.S.; on Panton, see *DNB*: on this occasion Hooke notes that Panton gave him his 'designe for Academy'; on Axe, see also below, p. 308, and note the diary entry under 24.1.78: 'Axe here, perswaded me to put bond in Suit' (p. 341). For other 'spies', see ibid., pp. 360 (30.5.78), 382 (31.10.78), 457 (4.11.80).

[99] Ibid., p. 49, and see above, 'Towards Solomon's House', p. 193, including the references cited in n. 31. The figure of £1000 mentioned there clearly makes sense in the light of Cutler's bond, of which I was ignorant when that passage was written.

already been noted.[100] If so, such an arrangement would have raised in acute form the ambivalence which has already been noted as to whether Cutler's endowment was intended for Hooke or the Society: for, whereas it was to Hooke that the arrears of the annuity were due, the Society appears to have presumed that it would be the corporate body which would pocket the penalty that Cutler paid.

In any case, nothing came of this, and for the remainder of the 1670s Hooke's diary records a series of promises on Cutler's part — often made in public, like that in June 1673 — which never seem to have come to anything. Many of these involved such intermediaries as Thomas Axe, Edmund Boulter — Cutler's nephew and a fellow Grocer, who was ultimately to become the beneficiary of Cutler's will — or Dr Daniel Whistler, whose involvement with Cutler's original gift to the Society has already been indicated. Thus on 15 January 1674, Cutler promised a 'final determination' before 1 March that year, agreeing in front of Boulter what he owed Hooke; later in 1674, however, Hooke complained on more than one occasion how Cutler 'playd the foole with me'.[101] Moreover, this pattern was to repeat itself over the subsequent years, while Hooke continued doggedly to record the debt in his lists of moneys owing to him: on 21 June 1677, for instance, when Hooke bumped into him in the City, Cutler was still promising 'that he would speedily pay me', while a year later, on 30 November 1678, Hooke records how Cutler again publicly acknowledged his debt and 'promised to pay me every farthing soe soon as ever I would call on him with Dr Whistler'.[102]

Clearly, Hooke was getting increasingly frustrated, and it was only a matter of time before he sought some other means to extract the money from Cutler. Nothing seems to have come of his appeal to the Royal Society on 25 February 1675, when he had 'Moved the Councill about Sir J.Cutler'.[103] But a show-down apparently occurred on 27 July 1680, when Hooke went with his lawyer friend, Sir John Hoskins, and Sir Christopher Wren to Cutler's house, where 'Sir J.Hoskins spake to him much to the purpose, he refused payment, said he designed to arrest me, etc.'[104] Perhaps Hooke was deterred by this, for matters then simmered for a further two years. But in the summer of 1682 the Royal Society was at last persuaded to take the initiative in the matter, responsibility being deputed to Sir William Petty, who, as a friend of John Graunt's and hence of Cutler's, had been closely involved in the foundation when it was originally set up. On 12 July that year, Petty was 'desired to accompany those members of the council, who by a former order were desired to speak to Sir John Cutler for

[100] See particularly Ward to Sheldon, 19 Aug. 1673, Bodleian Library MS Tanner 42, fol. 29.

[101] Hooke, *Diary*, pp. 80, 117 (13.8.74), 136 (17.12.74).

[102] Hooke, *Diary*, e.g. pp. 131 (20.11.74), 161 (21.5.75), 265 (31.12.76); pp. 297, 386.

[103] Hooke, *Diary*, p. 149: this is not recorded in the minutes of the meeting on that date published in Birch, *History*, iii, 190 – 1.

[104] Hooke, *Diary*, pp. 449 – 50.

obtaining Mr Hooke's arrears'.[105] Unfortunately, no record survives of the 'former order', so it is not known when the first official step was taken to which this was a sequel.

In any case, this initiative evidently achieved nothing, and it was at this point that Hooke resorted to legal action, taking the case to the Court of King's Bench. There, on 23 October 1682, Hooke's lawyer, Charles Ballett, displayed Cutler's bond of 29 June 1665 and applied for the payment of £1000 as guaranteed in that in the event of Cutler's failing to fulfil his obligation to pay Hooke his annuity, together with costs of £10. Cutler's response was to make the standard plea 'non est factum meum' (it is not my deed), and he thereupon 'put himself on the country' — in other words, applied for the matter to be tried by jury.[106] It must have been in connection with this that — in one of the increasingly fragmentary diary entries which Hooke kept between 1681 and 1683, which were omitted from the published edition — he records that on 6 November 1682 he was 'with Ballet[.] Gave coppy of Sir J.Cutlers bond'.[107]

In addition, Hooke's case was pursued on his behalf by the Royal Society, and a letter survives to Cutler, dated 19 December 1682, from Sir John Hoskins and Daniel Colwall, another prominent member of the Royal Society who had been Treasurer until 1679.[108] Cutler had apparently impowered Hoskins and Colwall to deal with Daniel Whistler about the affair, but, after he had broken two engagements, they had enquired about the matter for themselves. Stressing how Cutler could not have found a worthier object for his charity, they reported that they 'cannot find any footsteps of any other Consideration or Agreement then what is taken notice of in the Condition of the bond, But many proofs of a more unrestrained Liberality'. What they were evidently trying to do was to disprove the existence of any contract independent of the bond, in the absence of which they would presume gratuitous performance on Hooke's part. As Hooke's close friend as well as a lawyer, Hoskins was perhaps likely to take Hooke's part in the matter, and he and Colwall added: 'Our own memorys satisfy us that Mr Hooke has done much more then he is ever sayd to have been obliged to, and even in the same kind, at least till stoppage of Payment began'. Moreover, after noting how Hooke had had to 'foregoe other more profitable Employments, and Relinquish a Considerable and aequivalent demand to what he hopes for from you', they continued 'that any Guift be it as free and voluntary and without Consideration as is

105 Birch, *History*, iv, 155. On Petty's earlier role, see above, pp. 284f.

106 PRO KB27/2023, rot. 461. Ballett may have been Charles Ballett of Clement's Inn: cf. H. F. MacGeagh and H. A. Sturgess, *Register of Admissions to the Honourable Society of the Middle Temple* (London, 1949), i, 224. For an earlier reference to him in connection with Cutler, see Hooke, *Diary*, p. 261 (8.12.76).

107 City of London Guildhall Library MS 1758, fol. 77.

108 Bowood House, Petty MSS H[8]15; endorsed 'to Sir John Cutler', 'Mr Hook' and 'Mr Hooke & Sir John Cutler'. I have consulted the Bowood material through the microfilm copies of it on deposit at the Bodleian Library, Oxford, by courtesy of the Earl of Shelburne.

Imaginable yet when once Bestowed there is as firm a right transferred as in the hardest Smithfeild Bargain'. They therefore requested Cutler to pay Hooke 'what your Bond to him has Given him and made justly his'.

The jury trial for which Cutler had applied in October was appointed to occur on 23 January 1683, but it apparently failed to take place then because, instead, negotiations took place between the two parties on the previous day, as is revealed by an elaborate, if laconic, entry for that date in Hooke's diary: 'Sir W Petty about Sir J.C. with Lord Russell — Sir J. Hosk[ins]. met Sir J.Cutler with D.Whistler at Sir W.Petty. Sir J.C. Left all to Sir W.Petty. I Refusd[.] Dr Whistler[.] Sir J.C. gave me a note to Receive £200 from Bolter. We parted.' As a sequel, on 23 January 1683, Hooke notes how Boulter 'accep[ted] £200 bill' and on the 25th he records: 'Re[ceived] of Edmund Bolter. 200 on Sir J.Cutlers note'.[109] Thus the threat of legal action had finally galvanized Cutler into making an interim payment of part of the arrears, while, equally significantly, he had agreed to arbitration at the hands of Petty. It was this that now ensued, despite Hooke's claim that he 'Refusd': arbitration was a standard way of attempting to deal with legal disputes like this, and in this case we are very fortunate that a series of papers relating to the matter survives among Petty's papers at Bowood House. One of these is the letter to Cutler from Hoskins and Colwall of 19 December 1682 which has already been referred to, which is in fact in Hooke's hand, either because he drafted it in the first place, or — more likely — because this is a copy that he made.[110] The remaining papers relate to the subsequent arbitration. They include a draft agreement signed by Petty and dated 14 May 1683, accompanied by a note to Hooke from Petty of the same date, and various overlapping documents which must have been compiled at about the same time. The date of the agreement was just before that set by the Court of King's Bench for the case to return to that court, thus explaining why Petty reached his final judgment when he did.[111]

Though the documents do not include a submission from Cutler, his point of view can be deduced from the extant papers, supplemented by what may be learnt from his submission to the Court of Chancery in June 1683. Clearly he must have stated that he was prepared to pay Hooke only on condition that the arrangement concerning the lectureship was revised so as to be more satisfactory in his eyes. For, although he had paid some of Hooke's outstanding arrears, Cutler could himself claim a sense of grievance. His fundamental complaint was obviously that Hooke had failed in his obligation to give sixteen lectures a year which were clearly distinguishable as 'Cutlerian' orations and from which credit therefore accrued to Sir John; but Cutler evidently also claimed that the lectures had

109 PRO KB27/2023, rot. 461 (there is an illegible erasure in the MS under the January date); City of London Guildhall Library MS 1758, fol. 77v. For William, Lord Russell, see ibid., fol. 73v (4.2.81), Hooke, *Diary*, p. 457 and *DNB*. On Boulter, see above, p. 308.

110 The neatness and lack of alterations suggest this. For the other documents, see Bowood House Petty MSS H[8]16 – 19.

111 PRO KB27/2023, rot. 461.

not dealt with the subject that he had intended. Probably he suggested a once and for all settlement, the exact details of which we unfortunately do not know, which would have rectified these defects in the original agreement from his point of view in return for payment: it may well be that he argued that it was not altogether fair that he should be legally obligated while Hooke was not, and that Hooke should sign as strongly binding an agreement as he had. Hooke, however, evidently obstructed this, seeing Cutler's alterations as a restriction of the freedom of action that he had enjoyed under the original gift.[112]

Though Hooke himself felt strongly about this, however, Cutler's case was arguably quite a powerful one, telling enough (and embarrassing enough to the Royal Society, which had presided over the unsatisfactory original agreement) to make some members of that body feel that Hooke should be prepared to make more concessions than he was evidently prepared to, and perhaps to accept the suggested compromise without more ado. This is implied by Petty's exchange with Hooke in May 1683, as we shall see, while it is further suggested by Hooke's comments in a letter of 31 March that year to Sir John Hoskins: 'I am extremely Disatisfyed with the proceedings concerning Sir J.C. businisse', he wrote, 'and Doe exceedingly want the Comfort & councell of a true freind who is not to be known but at such a time for I find that many of those I thought I might have Relyed upon are nothing soe and I am like to Reap noe benefit by my delay but rather much detriment. Besides I find others very active in indeavouring to find faults and defame me to make way for their own designes, and what ever I have done though to the best of my abilitys are — and at best but = 0. Soe that I find I must study other principles of Algebra then what I have Hitherto been acquainted with.'[113]

The areas in dispute over which negotiation evidently took place are apparent from a lengthy statement in the hand of an amanuensis who can be associated with Hooke, which Hooke must have prepared for Petty.[114] Part of this comprised six arguments concerning the subject-matter of the lectures, which we will come to shortly, but before that Hooke made a series of more general points about the terms of the gift.

> Mr Hooke desires to bee but in the same condition[,] as if hee never had bin in Sir Johns disfavour and not to have any new obligation imposed upon him For that would bee a diminution to Sir Johns generosity and make that seeme a bargaine which was a guift.

112 For his advice on the significance of the negotiations from the point of view of the law of contract, I am indebted to Christopher Whittick.

113 E.L. H.3.70 (Hooke's draft). In the last sentence, 'that' has been repeated and the second erased.

114 Bowood House Petty MSS H[8]16. This document is endorsed at the top in pencil in a neat italic hand, 'from Mr Hook'; the verso is blank. Cf. Cl.P. xx.50, fol. 97, and above, 'Towards Solomon's House', pp. 242 – 3.

2 Hee has bin trusted hitherto & notwithstanding failer [sic] of payment has performed new and laborious exercises, as appeares by the catalogue

3 The condition of the bond, the first profession of both sides and the constant practice since, interest the R S in the minding this lecture and they will still so bee the R S having[115] an endowment though not sufficient for their occasions, yet sufficient to warrant their duration equal to that of any other corporation, at least they will not bee extinct while Mr H lives

4 Mr Hookes honour & gratitude will oblige him, which are weakned by being stinted or limited, & hee would receive an honorarium but scornes hire

5 It is impossible but hee must bee trusted for hee may else degenerate into perfunctory wall lectures

6 Sir John has often expresst his meaning to bee free, and liberall and that hee would not have Mr H bee bound & cared not whether he read or no

7 Mr Hooke's present circumstances assure against a change of studyes.[116]

As this shows, Cutler was clearly requesting a much more formal arrangement than hitherto, which Hooke resisted on the grounds, as he said, that this would 'make that seeme a bargaine which was a guift'. To illustrate his case, Hooke provided a 'catalogue' of what he had done by way of lecturing in the past, though this does not survive. More specifically, Cutler was evidently insisting that what had originally been arranged *was* contractual, and what is more that it bound Hooke rather than the Royal Society: indeed, Cutler may have suggested bypassing the Society, whose role in supervising the lectures he had perhaps realised had turned out to be extremely perfunctory. Possibly, he hinted that the Royal Society might not outlive Hooke, and Hooke's third clause was obviously intended to counter this, stressing the Society's corporate role and alluding to the royal endowment that it had just acquired at this time:[117] the implication of this from Hooke's point of view was that he was under no obligation because the bond was between Cutler and the Royal Society. It is also interesting that he asserted his 'honour' could be trusted and that an honorarium, rather than 'hire', was appropriate: primarily, he was denying the existence of a contract, but did he also feel that Cutler was treating him like a menial?

More interesting still, however, is the second part of this document, which makes it clear that one of the issues in dispute at this stage was the

115 Before 'having', 'being' has been deleted.
116 Item 7 is inserted at the end of the line after item 5.
117 See Hunter, *Royal Society*, p. 40.

subjectmatter of the lectures. The result was that Hooke was inspired to make a clear statement of his priorities as to research and exposition in science and technology. He continued as follows:

> As to the subject of these lectures, Mr H. thinkes it inconvenient to bee put immediately upon the Mechanicke part of trades For
>
> 1 The reputation of Sir J^{n} Mr H. the R S. & England will bee most advanc't by teaching perpetuall & universall knowledge whereas the mechanicke is mostly pro hic & nunc & will not as the other, bee a part of naturall history
>
> 2 All tradesmen know the operative part and that cannot perfectly bee taught by words onely but the speculative & rationall part may
>
> 3 There is nothing properly taught to bee knowne but where the reason and cause is explained which will also give light, rules, & hints for further inventions and it is better doe this at first altogether, that is, read lectures of principles than upon each severall subject begin & demonstrate from first principles, and trades or parts of them may bee instanced as experiments to prove theoryes[118]
>
> 4 Jealousyes will bee put in tradesmens minds that this may prejudice them, unlesse some philosophicall introduction vindicate it from contempt of gentlemen & contumely of the vulgar, and this suits with promoting naturall science, the worke of R S
>
> 5 Mr H. is prepared with unheard of discoveryes & demonstrations in physicks but must compile mechanicks though already hee know them[119]
>
> 6 The first and continued understanding of the designe was to explaine nature first, then art and so hee intends, and fancy can never be forc't & gratifying it is possibly, what most inclines Mr Hooke to these studyes, who could gaine much more by other imployments

It is perfectly clear from this that one of Cutler's justifications for failing to maintain his side of the bargain after the first few years must have been that Hooke was not lecturing on the topic that Cutler had initially intended the lectureship to deal with, namely the history of technical practices. Indeed, if this is read in the light of the negotiations of 1664 – 5, one wonders if Cutler had always really hankered after a lectureship on technology: though he had agreed to the Royal Society's broadening of its terms of reference, it may be that he had not realised quite what was

[118] 'and trades . . . theoryes' has been added in a smaller hand, as has 'and . . . R S' in item 4 and 'and fancy . . . imployments' in item 6.

[119] after 'discoveryes' 'in' has been deleted, and '&' has apparently been altered from something else. In item 6, 's' has been deleted at the end of 'other'.

involved in the altered formula, and had presumed that it was closer to his actual intention than the Society, for its part, intended. Hence this renegotiation must have seemed an ideal opportunity to sort the matter out.

Hooke's views, on the other hand, seem if anything to have hardened. Whereas in the mid-1660s, as we have seen, he had been happy to lecture on the History of Trades and to argue that it formed a valuable portion of the broader project of a history of nature, now he categorically argued that mechanical knowledge was too transitory to be part of natural history, though his third point arguably slightly modified this in acknowledging that trades could be instanced in connection with theories. In addition, his comments, and particularly the second, may reflect the discovery by the Royal Society as a whole of how much more difficult it was to succeed in its technological ventures than had originally been expected, particularly when it came to trying to teach 'the operative part' of techniques.[120] Hooke also implies a change of his own views in complaining how he 'must compile mechanicks though already hee know them', while particularly interesting is his claim that he only engaged in such studies out of 'fancy', and that he could gain much more by 'other imployments' — doubtless a reference to Hooke's architectural and surveying practice, though he may also have been wistfully considering the potential of using his aptitude for invention to reap the benefits available to projectors.[121]

Lastly, it is interesting how Hooke claims that a philosophical veneer would reduce the 'jealousyes' of tradesmen — though Hooke invokes so many arguments that one suspects that he is indulging in special pleading in order to make his case — while his feeling that without 'some philosophicall introduction' such studies might be subject to the 'contempt of gentlemen & contumely of the vulgar' is also a telling comment on the way in which the Royal Society's priorities tended to shift from applied to pure science as it became aware of the superior prestige attached to that in the intellectual milieux of the later seventeenth century.[122] Indeed, Hooke's statement is worth pausing over, for the clarity with which he expounds the relative value that he placed upon pure and applied knowledge contradicts what some modern scholars have asserted — namely that 'there was no sense at the time that the utilitarian aspect of science was any less prestigious, or of any lower priority than the better-known experimental case studies not obviously related to practical affairs'.[123]

This is not to say that at this point in his career or at any other Hooke was uninterested in utility: far from it, for this is a motif that runs through his lectures and other writings, particularly the extent to which he and other

[120] Cf. Hunter, *Science and Society*, p. 104 and ch. 4, passim.

[121] For Hooke and projects, see above, 'An Experiment in Corporate Enterprise', pp. 89 – 90, and Christine MacLeod, 'The 1690s Patent Boom: Invention or Stock-jobbing?', *Economic History Review*, 2nd series 39 (1986), 566 n. 96.

[122] See Hunter, *Science and Society*, pp. 110 – 11.

[123] Charles Webster, *From Paracelsus to Newton* (Cambridge, 1982), p. 63.

natural philosophers might devise inventions of wide public benefit.[124] It was simply a matter of priorities and of the best means by which technological improvement might be brought about, and he had no doubt whatever that the establishment of general principles for the understanding of nature was paramount and that it was to the exposition of these that lectures were best devoted. Moreover, he evidently felt that, whatever he could achieve through inventions, he would be wasting his time if he devoted himself, as Cutler wanted him to, to the History of Trades.

Hooke's paper must have been relayed to Petty, who employed it in attempting to draw up a mutually agreeable statement of principle which would ensure 'a Freindly composure of the said Defferences, without proceeding to the extreamitys of Law'.[125] This 'Award' deemed it 'Reasonable' that Cutler should pay the arrears due to Hooke at Midsummer that year and support him thereafter according to his bond, and that Hooke should 'proceed to the History of Nature & Art and therein comprise the History of trades and Manufactures'.[126] Petty continued how 'from the Inclination and temper of Mr Hooke, his continued course of Studyes, the Obligation he hath of Honor & Reputation, and his serious Declaration to me of such his Resolutions, I doe Beleive he will perform, (to the Best of his abilitys) for the Honor of Sir John Cutler, and to the Satisfaction of the world: And will doe more then what he did at first promise or undertake, or can any other ways be Obliged unto'.

The degree of verbal overlap between this and the document just quoted makes it clear that Petty must have had that in front of him when drawing it up. As for the subjectmatter suggested in this attempted compromise, this echoes a further document among the Bowood Manuscripts, an undated note from Hooke to Petty endorsed 'Mr Hookes promise to Sir W^{m}. Petty'.[127] In it, Hooke guaranteed that, if his arrears were paid and the annuity paid for the rest of his life 'without further suit or trouble', he would 'proceed to the best of my abilitys in Compiling the History of Nature & Art[.] And in it to Comprise all such Physicall and Mechanicall Operations as are Curious in Trades and Manufactures, which I Resolve as Desired either to Read as Lectures at the Meetings of the Royall Society, or Annually to Print and Publish therein making honourable mention of Sir John Cutler. In the Performances of which I shall free and voluntary (as now I am) Doe much more then any Legall Bonds can Obleige me to'. A postscript reiterated how 'I desire that noe construction may be made of this to Invalidate the Bond, which is Absolute and was a free donation without Conditions to be performed on my part, As you yourself can testify'.

124 See, for instance, Hooke, *Posthumous Works*, p. 555; or above, 'Towards Solomon's House', p. 213.

125 Bowood House Petty MSS H[8]17, in what is probably a scribal hand, endorsed by Petty 'Award betwe[en] Sir J: Cuttler & Mr Hooke'.

126 'Nature' has been altered from 'trades'; 'Art' has also been altered, but it is not clear from what.

127 Bowood House Petty MSS H[8]19. 'Resolve' is altered from something else.

On the issue of Hooke's subjectmatter, this 'promise' like — Petty's putative award — appears to represent a concession by Hooke in the direction of Cutler's wishes, possibly imposed on him by Petty or others. One may speculate that they either felt themselves, or were told by Cutler when he was consulted about the proposal, that Hooke could not have it both ways: if he wanted his arrears and annuity, he would have to move further towards Cutler's demands concerning the topic of his lectures than he himself would have liked. Probably this was Petty's first attempt at a settlement, but it evidently proved unacceptable and was replaced by a different version, dated 14 May 1683.[128] The latter states that Cutler was to pay the arrears (without costs or damages), but no mention is made of future support for Hooke. Hooke, on the other hand, was to withdraw his action at law and to 'performe what is mentioned in the preamble of the Bond which Sir John gave him, with Honourable mention of Sir John Cuttler in what he shall read or Print' — in other words, a far vaguer statement of Hooke's responsibilities than that in the undated document.[129] The alteration concerning Cutler's obligation was presumably a concession to him, whereas it is quite likely that the latter change was made at the behest of Hooke, since attached to the award of 14 May 1683 is a note of the same date from Petty to Hooke responding to a lost letter from him and offering to withdraw one clause. Petty wrote — with an evident degree of annoyance — 'Mr Hooke, I am sorry you cannot trust the Society in such a matter as I propound. Wherefore to prevent publick noise, I am content to withdraw that Clause and have sent another award without it'.[130]

The Hooke-Cutler Dispute, Round 2: Litigation

It must be to his receipt of this letter that Hooke refers in the final entry of the diary that he had kept since 1672, that for 15 May 1683: 'At Ballets. opend Sir W Pettys Letter. at Sir W Pettys. hufft. Westminster hall. V.Many[?] Cause tryd at Guildhall[.] Verdict on Bond. Sir J.Hoskins Mr Colwall Lodowick [H]unt.'[131] Clearly Hooke was not prepared to accept Petty's offer, presumably because of its failure to ensure Cutler's future

[128] Bowood House Petty MSS H[8]18 (first leaf), signed by Petty and in his hand; verso blank.

[129] For what is known of the wording of the bond, see above, pp. 295 – 6.

[130] This is a conjugate leaf, the verso of which is endorsed: '14 May 1683/ Sir W Pettyes award between Sir John Cuttler & Mr Hooke'.

[131] Guildhall Library MS 1758, fol. 78v. Francis Lodwick, like Hoskins and Colwall, was a close contact of Hooke's; Henry Hunt was his assistant and Operator to the Royal Society. Hooke's laconic entry presumably implies that he spent time discussing the matter with these friends.

payment to him as well as the settlement of arrears, despite the fact that Petty had toned down the requirement concerning Hooke's subjectmatter. The case therefore returned to the Court of King's Bench as scheduled. A jury was appointed to hear it on 16 May, who on the 22nd of that month decided that the bond of 29 July 1665 *was* Cutler's and, on this basis, adjudicated that Cutler owed Hooke the £1000 referred to in it, awarding Hooke a total of £9 costs, which included one shilling damages.[132]

Cutler countered by instituting a process of appeal by Writ of Error against the King's Bench judgment, which was heard in the Court of Exchequer Chamber. Since that court did not keep records of its own, all we know about the progress of the case is what was said about it in the records of the other courts to which the matter came, which indicate that a decision was reached in Hilary Term 1684.[133] In addition, Cutler turned to a court of equity in the form of the Court of Chancery, where he was able to invoke the broader considerations concerning the arrangement which had come up in Petty's arbitration but which had been beyond the remit of the Common Law Court of King's Bench, which could simply handle the evidence of the bond itself. It is perhaps significant that at this time Chancery was taking an interest in offering relief from the inequitable insistence on the performance of terms of legal bonds, which may well explain why this seemed to Cutler the obvious step to take. Be that as it may, Cutler's Bill of Complaint to Sir Francis North, Lord Keeper of the Great Seal, dated 13 June 1683, still survives among the Chancery records at the Public Record Office, though it has hitherto been overlooked.[134]

In this document (in which he conventionally described himself as 'Orator'), Cutler gave his version of the history of the case. He described his motives in words that have already been quoted, explaining how it was agreed that Hooke should read sixteen lectures a year 'in the History of Trade and Mechanicks as it Concerned Naturall Philosophye'. Details of the elaborate arrangement in relation to the law terms was given, and Cutler then recapitulated how, in return for the performance of the lectures, he agreed to pay Hooke £50 a year in two equal instalments, referring to the bond in which he promised to forfeit £1000 in the event of his failing to honour the agreement. He went on to explain how for some years he had made the payments, rather disingenuously claiming that in total he had paid Hooke 'One thousand Pounds or thereabouts' (in fact, it seems likelier that — even allowing for the £200 that had changed hands in January 1683 — the total paid up to that date was less than half this).[135] But 'the sayd Robert

132 PRO KB27/2023, rot. 461.

133 PRO KB27/2105, rot. 226. Cf. C33/261, fol. 234, and see below, p. 324.

134 PRO C7/564/29, item 1. In C33/261, fol. 613v, it is stated that Cutler's bill was exhibited in Easter Term 1683, and Hooke's reply in Trinity Term that year. On Chancery's interest in inequitable bonds, see D. E. C. Yale, ed., *Lord Nottingham's Chancery Cases*, ii (Selden Society, 79, 1961), Introduction. I am indebted to Christopher Whittick for this reference.

135 Payments at £50 a year from 1665 (when the bond was signed) to 1670 would have come to £250. See also above, pp. 306, 310, and below, pp. 322 – 3.

Hooke afterwards (as hee had in truth much done before) Neglecting to Read the sayd Lectures and to performe his Agreement on his part your Orator Did thereupon withold the Payment of the sayd halfe yearely Payment of fifty Pounds as hee was advised hee might Justly Doe . . . the sayd Robert Hooke haveing not for many yeares last ever read any Lectures according as by this Agreement hee was to Doe and Designeing to Read noe more thereof and soe Declareing but waveing the sayd Imployment and undertaking'.

In Cutler's view, since he had already paid more than was owing to Hooke according to the agreement and for more years than lectures had actually been given, Hooke should stop pestering him and 'the sayd Bond ought to bee Delivered up to bee Cancelled or the Originall Agreement mutually performed'. He stressed how Hooke should 'very well know that the sayd Designe was Purely Voluntary in your Orator and for the Advancement of Trade and Arts and the Learning in Reference thereunto and to Promote and Incourage the same and that the sayd Robert Hooke soe farre as his Lectures might Availe might bee Incouraged to give his Assistance'; he also argued that he had paid as much as was due to Hooke 'in Justice and Equity', 'or that hee ought to have had according to the true Intent and meaning of the sayd Agreement and Security', particularly in the light of his non-performance of his duties. Yet Hooke had 'gott the Condition of the sayd Bond soe worded as that hee thinks hee may by Law Obteyne the money without doeing any thing for it', and, having 'Conceived some Causelesse Displeasure against your Orator and entred into A Combination with severall persons unknowne unto your Orator how to gett from your Orator some grate summe of money and to share and Devide the same among themselves upon that Designe and Intention onely and not out of any sense or sattisfaction of the money being Really Due to him, has of late Comenced A Suite att Comon Law upon the sayd Bond or Obligation of such Penalty as aforesayd against your Orator', claiming several years' arrears: moreover he had obtained a judgment in his favour by default for the penalty of the bond, with no regard of the money received and despite his failure to perform his side of the agreement. Now, he threatened to proceed against Cutler 'with all vyolence' and to take out execution on the penalty, although 'in Equity and good Conscience there is noe thing really Due to him'.

Cutler then elaborated further on his allegation about Hooke's dereliction of duty: 'And although the sayd Robert Hooke Doth some tymes pretend that hee did Duly and Dilligently attend and performe the sayd Lectures Yearely ever since the sayd Agreement was made as aforesayd yett hee knowes the truth to bee otherwise and that hee did neglect waive and Desist from the Performance thereof very many weeks hee Reading noe other Lectures then were Incumbent on him to Doe and performe as Geometry professor to Gresham Colledge'. Cutler claimed that Hooke had given no Cutlerian Lectures 'for severall yeares', 'nor did hee applye to the said [Royal] society for Direction nor did they Direct the matters of the sayd Lectures as they ought to have Done and as hee ought to have Procured them to have Done according to the true Intent and meaning (if not the Expresse words) of the sayd Agreement and Condition of the sayd Bond or

Obligation and as hee Agreed and ought to have Done in case hee would have the benefitt thereof or if hee Did your orator ought to have A Discovery thereof'.

Cutler claimed that he had applied to Hooke to find out exactly 'what Lectures hee read and when in particular', and he now requested the court to demand that Hooke should provide evidence on this point, which he had so far refused, namely 'how long hee did truely and Duely Read the Lectures agreed upon in Respect of and which he ought by the sayd Agreement and Condition of the sayd Bond to have read and what Lectures hee from tyme to tyme Read and where particularly Yearely the sayd Lectures were Read And whether in Respect of the sayd Agreement and halfe yearely payment agreed to bee made by your Orator hee did Read Different and other Lectures then were Incumbent upon him by reason of his being Geometry Professor as aforesayd or of his haveing any other Imployment in the sayd Society and what such Lectures were and when read particularly and whether any and which of them were Directed by the sayd Society and when the same particularly were soe Directed'. In addition, only by a sworn statement from Hooke would it be possible to establish what had been paid since 29 June 1665, since on this point too witnesses that Cutler might have called were 'either Dead or removed into Parts Remote or beyond the Seas and unknowne to your Orator': he requested that Hooke be subpoenaed to provide this information, 'And that hee may shew cause (if hee can) whye your Orators sayd Bond or Obligation should not bee Delivered up to bee Cancelled and the Iudgement thereupon vacated'.

This document has been quoted at length so that it may speak for itself, but a few points may be made in relation to it. One is that, in contrast to the negotiations conducted earlier in the year through Petty, Cutler here complained almost exclusively about the actual delivery of the lectures rather than their subjectmatter: possibly he was advised by his lawyers that this was a more satisfactory basis on which to bring a legal case, but, in addition, since he *had* agreed in his bond to the initiative being taken by the Royal Society as far as the topics of the lectures were concerned, he hardly had a case at law for claiming direct supervision, in the absence of privity of contract between himself and Hooke. It may be noted, however, that in each of the places where Cutler referred to the subject of the lectures he stressed technical matters, and it may be that this was a secondary line of attack that he was holding in reserve. On the other hand, there is more than a hint of Cutler's dissatisfaction with the Royal Society's management of the Lectureship, though this is something for which he seems to have been inclined to blame Hooke: perhaps because it was against Hooke's suit that he was appealing, it never seems to have occurred to him to sue the Royal Society.

Though Cutler may have had some justification in his claim that Hooke had not given his full complement of lectures, and that it had become hard to distinguish his lectures on the Cutlerian foundation from those given in his role as Gresham Professor or as Curator to the Royal Society — something which it is now almost possible to assess — the financial side of his case was palpaby disingenuous. Thus his claim that he had paid Hooke over £1000 was certainly false: if he *had*, there would have been no arrears at all in 1683. It seems likelier that Cutler claimed that he had expended exactly the sum mentioned in the penalty clause in the bond so that he could be released from his obligation. It also seems hardly equitable that it should have been on Hooke, rather than Cutler, that the burden of proof concerning the money actually paid should fall, and highly suspicious that all Cutler's potential witnesses should be either dead or otherwise unavailable. Cutler's dark implication that Hooke and some unknown conspirators were out to extract 'some grate summe of Money' also seems rather fanciful. Indeed, one might argue that by overstating his case, Cutler had actually slightly damaged his chances of winning in a court of equity and made it easier for Hooke to frame a convincing reply than would otherwise have been the case.

Hooke must have set to work on his response almost immediately. At a meeting of the Council of the Royal Society on 20 June 1683, he was given leave 'to have access to the journals or council-books upon occasion of his business with Sir John Cutler', and his answer to Cutler's Bill was evidently submitted on 26 June.[136] By this time the Chancery records show that Cutler's complaint had been registered, Hooke's Common Law case suspended and the case postponed awaiting an answer from Hooke.[137]

In his Answer (in which he describes himself throughout as 'Defendant' and Cutler as the 'Complainant'), Hooke gave his own lengthy account of the case. He reported on the circumstances in which the endowment was first mooted in words that have already been quoted, while he slightly manipulated the actual course of events in his favour by claiming that he had already been receiving a salary of £80 a year as Curator of Experiments to the Royal Society 'for some time' before Cutler came forward with his proposal (it was presumably because the judge would not understand the difference that he claimed that his task as Curator had been to 'read lectures to them about naturall and experimentall Philosophy besides such other Subjects and matters as they did or should desire and direct this defendant to read upon').[138]

He also made the best of the case from his point of view on the question of subjectmatter, claiming, hardly surprisingly, that Cutler had allowed him a free hand. Cutler, he said, had told him to go on doing what he was doing already, 'which said promise the said Complainant did afterwards several times renew to this Defendant but concluded not upon any other particular business to the best of his remembrance or that he would have this Defendant to doe or performe other matters or business then what this defendant was then engaged and had proceeded in by direction of the Royall Society which was in the improvement of art and nature'. In the light of the negotiations with Petty over the lectures' proper subjectmatter that have already been surveyed, one should almost certainly interpret such statements on Hooke's part with a pinch of salt (though it is interesting firstly

136 Birch, *History*, iv, 210; PRO C7/564/29, item 2.
137 PRO C33/259, fols. 616v, 860.
138 PRO C7/564/29, item 2. See above, pp. 287 and 289 – 90.

that he here placed 'art' before 'nature', and secondly that he substituted the word 'improvement' for 'history'). Indeed, one might feel that Hooke protested suspiciously much, reiterating later in his submission how 'the said Complainant did both before and since the executeing the said Bond often declare to this Defendant that he left it wholy to this Defendant to doe what he this Defendant thought fitt in the management of his said Studyes and declared that his the said Complainant's donation was Free and as a Particular kindness to this Defendant and that the fifety pounds per Annum should duely be paid to this Defendant by him whether he read any lectures or not'. Subsequently, he repeated how Cutler had entered the agreement 'freely and voluntarily', 'without any Sollicitation from him this Defendant': Hooke was desperate to prevent any contract from being construed.

In addition, Hooke argued that, as he had understood it, Cutler was simply paying part of the salary that he would otherwise have received from the Royal Society for his responsibilities as Curator of Experiments, the very interpretation which the Royal Society had encouraged by its ambiguous statements in 1664 and of which Hooke was arguably to an extent an innocent victim:[139] 'the Defendant tooke & understood the said Complainant's generous and kind proffer of fifety pound per annum as an additionall reward and encourragement for his the Defendants laboure and paines therein', namely his tasks as Curator. Moreover, having asserted that the gift was without conditions, he added: 'which declaration made in presence of severall of the Society was the reason why the said Defendant relying upon the said Promise and settlement made by the said Complainant did relinquish the said fifty pounds per annum to the said Society'.

He elaborated on this in connection with the negotiations that, as we have seen, had occurred between Cutler and Petty, Wilkins and other members of the Royal Society, rather than Hooke himself. Hooke claimed that during these 'this Defendant was treated with by the said Society and acquainted by them that the said Complainant had left it to them to direct this Defendant what to doe & how to prosecute these his Studyes and that the said fifety pound per Annum so given by the Complainant was to be accounted[?] as part of the eighty pound per Annum this Defendant was to have from the said Society and that the Defendant was to continue to doe what he had before done and accordingly about the time of the entring into the said bond they tooke off fifety pounds per Annum of the said eighty pound per Annum and allowed him only thirty pound per Annum for the future', which was what he had had ever since. All this, Hooke asserted, Cutler knew perfectly well, part of the bargain being his election as Honorary Fellow of the Society, and he added: 'nor was there any other agreement made with the said Complainant by the said Defendant'. Interestingly, at this point the words 'to the best of his remembrance' are inserted above the line in what appears to be the handwriting of Hooke himself: the rest of the document is written in a clerical hand.

Hooke then gave the exact terms of the bond of 29 June 1665, as quoted above, claiming that he had not even been privy to the final wording of it:

139 See above, esp. p. 292.

'this defendant sayth that the recitall Preceeding the Condition of the said bond as if he this defendant should agree or promise to read sixteen lectures in the vacation to the Royall Society as is therein mentioned, was inserted without this defendants direction privity or agreement nor did this Defendant know any thing of the Condition of the said Bond nor of the Complainant's executeing of the said Bond untill he this defendant had the said Bond delivered to him ready executed as aforesaid'.[140] As we have seen, it is all too likely that the grandees of the Society should have treated their employee in this high-handed manner eighteen years earlier, though Hooke had obvious ulterior motives for minimizing the extent to which he was a direct party to the agreement. For this meant that in retrospect Hooke could claim that he had made no promises to Cutler 'otherwise than that he agreed with the said Society touching his lectures': to earn his £30 from the Society and £50 from Cutler, he 'did from time to time read such lectures and upon such subjects and observed to doe such things as the said Society enjoined or directed him which was as much as ever this Defendant undertooke that he would doe in respect and consideration of such sallary gratuity or reward of eighty pound per Annum Promised and agreed to be given and allowed him'.

Hooke then stated the exact amount paid by Cutler — in other words, a total of £450 since the entering of the bond, including the £200 arrears paid on 25 January 1683 — 'for all which said sumes this Defendant gave receipts and which the said Complainant as he beleives now hath or can come by and to which this Defendant doth referr himselfe and the Complainant for more certainty therein'. He also defended his record: 'And this Defendant hopeing to have the said Annuityes continued to him did not only read such lectures as the Society desired and directed him which were more in number as hee believes then the Complainant by his Bill pretends this Defendant was obliged to but he this Defendant hath comprised several of his said lectures in discourses and hath printed severall of them under the generall title of Cutlerian lectures'.

Continuing, Hooke arguably went a little over the top in the presentation of his case, albeit in a characteristically Hookeian manner: here his target was less Cutler than the Royal Society itself, as his employer, Hooke's relations with which were at a low ebb at this juncture, quite apart from the grievance he may have felt about the events of the 1660s that he had recounted.[141] 'All which but more especially the first of them were done by the directions and comands of the said Society which did cause this Defendant to sitt up and watch for the most part of the night for neere a whole yeare together for makeing of diverse observations without which the directions of the said Society could not be performed which was extreamely prejudiciall to his this Defendants health and had like to have cost the defendant his life'. He also claimed — as in his submission to Petty earlier in the year — to have sacrificed more lucrative sources of income for

[140] For the bond, see above, pp. 295 – 6.

[141] For a parallel, see, e.g. the letter to Hoskins of 31 March 1683 quoted on p. 311, above. See further below, p. 330.

the sake of the job: 'And this defendant further sayth that he this defendant so Farr depended and relyed upon the due payment of the said Annuitye or reward agreed to be paid him by the said Society and the Complainant as aforesaid that from the time of his the said Complainant entring into the said bond and delivery thereof to him this Defendant he this Defendant wholy left off Such other Studyes which before he had prosecuted and intended to have proceeded in which would have byn farr easier & much more profitable as he beleives unto him and from that time directed his Studies and endeavours to the Performance of such things as tended to [the] improvement of the knowledge of art and nature and in makeing such experiments as might most conduce thereunto and were directed by the said Society and had byn allwayes ready to explaine and elucidate the same by discourses or lectures and at such times and in such places as the said Society pleased to meet in and direct this Defendant and this Defendant performed the same accordingly to the best of his Abilities'.

Turning to the period since 1670, when Cutler had ceased to pay his annuity despite promising both to Hooke and others to do so, Hooke claimed that 'haveing allwayes had a great esteeme and honour for and beleife of the said Complainant and so relying upon his promises did not make use of the law for the recovery of the same before the said Complainant had detained full twenty four halfe yearly payments and untill the said Complainant absolutely denyed the payment of his arreares with threatening and approbrious language which was very differing from what the said Complainant not long before had promised the defendant'. This was the figure owing when Hooke had first taken legal action in October 1682; he now claimed that, allowing for the arrears paid in January and the two further instalments due at Christmas 1682 and Midsummer 1683, the total due amounted to £450 or eighteen half-yearly payments, 'besides interest'.[142]

Then, Hooke dealt with his performances as lecturer. 'And this Defendant further sayth that albeit he cannott sett forth the severall lectures or discourses he hath made to the said Society nor the particular subjects times and places by reason of the length of time yet he this Defendant doth deny it to be true that he hath not read lectures for many or any yeares last past as in the Bill is Suggested or that he refused or neglected to performe such lectures readings or discourses as the said Royal Society thought fitt from time to time to direct this Defendant to performe[,] Allthough this Defendant did always[?] looke upon the said Performances not as a Duty incumbent upon him by reason of any contract or agreement concerning the aforesaid Donation of fifety pounds but as a thankfull acknowledgement to the Complainant for his free Donation, and to the said Society for their favourable acceptance of what he the said Defendant was able to doe and performe'. He also stressed how he 'did read to or in the said Society differing and other lectures and at other times then were incumbent upon him by reason of his being a Geometry professor in Gresham Colledge [or] reason of his receiving thirty pound per Annum from the said Society' (the

142 See above, pp. 306, 310, 317, 322.

latter point seems mildly inconsistent with his earlier insistence that Cutler's benefaction had been intended to pay for his work at the behest of the Society). Lastly, Hooke acknowledged how, since Cutler had refused to pay his arrears, he had put Cutler's bond in suit and obtained a judgment, thus stimulating Cutler to bring his Writ of Error against him. But he stated that if Cutler would pay his arrears 'together with damages and this Defendant's costs at Law and in this Court (and continue payment of the said Annuity for the future)', he would withdraw his Common Law proceedings.[143]

These, then, were the conflicting claims of the two contestants, as laid before the Court of Chancery in the summer of 1683. For the remainder of that year nothing seems to have happened in connection with the case, though on 24 November Hooke had the satisfaction of gaining a statement of support from the Council of Royal Society, who opined at a meeting on that date 'that Mr Hooke should be owned and assisted by the Society as far as lawfully they may, and is warrantable by the orders and journals of the Society'.[144] The case came up again in Chancery on 1 February 1684, when the positions of the two parties were expounded (in Hooke's case by his counsel, Owen Feltham and Mr Hutching).[145] On the whole the arguments were identical to those expounded in the documents already summarized, though on Hooke's side it was explicitly claimed that the Writ of Error — which was said to be due to be argued that term — had been a tactic to cause delay. The judgment that ensued was in Hooke's favour, vindicating his confidence in his case against the fainthearted Royal Society members who had advocated compromise the previous year. Since no detailed record survives of what was said in court, one can only speculate on the grounds on which the decision was made, but one presumes that the judges took the view that Hooke *had* satisfied his side of the agreement and that it was Cutler who was at fault.

It was thereupon agreed that Sir Robert Legard, Master in Chancery, should investigate the arrears and that Cutler should pay them within a week of Legard's reaching a decision. Hooke was told that in the meantime he was 'att liberty to proceed to the affirmation of the said Judgment upon the Writt of Error but all further proceedings are hereby stayed until a Week after the Master's Report made as aforesaid'. Legard made his report on 15 February, computing the arrears due at Christmas 1683 as £475 (interestingly, he described the annuity as a 'Pension'). These were therefore paid at the Rolls Chapel at 10 a.m. on 22 February 1684, in return for Hooke

143 The document ends, as it began, with formal legal disavowals. It is signed 'Owen Feltham' at the end, and 'John Hoskyns' at the start. For Owen Feltham, admitted to Gray's Inn in 1654, see Joseph Foster, ed., *Register of Admissions to Gray's Inn, 1521 – 1889* (London, 1889), i, 268.

144 Birch, *History*, iv, 229.

145 PRO C33/261, fol. 234. For lawyers called Hutchins, see MacGeagh and Sturgess, *Register* (n. 106), i, 154, and Foster, *Register* (n. 143), i, 300, 337. On Feltham, see n. 143.

giving a recognizance ackowledging the authority of the court and its decision.[146]

The arrears were therefore now dealt with, but this left the question of continuing payment outstanding, and the case was therefore brought up again on Hooke's behalf on the following 19 June. On this occasion his counsel was a Mr Bretland, who claimed that, although Hooke had submitted his response a year earlier, 'the plaintiff hath not Replyed or anywayes proceeded in the Cause'; in a subsequent report of the matter, Cutler was accused of 'not bringing his owne Cause to a heareing, as by an order he was to have done'. At the defendant's request, Cutler's bill was therefore dismissed, and, as we learn from the same later report on the matter, Hooke was also awarded costs to the value of seven nobles.[147] It was also later alleged that, having obtained this order, Hooke 'kept [it] private and never gave any notice thereof': knowing that he had a Chancery judgment in his favour which would enable him to collect fresh arrears as they arose whenever he wanted them, he may have decided at this point to bide his time, awaiting Cutler's death in the hope that his heirs might be more amenable than the 'old curmudgeon' himself. That he was prepared for further trouble is suggested by the fact that 25 June 1684 is the date of the earliest surviving example of a practice which he was repeatedly to employ thereafter, of endorsing the manuscript of a lecture that he gave with the names of those in attendance, presumably so that he could call on them to certify his performance of his duties should need arise.[148]

What is clear is that Cutler did *not* revert to paying Hooke his annuity, for by the beginning of 1695 Hooke was to claim further arrears of £550 — in other words, eleven years, taking us back to the point at Christmas 1683 up to which Cutler's debt had been computed.[149] Once again, however, negotiations appear to have taken place: it was later claimed by Cutler's executors that he and Hooke 'had severall treatyes in order to an accomodation and the Testator [Cutler] made some agreement with him touching the same and the defendant [Hooke] accepted of what was then agreed on and released the Testator or agreed soe to doe and that therefore the Testator did not proceed in the Cause and the defendant did not afterwards read any Lectures as by the condition of the bond he ought to have done and the said defendant did not dureing the Testators life claime any more of the said Annuity'. That discussions of some kind occurred seems plausible enough, although Hooke countered any suggestion that he had agreed to terminate the arrangement in this way by denying 'that att any tyme afterwards he had any discourse with the Testator in order to an accomodation or that he made any agreement touching the matters in

146 PRO C38/114 (unfoliated); C33/261, fol. 234. On Sir Robert Legard, see W. A. Shaw, *The Knights of England* (London, 1906), ii, 258.

147 PRO C33/261, fol. 613v; C33/285, fol. 834v. Mr Bretland was perhaps Reginald Bretland, admitted to Gray's Inn in 1661: Foster, *Register* (n. 143), i, 291.

148 PRO C33/285, fol. 834; Cl.P. xx.62 (cf. Birch, *History*, iv, 308). For later examples, see below, n. 153.

149 PRO C33/283, fol. 416; 285, fol. 834v.

difference or agreed to release the Testator of any of these arreares or that he made noe further demands from him'.[150]

Certainly, on 12 January 1687, the Royal Society promised Hooke their 'lawful assistance and recommendations towards his recovery of the £50 a year, which Sir John Cutler stood obliged to pay him during his life', while it is possibly also relevant that Hooke recorded in his diary on 15 May 1689 that he 'Read Cutler Letter to the Royal Society'.[151] More significantly, we learn from the records of the legal case that followed Cutler's death that, in about 1691, 'the late Archbishop of Canterbury [in other words, Hooke's friend John Tillotson, who died in 1694] and the now Bishopp of Rochester [Thomas Sprat, erstwhile historian of the Royal Society] did att the defendants [Hooke's] request endeavour what they could in a freindly manner to perswade the Testator [Cutler] to pay what was due to him for further arreares but could not prevaile therein'. Hooke admitted, however, that 'hopeing that the said Testator [Cutler] would neverthelesse have bin perswaded thereto did for that reason forbeare to proceed against him dureing his life'.[152]

On the other hand, contrary to what was to be claimed by his opponents in the later legal case, Hooke *did* continue his Cutlerian Lectures. In his surviving diary for 1688 to 1693, he refers specifically on two occasions to giving a 'Cutlerian Lecture', on 10 July 1689 and 19 July 1693. Moreover both these lectures survive, as do a number of others that were clearly given under the terms of the Cutlerian benefaction, whose content will be considered shortly; on a number of these, Hooke continued the practice that he had apparently begun in June 1684, of recording the names of those present.[153]

150 PRO C33/285, fols. 834, 834v.

151 Birch, *History*, iv, 518; Robert Hooke, *Diary 1688 – 93*, in R. T. Gunther, ed., *Early Science in Oxford*, x (Oxford, 1935), 121 (I have collated this with the original in British Library MS Sloane 4024, fol. 14v, which *does* appear to say 'Letter', though it could conceivably be intended to be 'Lecture'). Ibid., p. 181 (22 Jan. 1690) may also be relevant. See also a letter from Hooke to James Yonge of Plymouth of January 1685, E.L. H.3.71 (a scribal copy), in which he refers to 'my troblesome affaires', which he hoped to finish 'this terme'. It was perhaps because the matter remained pending that Richard Waller inserted it in his narrative of Hooke's life under 1688 (*Posthumous Works*, pp. xxiv – v), thereby misleading the editors of Hooke, *Diary*, p. xxiii, and 'Espinasse, *Robert Hooke* (n. 91), p. 83n., into believing that the case began in that year.

152 PRO C33/285, fol. 834v. Before 'due', 'justly' has been deleted. On Hooke's friendship with Tillotson, see Hooke, *Diary*, passim.

153 Hooke, *Diary 1688 – 93* (n. 151), pp. 135, 260. For other lectures of which the text survives, see below, n. 181; for others, see ibid., pp. 84, 125, 169, 171, 173, 191, 245, 258, 264. For Hooke's notes of those in attendance, see Hunter, *Royal Society*, ch. 1, n. 77: to the list given there should be added Cl.P. xx.62, 73, 85 and 86. It should also be noted that I then mistakenly presumed these to be lists of those in attendance at meetings of the Society: in fact, one or two non-members were present on a number of occasions, and it is possible that some Fellows arrived for the meeting proper after the lecture was over.

Whatever had been going on during the later years of Cutler's life, it is not surprising to find that Hooke's hopes were raised once again by Cutler's death on 15 April 1693.[154] A few days earlier, when it was clear that Cutler was on his deathbed, a dramatic reconciliation had been effected between him and his daughter, who had married the Earl of Radnor without his consent: recording this in his diary, Hooke also hopefully noted 'that Lord Radnor had promisd him for me'.[155] In the immediate aftermath of Cutler's death, Hooke showed an interest in his will and the various legacies that he was said to have left, possibly hoping that there might be something in it for him, though in this he was disappointed.[156] Instead, Hooke turned his attentions to Cutler's nephew, Edmund Boulter, whose earlier dealings with Hooke in this connection have already been referred to, enlisting the help in negotiations of Sir Richard Levett, a prominent alderman and future Lord Mayor of London.[157]

In May 1693 Hooke seems to have been quite hopeful that Boulter would pay, since Boulter told him on the 20th that 'he would consider of it soe soone as some other Acc[ounts] were past with Lord Radnor, to whom he was going, that he knew of it, and had payd me the last £200'.[158] On 7 June, Levett seems to have reported that Boulter was stalling — 'he would shortly consider of it' — and though Hooke met Boulter on the 13th and again on the 21st, when he 'spak kindly', evidently nothing of value transpired.[159] July saw a similar pattern: on the 6th, Boulter told Hooke 'that he had not yet had time to settle it with Lord Radnor, but would speedily', while on the 8th Levett 'promised to urge Mr Bolter to speed my Arrears'; but then a series of further encounters had produced nothing by 8 August, when the diary frustratingly ends.[160]

One can only presume that, as in the late 1670s and 1680s, Hooke remained hopeful for a negotiated settlement as long as possible, but was

154 Hooke, *Diary 1688 – 93* (n. 151), p. 231 (for an earlier false alarm on 16 January 1690, see ibid., p. 180).

155 Ibid., p. 230.

156 Ibid., pp. 232 – 3.

157 Ibid., pp. 234, 239. On Levett, see A. B. Beavan, *The Aldermen of London* (London, 1908 – 13), ii, 118; Shaw, *Knights* (n. 146), ii, 266; see also Hooke to Levett, 20 June 1691, British Library MS Sloane 1039, fol. 131, which shows that Hooke had been associated with Levett in his capacity as architect. Hooke also appears to have enlisted the help of Sir Robert Southwell, President of the Royal Society: *Diary 1688 – 93* (n. 151), pp. 248, 251, 252. On Boulter, see above, pp. 308, 310.

158 Hooke, *Diary 1688 – 93* (n. 151), p. 241. See also ibid., pp. 233, 237, 244. The £200 was possibly that paid in January 1683: see above, p. 310.

159 Ibid., pp. 247, 249, 252. See also ibid., pp. 246, 248, 256.

160 Ibid., pp. 256, 257. See also pp. 260, 262, 263. The entry for 31 July refers to the fact that Boulter and Radnor 'were to receive an account from Joderell of my affair'. Joderell was a solicitor in Chancery: F. C. Inderwick, ed., *Calendar of Inner Temple Records* (London, 1901), iii, 199; this may have been the same man as the Edmund Jodrell admitted to the Middle Temple in 1680: McGeagh and Sturgess, *Register* (n. 106), i, 201 (see also ibid., i, 156, for an earlier entrant of this name).

finally forced to resort to law, this time deploying both of the weapons that he now had available to him — Cutler's original bond of 1665 and the Chancery judgment of 1684, an equally valuable instrument for extracting the whole of what he was entitled to. Hence, on 1 May 1694, Hooke brought a writ of *scire facias* to have execution of the Chancery judgment: in response Boulter was said — 'for delay' — to have made what Hooke's counsel considered 'a very frivolous plea'.[161] Hooke's next strategy, in Michaelmas Term 1694, was to return to the Court of King's Bench and obtain an order for the execution of the judgment on the 1665 bond that had been made in May 1683: there, on 25 October, Boulter tried to argue for a stay of judgment on the grounds that the arrears of £475 that Cutler had paid in 1684 had in fact been a partial payment of the £1000 guaranteed in the bond, and that, since Cutler's death, he had himself promised to pay a further £543, which Hooke had refused to accept. On 26 November, however, the court quashed Boulter's plea and allowed Hooke to proceed.[162]

As a result, the case reappeared in Chancery on 23 February 1695, with Hooke invoking the 1665 bond and Boulter's counsel arguing that 'according to the true meaneing of the agreement', nothing was due to be paid; that the bond should be delivered up; and that an injunction should be granted to stop Hooke's proceedings. In response, Hooke's counsel 'insisted that the said defendant hath duely read the Lectures and performed his part of the said agreement and that there is now due to him from the plaintiff £550 over and above what he hath received besides Costs att Law and in this Court'. The court decided that Boulter should pay Hooke £100 within a week 'subject to such order as this Court shall make upon heareing of the Cause', while also granting an injunction as requested, on condition that the case was set down for the next (i.e., Easter) term.[163] Meanwhile, on 27 March 1695, the Council of the Royal Society ordered 'That the Common Seal of the Society be affixed to a Certificate for Dr Hooke testifying his performances in relation to his present Law Suit with the Executors of Sir John Cutler', and what is evidently the draft of this certificate survives among Hooke's papers: it stresses how he had carried out his duties 'to the full of his undertaking'.[164]

161 PRO C33/285, fol. 834v.

162 PRO KB27/2105, rot. 226. This also records that, when the King's Bench judgment was affirmed against the Writ of Error, Hooke was paid a further £9.

163 PRO C33/283, fol. 416 (after '£550', 'ob. . .' has been deleted).

164 Royal Society Copy Council Minutes, ii, 121; British Library MS Sloane 1039, fol. 95. The date of the latter is clear from the fact that Hooke is referred to as 'D^{r}. of Physick'; the state of affairs of the 1690s is arguably also in evidence in the fact that his post is described as 'Director & Curator of Experiments'; in addition, it is perhaps worth noting that 'to the full of his undertaking' is inserted in Hooke's hand.

On 10 May that year the case came up again at Chancery, when it transpired that the £100 had been paid as ordered. However, Boulter claimed that the cause had not been entered as requested, since 'the matters in difference have been under refference' which had proved ineffectual. Hooke's counsel denied that any such arbitration had taken place, insisting that this was merely a delaying tactic and requesting that the injunction should not be renewed unless the money was actually brought into court. But the injunction *was* renewed and the case was postponed till Trinity Term, when it was arranged that it should come up on 10 June.[165]

Instead, however, on 7 June, Boulter's counsel laid out a new strategy on Boulter's part, alleging that, since the point at issue concerned Hooke's performance of lectures to the Royal Society, it was 'very necessary' for him to inspect the Society's records 'and take notes of what relates to this matter before the Cause comes on': the case was therefore postponed so that Boulter could 'have a sight of the said bookes and take notes thereout as he shall think fitt'. This was agreed to despite objections from Hooke's counsel, who by now included two new lawyers, Mr Dobbins and Sir Thomas Powys, the latter the Attorney General, and a real 'big gun' for Hooke to have on his side;[166] there is, however, no evidence in the Royal Society's records that Boulter did in fact contact them in this connection. After postponement, the case came up again on 22 November, when it was apparently alleged that Boulter's counsel had not been properly notified concerning depositions by the defence, and a further delay was agreed to the next term, when — on 24 January — yet a further postponement was allowed.[167]

The matter was at last dealt with on 18 July 1696, and a full report of what transpired on this occasion survives in the Entry Book of Decrees and Orders of Chancery.[168] The cases made on each side in 1683 were summarized and the developments since described; it was then ordered — presumably on the grounds that nothing had been said to necessitate any alteration to the decision of 1684 — that Sir Robert Legard should compute what was due to Hooke and order its payment, though on this occasion his report does not survive. In addition, it was ordered that the Royal Society 'doe see that the Lectures be performed by the said defendant according to the condition of the said bond and that the said society doe every five yeares certifie to this Court that the said Lectures are performed'. It was also specifically ordered that the annuity *should* be paid in future, something which had been overlooked in 1684, thus providing the loophole which had

165 PRO C33/283, fols. 512v – 3.

166 PRO C33/283, fol. 670. *DNB*; Dobbins may have been Robert Dobbins, one of the Masters of the Utter Bar (MacGeagh and Sturgess, *Register* (n. 106), i, 208 and evidently also 150).

167 PRO C33/285, fols. 87v, 112v.

168 PRO C33/285, fols. 833v – 5.

made this second case necessary. Hence, coincidentally on his sixty-first birthday, Hooke had at last obtained the settlement he wanted and Richard Waller, in his life of Hooke, states how this outcome to the case was 'to his great satisfaction, which had made him very uneasy for several Years'. Waller proceeded to quote the effusive entry that Hooke inserted in his (now lost) diary: '*God* has given me a new Birth, may I never forget his Mercies to me; whilst he gives me Breath may I praise him'.[169]

The Royal Society and the Later Cutlerian Lectures

By this time, Hooke's post as Cutlerian Lecturer had become separated from his office as Curator of Experiments, which had terminated in a rather confused and unsatisfactory manner in the 1680s. The background is provided by events in the Royal Society following the death of Oldenburg in 1677. At that point, Hooke had acquired what virtually amounted to a new status within the Society, being elected to the Council for the first time and from 1677 to 1682 taking Oldenburg's post as Secretary. In 1681, Hooke was even paid an extra £40 salary 'as an encouragement' when he proposed 'a more sedulous prosecution of the experiments for the service of the Society, and particularly the drawing up into treatises several excellent things, which he had formerly promised the world'.[170] In fact, however, all this represented an overcommitment on Hooke's part, and he was unable to fulfill the obligations that he had undertaken. The result was that mounting complaints about his performance of his secretarial duties led to a fresh coup on 25 November 1682, when he was replaced as Secretary by Robert Plot and was not re-elected to the Council for the following year — for the first time since 1677. The existence of continuing bad blood between Hooke and the Society the following year is evidenced by his claim that he 'could not be admitted' to show experiments that he had devised at meetings.[171]

Indeed, it is rather ironic that when Hooke's struggle with Cutler reached its climax, his relations with the Society seem to have reached an all-time low. Moreover it was just at this time that the state of affairs changed concerning Hooke's post as Curator of Experiments to the Society. From 1664 to Lady Day 1683, Hooke had been consistently — if sometimes

169 Hooke, *Posthumous Works*, p. xxv.
170 Birch, *History*, iv, 91. See also Royal Society accounts, 1682.
171 See below, 'Reconstructing Restoration Science', pp. 350 – 1; Hooke to Hoskins, 31 March 1683, E.L. H. 3. 70.

belatedly — paid £30 a year for his performance of his duties as Curator.[172] On 6 June 1683, however, it was resolved that Hooke should be more closely supervised than hitherto in his experimental work: it was ordered that his performance should be assessed quarterly 'and a gratuity ordered him accordingly; and that from this time he have no other salary'.[173] The following November, this arrangement was reiterated, and, though Hooke told the Council 'that he intended to write an historical account of the experiments, which he had shewed before the Society, together with a declaration of the use and consectaries of each, and an idea of natural philosophy built upon them', and to repeat those hitherto imperfectly executed or recorded, the Council seem to have been unimpressed: at a subsequent meeting in December it insisted on a record of his performances during the previous half year, while also asking questions about the recording of experiments over the previous seven years — in other words, since 1677.[174]

Though Hooke was paid a total of £32.10 in 1684, there is then a cessation of payments to him in the accounts. In fact, in 1684, the Society seems to have tried to find an alternative Curator of Experiments, first requesting Edmond Halley to carry out the job and then appointing Denis Papin as 'temporary curator' on 2 April. The fact that Papin inherited Hooke's salary of £30 per annum presumably means that he was intended as a replacement for him. This was in contrast to the state of affairs concerning two supplementary curators appointed earlier that year to deal with experiments in anatomy and chemistry: for Edward Tyson and Frederick Slare, whose employment is to be seen in the context of the Society's expansiveness at this time as a result of its belated endowment by Charles II, were paid only £20 a piece.[175] If Papin was intended as a replacement for Hooke during the time when he held the post from 1684 to 1687, however, Hooke does not seem to have accepted that the arrangement for his Curatorship had come to an end. In March 1686 there was discussion of an experimental programme which Hooke might carry out, while in June that year he persuaded the Council that he was entitled to a Curator's salary since 1684, being paid £60 arrears for the years 1684 and 1685, and then a further £60.[176]

172 See Royal Society accounts, 1664 – 84, passim (the accounts for 1674 do not survive, but they presumably included payment of a year and a half's salary). See also above, p. 292 n. 41.

173 Birch, *History*, iv, 207 – 8. Cf. ibid., iv, 210.

174 Birch, *History*, iv, 229 – 30, 237.

175 Birch, *History*, iv, 187 – 8, 260 – 1, 277, 284, 289; Royal Society accounts, 1684; Hunter, *Royal Society*, pp. 40 – 2.

176 Birch, *History*, iv, 462, 464 (cf. 468), 486, 489, 542, 553, 554, 557; Royal Society accounts, 1688: the £22.10 paid on 7 May certainly related to a £30 salary, since it is described as being for three quarters.

Then, in January 1687, Hooke tried to establish a new bargain with the Society, proposing to 'produce one or two experiments and a discourse at every meeting, providing his salary be made up £100 *per ann.*'. This unprecedented proposal stimulated 'much debate' in the Society; the minutes record that it was agreed that his offer be accepted (with the proviso that 'the said experiments should proceed in a natural method'), and that he should receive £50 from the Society, which would also help him to obtain the £50 a year that he was owed by Cutler.[177] But the Society's accounts contain no payments which could be construed to fall under this rubric as against the arrears of Hooke's £30 salary which have already been mentioned, and, after the last of these had been paid in 1688, Hooke received no further stipend from the Society (in fact, he very nearly failed to receive the last of these, it being threatened that he might be paid in copies of Francis Willughby's *Historia Piscium*, which the Royal Society had jointly published in 1686 and with copies of which it was oversupplied).[178]

In 1686, Edmond Halley was appointed Clerk to the Society, and from this point until the early eighteenth century the Society survived with a Clerk and an Operator, Henry Hunt, the office of Curator becoming extinct. This was due to financial stringency, of which Hooke was evidently the victim.[179] He did attempt to argue that his Curatorship should continue: the minutes of a Council meeting on 13 March 1691 record that 'Mr Hook was desired to State his Case in relation to his Curatorship in Writing that it might be considered in the next Council', but no further discussion of the matter ensued, and neither do the accounts record any further salary payments.[180]

On the other hand, Hooke did remain very active in the Society, as even a cursory glance at its journal books will reveal. To some extent, he may have participated as a volunteer, like other active members of the Society in these years, and many of his less elaborate contributions to the proceedings are to be seen in this context. But, in addition, he also continued to make a major contribution to the Society's business by giving formal lectures. Of such presentations, those that have received most attention are the ones that he gave on geology, but these are only part of a larger corpus, only part of which has been published. Moreover, if the dates of these lectures are examined, a high proportion of them appear to fall on dates when Cutlerian Lectures were due — in other words, the first four weeks of each vacation,

[177] Birch, *History*, iv, 518. Cf. ibid., 516.
[178] Birch, *History*, iv, 545.
[179] Hunter, *Royal Society*, p. 44. See Royal Society accounts, passim, and, on Hunt, H. W. Robinson, 'The Administrative Staff of the Royal Society, 1663 – 1861', *Notes and Records*, 4 (1946), 196 – 7. On the early eighteenth century, see J. L. Heilbron, *Physics at the Royal Society during Newton's Presidency* (Los Angeles, 1983), pp. 30 – 1.
[180] Royal Society Copy Council Minutes, ii, 111. Cf. Hooke, *Diary 1688 – 93* (n. 151), pp. 210 and 214. For a further possible arrangement of June 1696, not recorded in the Council minutes, see Hooke, *Posthumous Works*, p. xxvi.

the presentations often forming a sequence.[181] In contrast to the practice of the 1660s, these were now almost invariably seen as part of the Royal Society's meetings rather than a separate activity and are hence briefly recorded in the minutes: but that at least some of them are Cutlerian Lectures is confirmed by Hooke's description of them as such in his diary for 1688 – 93, while a number of them are endorsed with the names of those present, a tell-tale sign that these were lectures which Hooke was delivering according to the terms of his obligation, something for which he wanted to be sure that he had witnesses.[182]

Equally interesting is the fact that — though they cover a wide range of topics, from magnetism to hydraulics — many of them include general reflections on scientific research and its implications and not least methodological prescriptions similar to those which, as we have already seen, had been associated with Hooke's principal surviving Cutlerian

[181] For geological lectures, see Rhoda Rappaport, 'Hooke on Earthquakes: Lectures, Strategy and Audience', *British Journal for the History of Science*, 19 (1986), 129 – 46, esp. pp. 143 – 6, with the handful of addenda noted in Yushi Ito, 'Hooke's Cyclic Theory of the Earth in the Context of Seventeenth-century England', *ibid.*, 21 (1988), 295 – 314. See also David Oldroyd, 'Geological Controversy in the Seventeenth Century: "Hooke vs. Wallis" and its Aftermath', in Hunter and Schaffer, *Robert Hooke* (n. 67). For printed lectures see esp. Hooke, *Posthumous Works*, passim; see also Robert Hooke, *Philosophical Experiments*, ed. William Derham (London, 1723), esp. pp. 304f. For unpublished lectures either certainly or possibly delivered on the Cutlerian foundation, see Cl.P. xx.62 (25.6.84: cf. Birch, *History*, iv, 308), 63 (17.12.84: Birch, *History*, iv, 348), 68 (2.12.85: Birch, *History*, iv, 447); 76 (20.7.87: Birch, *History*, iv, 549 – 50), 77 – 8 (26.6 and 3, 10, 24.7.89: cf. Royal Society Copy Journal Book, vii, 213, 214, 216, 219; Hooke, *Diary 1688 – 93* (n. 151), pp. 132, 135, 138), 79 – 80 (4, 18.12.89: cf. City of London Guildhall Library MS 1757, item 10, another lecture from the same series delivered on 11 Dec.; Journal Book, vii, 244 – 5, 245 – 6, 247 – 8; Hooke, *Diary 1688 – 93*, pp. 169, 171, 173), 82 (19.7.93: Journal Book, viii, 192; Hooke, *Diary 1688 – 93*, p. 260), 84 (29.11.93: Journal Book, viii, 206), 85 (8.7.96: Journal Book, viii, 364), 86 (3.6.96: Journal Book, viii, 356), 89 (7, 14.7.97 — perhaps for 6 and 13: Journal Book, ix, 58, 60), 91 – 2 (1, 8, 15.12.97: Journal Book, ix, 74, 75, 77) and 93 – 4 (24,31.5.99: Journal Book, ix, 152, 154). Other undated items in this collection might conceivably fall into this category.

Further items which could be Cutlerian Lectures comprise Trinity College, Cambridge, MS 0.11a.1[8], a lecture on longitude, in the course of which 'Gresham Colledge' is altered to 'this place', though this could be a Gresham Lecture related to those in Hooke, *Posthumous Works*, pp. 510f.; 0.11a.1[14A-D], lectures on Huygens' *Traité de Lumière* delivered to the Royal Society on 19 and 26.2.90 (see Journal Book, vii, 266, 268; Hooke, *Diary 1688 – 93*, p. 189 and perhaps also p. 190; printed in A. R. Hall, 'Two Unpublished Lectures of Robert Hooke', *Isis*, 42 (1951), 219 – 30); and O.11a.1[22] (two sides on geology: n.d.). It is perhaps worth noting here that 0.11a.1[14E-I] (like a number of lectures in Hooke, *Posthumous Works*, e.g. pp. 71f., 451f.) are clearly Gresham Geometry Lectures: interestingly, those at Trinity (dated 18, 25.6.91, 4.2 and 21.4.92) have notes on attendance. See also below, n. 184.

[182] See above, nn. 148, 153.

Lectures of the 1660s and his overlapping *General Scheme*. This is true both of some of the lectures on geology and other topics that Richard Waller included Hooke's *Posthumous Works*, and of various unpublished lectures, the bulk of which survive in the Royal Society's Classified Papers.[183]

Thus in the course of a lecture series in July 1697 which he mainly devoted to an account of recent work by Van Leeuwenhoek, Hooke took the opportunity to use the microscopist's findings to deliver a swingeing attack on theories of spontaneous generation and their dubious religious implications, which evidently so much impressed its audience that Hooke was prevailed upon to make a fair copy of it for the Vice-President of the Society, Sir Robert Southwell.[184] Earlier, a course of lectures given in June and July 1689 had contained extensive comments on the proper programme of natural philosophy and the qualifications needed in its devotees reminiscent of Hooke' sentiments in his *General Scheme*; he also stressed how, through apparatus, things could be made sensible which would otherwise not have been, and he echoed the Cutlerian rubric in emphasizing the need for studies of art and nature to be interrelated.[185] Then, in a further interesting series of lectures delivered in December of the same year, he devoted attention to what could be learnt from travel books about art and nature against a background of the 'Deficients' of natural history and the need for collaboration in promoting it, at the same time deprecating a narrow definition of utility.[186]

In this period, as earlier, it may therefore be argued that the Cutlerian endowment had a significant impact on Hooke's output. Though Hooke stuck to his guns concerning his rejection of the concern with applied knowledge which — as we have seen — Cutler would have preferred, the formula adopted in the 1660s as a compromise between Cutler's rubric and the desiderata of Hooke and the Royal Society had a genuine influence on the subjectmatter of the lectures. In addition, the task of giving formal lectures inspired Hooke to a degree of reflection which is not in evidence, for instance, in his accounts of experiments displayed at the Royal Society, and this stemmed directly from the idea of lectures on natural philosophy which Cutler and the Royal Society had agreed was desirable in 1664. In

183 Hooke, *Posthumous Works*, e.g. pp. 329f., 473 – 4, 553; for the unpublished lectures, see above, n. 181.

184 Cl.P. xx.89; see the appreciative comment in Royal Society Copy Journal Book, ix, 58. For the copy, see Royal Society 'Newtoniana', 3.137, item 1, a copy in different format of part of the commentary on Van Leeuwenhoek to be found in Cl.P. xx.89 (fols. 203 – 4, 205 – 6). It is endorsed as having come from the Southwell sale. Item 2 is a section of a lecture of June 1698 which does not otherwise survive (see Copy Journal Book, ix, 98): this again deals with the religious implications of natural philosophical findings, and may also have been copied for Southwell.

185 Cl.P. xx.77 – 8.

186 Cl.P. xx.79 – 80 and City of London Guildhall Library MS 1757, item 10.

particular, Hooke was often inspired to methodological reflections in which he might not otherwise have indulged, which are of great interest in themselves, and may well have had a significant influence on the band of listeners to whom they were addressed.

Cutler may also claim some credit for the fact that Hooke brought out in published form his more famous *Lecturae Cutlerianae* of the 1670s, for it is notable that, lacking any similar incentive at a later date, and in spite of the best of intentions, Hooke never again engaged in a programme of publishing his own work, a task left for Waller, Derham and their successors.[187] It is ironic that it took the pursuit of arrears to achieve this; and it is still more ironic that Hooke did not really need the money, in that it simply went to swell the immense accumulation of capital that he left when he died.[188] But the importance of the stimulus of Cutler's endowment for various aspects of Hooke's career is undoubted, and Hooke's own awareness of this is suggested by the fact that, if he had made a will, the purposes to which he would have devoted his fortune would have included 'a perpetual *Physico-Mechanick Lecture* of the Nature of what himself read' — though, since he died intestate, this did not materialize.[189]

The Royal Society also benefitted from Cutler's gift. Quite apart from the extent to which Cutler effectively subsidized Hooke's work as Curator of Experiments for twenty years, the Society's activities were undoubtedly enriched by the actual presentations that Hooke gave under the terms of the benefaction, first in addition to the Society's regular activity — in the heady days of the 1660s, when these were regarded as separate from its meetings — and then as an essential part of the meetings themselves. Moreover, by the 1690s, when the Royal Society had ceased to supplement the Cutlerian stipend as it had earlier, the Society arguably gained more than ever from Sir John's generosity, reluctantly as it may have been continued towards the end.

On balance, therefore, both Hooke himself and the advancement of learning more generally may be argued to have gained significantly from the endowment, thanks not least to the firmness with which the Court of Chancery came down on Hooke's side on the argument over the contractual obligations that the parties had entered into, and to Hooke's own tenacity in pursuing his case — a symptom, perhaps, of that cantankerousness for which he was notorious, which meant that he did better than many faint hearts in the Royal Society thought possible in 1683. But in many respects what is most interesting about this case is the sheer complexity that is revealed by the full account that new archival discoveries have here made

187 I here ignore Hooke's publication of his *Philosophical Collections*, most of the content of which, after all, was not by him.

188 See the inventory of Hooke's posthumous possessions printed in Hunter and Schaffer, *Robert Hooke* (n. 67).

189 Hooke, *Posthumous Works*, p. xxvii.

possible. All the parties involved were guilty of somewhat doubtful practices, the Royal Society during the initial negotiations, Cutler and Hooke later on, and these hostages to fortune contributed to the convolution of the legal dispute and the attempted arbitration. Though Cutler's parsimony obviously did not help matters, in many ways he may be seen as the victim of manipulation by Hooke and the Royal Society, who indulged in deliberate vagueness so as to get funding for Hooke for purposes rather different from those that Cutler himself anticipated, purposes which he may not have fully understood until the matter reached arbitration and litigation. It is also ironic — though, of course, he had no reason to believe that Hooke would live as long as he did — that even in the early 1670s Cutler would have made a net gain by terminating the contract and coming to an arrangement with Hooke or the Royal Society on the basis of the penalty laid down in the 1665 bond.

As for the Royal Society, in many respects the affair reflects badly on that body. It was surely unfair — if symptomatic of the rather disdainful attitude of the Society's grandees towards their employee — that Hooke was not fully consulted concerning the arrangements when the Lectureship was first set up; in addition, the Society helped him less in his struggle with Cutler in the 1680s than it might have done. Hooke may have been a difficult colleague, but he did, after all, contribute more than almost anyone else to the Society's vitality: on the manuscript of one of the Cutlerian lectures that he delivered in 1689, Hooke recorded how his ally, Sir John Hoskins, told him after it 'that if I did not doe all I was able I should be the Occasion of the Societys Ruine', to which Hooke replied, 'I would doe as much as any to uphold it & more then I was obleigd to. &c.'[190] It is ironic that the Society was so offhand towards one of its chief pillars.

Yet it is possible to be too harsh, since, in this episode as in the others documented in this book, it is easy to forget the strict financial constraints under which the Society operated and its need to take advantage of opportunities that it was offered — if necessary slightly manipulating them — so as best to forward the ends to which it was devoted. The complications in evidence in the negotiations with Cutler stemmed from a genuine attempt to forward a wide range of topics rather than the rather narrow stress on technical practices which Cutler apparently always wanted. Even the attempt to interpose the Society between Cutler and Hooke did not represent simple meddlesomeness but a wish to achieve permanence which was itself part and parcel of the Society's institutional ambitions. Hooke may have been mistreated: but mistakes can be made with the best of intentions towards longterm goals, and the complications that are in evidence in this case are typical of the sheer intricacy of the problems associated with funding and organizing scientific research in this formative period.

[190] Cl.P. xx.78, fol. 172v.

DOCUMENT

An Unpublished Cutlerian Lecture by Robert Hooke

This lecture forms the first part of Royal Society Classified Papers xx.50b (fol. 110). It is written on a single leaf, and the text breaks off halfway down the verso. The remainder of Cl.P. xx.50b (fols. 111 – 2) is in different ink and deals with a different topic — the idea of 'inflection' as a property of the air, and how this might be measured. It is possible that this is part of a further Cutlerian Lecture, exemplifying an actual phenomenon and how it should properly be studied.[191] In terms of handwriting, the item printed here is closely similar to Cl.P. xx.50a (fols. 99 – 109), which immediately precedes it and which has recently been published by David Oldroyd:[192] the two are, however, clearly distinct, not least since Cl.P. xx.50a is written on one side of the page only, whereas this is a double-sided sheet. For further comments on its dating and significance, see above, pp. 299f.

I have formerly declard in this place the[193] Occasion of my appearing here[194] and therefore I suppose it will be needless to adde any thing concerning it. And indeed had I been silent[195] the Bounty & Noble Donation[196] of Sir John Cutler towards the advancement of Reall knowledge has been such, as did sufficiently spread its owne fame, and acquainted the world how much it stands ingaged[197] to soe Noble a Patron of Learning, [198]for setling a revenue for the promoting of Soe Excellent and Usefull a Designe as the Compiling of A Philosophicall history. And though I confess It cannot be expected [199]by the single indeavours of the most able[200]and Accomplisht person in the world that much should be done towards the compleating of soe noble a Designe, and much less[201] by my meane abilitys, yet tis my hope and It shall be my indeavour soe to labour in it that what I am not able to goe through

191 See above, p. 302. See also Pugliese, 'Scientific Achievement' (n. 24), p. 21, and Birch, *History*, ii, 91 – 2.
192 See Oldroyd, 'Some Writings' (n. 79).
193 After 'the', 'bounty of And Noble Donation of Sir J' has been deleted.
194 'here' is substituted for 'in this place', deleted.
195 The phrase, 'And. . . silent', is inserted above the line, replacing 'And indeed', deleted. The result of Hooke's rewriting is to leave a lacuna in the syntax at this point.
196 'Donation' is subtituted for 'Designe', deleted.
197 'stands ingaged' is substituted for 'was already bound', deleted.
198 'and soe' deleted before 'for'.
199 'from' deleted before 'by'.
200 'perso[n]' deleted after 'able'.
201 After 'less', 'fro[m]' has been deleted. In the previous line, 'that. . . Designe' is enclosed by a line.

with all my self I may receive assistances from[202] such whose leasure abilitys and opportunitys will much[203] better inable them to perfect[204] And what ever assistance or Information I shall Receive from any such person I shall most punctually[205] acknowledg by Entring their Description verbatim together with the time I receivd It and the Names of the Persons.[206] And As often as I have occasion of publishing a Classe of Observations they shall be in the same manner printed as registred unless the authours of them shall desire the contrary. And by this meanes I hope I may be able to rescue many excellent things from Oblivion each of which being perhaps not voluminous enough of it self has not been thought fit to be printed, or if soe by reason of its smallness[207] has been quickly lost. And that there may not be any fear of fowle dealing in the management of this business It shall be free for every such communicating person or any whom he shall appoint to search the Register to see whether there be any history[208] observation or conjecture the same with his already entred. And if Not then shall his be printed the first of any upon that head. and all others in the order they are Communicated. Now though perhaps the observations should not be soe full and compleat as to come up to the perfection of the Idea I shall by & by propound yet if they conteine any thing new /verso/ And considerable and Exprest in few words they shall be printed word for word as they are sent. Now After what method I shall epitomise & range them afterwards in tables to make them beneficiall[209] towards the advancement of this usefull part of Learning I shall on some other occasion shew.[210]

And as to my own particular I shall not be wanting in my indeavours to the utmost of my ability to make the severest[211] tryales & collect the most notable Observations I can meet with. all which I shall enter into the register in the same order & time I make them that soe I may not ascribe any observations or Experiments to my self that were anothers. By this method[212] of collecting & printing I hope some thing may be done towards the great Designe of Raising a reall & usefull Philosophy.

202 'receive from' replaces 'ingage', deleted.
203 After 'much', 'assist them' has been deleted.
204 After 'perfect', 'severall parts' is deleted, as is 'It shall' after 'And'.
205 An illegible deletion follows 'punctually' ('ascribe'?), and another follows 'by'.
206 Here and subsequently in the paper Hooke arguably shows a characteristic sensitivity about priority of discovery.
207 'by reason of its smallness' is substituted for 'quickly lost', deleted; 'being', following this phrase above the line, has also been deleted.
208 'history' follows 'thing', deleted; after 'history', 'or' has been deleted in light of the addition of 'or conjecture'.
209 'beneficiall' is substituted for 'usefull [to?] the', deleted.
210 After 'shew' and before 'And as. . .', Hooke began a fresh paragraph 'This course must be taken if there be any hopes', which he then deleted.
211 'severest' is substituted for 'severall', deleted.
212 After 'method', 'I hope' has been deleted.

APPENDIX

*Reconstructing Restoration Science: Problems and Pitfalls in Institutional History**

Lotte and Glenn Mulligan's recent paper[1] presents a statistical analysis of the Royal Society's membership from 1660 to 1700, with a commentary. It thus adds to an already sizeable literature inspired by the Royal Society's pioneering role in scientific organization, and by the profuse records that survive of its early aspirations and activity. The Mulligans' contribution offers information on some topics which had not previously been systematically scrutinized — notably the age, education and scientific interests of all Fellows elected during the late seventeenth century. (Previously, Frank had qualified such information only for active Fellows in sample years.[2]) The Mulligans' paper also contains a more elaborate analysis than has existed hitherto of data on the status and occupations of the total membership for the period 1660 – 1685 — a subject which I assessed in a paper published in 1976[3] — and for the further years 1686 – 1700. In addition, the Mulligans attempt to interpret recruitment in terms of successive secretarial regimes which, they argue, defined the pattern of the Society's early history.

In an appendix I offer a few detailed comments on the tables which form the central component of the Mulligans' paper, and on the value of the information that they provide to supplement what is already known about the Society's early membership. Here I wish to deal, not with the Mulligans' statistical data, but with their interpretation of it. This is because there is considerable agreement, on the topics on which they overlap, between the

* I am grateful to Annabel Gregory and Paul Wood for their comments on an earlier draft of this article.

1 Lotte Mulligan and Glenn Mulligan, 'Reconstructing Restoration Science: Styles of Leadership and Social Composition of the Early Royal Society', *Social Studies of Science*, 11 (1981), 327 – 64.

2 R. G. Frank, jr, 'Institutional Structure and Scientific Activity in the Early Royal Society', *Proceedings of the 14th Congress of the History of Science, 1974* (Tokyo, 1975), iv, 82 – 101.

3 Michael Hunter, 'The Social Basis and Changing Fortunes of an Early Scientific Institution: an Analysis of the Membership of the Royal Society, 1660 – 1685', *Notes and Records*, 31 (1976), 9 – 114.

Mulligans' analysis of recruitment and that in my 1976 paper (though it is disguised by a differing periodization and other complications discussed below in my 'Notes on the Mulligans' Statistics'). Moreover, though the period covered by my 1976 paper terminated in 1685, I have since extended my analysis of the membership to 1700.[4] I can therefore now collate my later findings with those of the Mulligans.

Not only does their analysis bear out my conclusion that the Society's early membership came predominantly from the professional and landed classes and from men associated with government and the court, with merchants and tradesmen proportionately rare: we also concur in our view of the changes that occurred between the 1660s and the 1680s, seeing a decline of aristocrats and others prominent in rank and office, and a slight rise in merchants, with the period around 1680 witnessing a high point in the recruitment of professional scholars (the Mulligans' 'dons' and 'teachers').[5] As I pointed out in my 1976 paper, there was reason to believe that this broadening (my word: not 'decline', the Mulligans' paraphrase)[6] of social basis was temporary and that by 1700 things were returning to a position close to that of the Society's earliest years. I can now confirm this on the basis of my new work, which shows the proportion of merchants and professional scholars falling off again in the 1690s, accompanied by a return of aristocrats, politicians and courtiers. In my 1976 paper I also, like the Mulligans, drew attention to the crucial influx of new blood into the Society in the late 1670s.[7]

The difference between the Mulligans and me is, therefore, primarily one of interpretation, centring around the distinct ways in which we have periodized recruitment around 1680. Whereas I distinguished the 1670s and 1680s and hence observed a general trend in the membership running through the two, the Mulligans have taken as a period the years 1677 – 1682 when Robert Hooke was Secretary of the Society, and they claim that the changes we agree took place were almost entirely confined to that era. They have chosen this period because they have divided up recruitment in the late seventeenth century as a whole by secretarial era, believing that they have found in these the clue to changes in the Society's character in its early years. They take me to task for offering no such unitary 'explanation' of the pattern of elections in my 1976 paper, since their article claims to vindicate, both through quotations from letters and the like and from the composition

[4] This extended version of my analysis is now published in the British Society for the History of Science's Monograph Series, under the title *The Royal Society and its Fellows 1660 – 1700: The Morphology of an Early Scientific Institution* (Chalfont St. Giles, 1982).

[5] Mulligan, 'Reconstructing' (n. 1), p. 346. I presume that these bear some relationship to the categories, 'Scholar', 'Science', 'Teacher', 'Secretary' and 'Academic' in Table 2, though the exact significance of these is nowhere explained.

[6] Hunter, 'Social Basis' (n. 3), pp. 36 – 37; Mulligan, 'Reconstructing' (n. 1), p. 352.

[7] Hunter, 'Social Basis' (n. 3), p. 30.

of the membership, the secretaries' overriding significance in the Society's organization.[8]

Who Ran the Society?

How convincing is this claim that contemporaries 'were aware of the importance of the secretaries in shaping the nature of the Royal Society'?[9] In fact, the evidence adduced falls far short of proving this point: it mainly illustrates the role of successive secretaries in carrying on the Society's correspondence and, in the 1690s, in editing the *Philosophical Transactions*. (One quotation has been misleadingly truncated to make it appear to support their case.[10]) Indeed, the weakness of this section makes one suspect that the hypothesis has been put forward to make sense of apparent contrasts observed when recruitment was divided up by secretarial era — contrasts which have then been taken (in somewhat circular fashion) as the chief evidence of the secretaries' decisive influence on the Society's early evolution.

It is far from clear why it is presumed that the influence of any one officer would always be significant in the Society to the exclusion of all others, or why the same office is presumed always to have been the crucial one. In fact, the relative importance of different offices fluctuated in a manner that was typical of the spontaneous way in which the Royal Society developed in its early years, while non-officers were also often highly influential. Thus I would agree with the Mulligans in discerning a period in the 1690s when the Secretary, Hans Sloane, seems to have been the most important officer. The presidency was so much an honorary position at this time that at least one President never attended a meeting during his term, while two others had to be elected Fellows on the day they became President.[11] But the Mulligans' period conveniently ends in 1700, so they have not had to consider why the President seems to have been so dominant in the Society under Newton, who missed only three Council meetings in the twenty years beginning in 1704.[12]

In the Society's early years there is good reason to believe that the President had a crucial role, if not the overbearing one occupied by Newton.

[8] Mulligan, 'Reconstructing' (n. 1), pp. 328, 348 – 49, 351 – 53.
[9] Ibid., p. 335.
[10] In the quotation from Ralph Thoresby in Mulligan, 'Reconstructing' (n. 1), p. 334, the words, '*the Incomes* of the R.S.' have been omitted after 'modelling' (my italics). Thoresby is referring to the financial retrenchment that the Society projected in 1698 – 99, discussed in Hunter, *Royal Society*, pp. 45 – 6, and in which the initiative was not taken by Sloane. The context of Richard Waller's letter to Sloane (which is ignored by the Mulligans) is also elucidated there, p. 50.
[11] Pembroke, Montagu, Somers.
[12] See Richard S. Westfall, *Never at Rest: A Biography of Issac Newton* (Cambridge, 1980), ch. 13, esp. p. 630.

Many contemporaries seem to have attributed to William, Viscount Brouncker, President from 1662 to 1677, much more than the titular headship attributed to him by the Mulligans, and this is borne out by one of the few constitutional analyses of the Society made by a contemporary. This appears in a letter from one 'A.B.' to the Council of the Royal Society, dated 19 October 1674, which has recently come to light and which will be printed and discussed at length in a forthcoming article by Paul Wood and myself.[13] Although we have been unable to precisely identify the author, the document is clearly the work of an established and well-informed Fellow. A.B. was analysing the current state of the Society and suggesting potential changes, and for him 'the President is our Sun, giving us light, heat & influence', whose role in supervising the Society's business he elaborated at length. All he said about the Secretary, on the other hand (clearly thinking of Oldenburg) was: 'As for the Secretary his business is, To take weekly Journals; to make Registers; & to keep a Philosophical Correspondence.' Moreover this estimate of the functions of the two officers was echoed by Edward Chamberlayne, FRS, in his *Angliae Notitia; or the Present State of England.*[14]

Indeed, Oldenburg was arguably less influential in the actual *direction*, of the Society — as against its administration and correspondence — than other figures besides Brouncker. One was Oldenburg's fellow Secretary from 1663 to 1668, John Wilkins, Bishop of Chester and formerly the centre of a seminal scientific circle in Interregnum Oxford. (It should be pointed out that throughout this period there were always two Secretaries, who sometimes shared the responsibilities of the job fairly equally — as with Hooke and Nehemiah Grew in the late 1670s — sometimes served complementary functions — as with Wilkins and Oldenburg — and were sometimes unbalanced, with one a virtual 'sleeper' — as with Sloane and Richard Waller from 1694 onwards.) Wilkins was responsible for major policy initiatives in the Society after 1668 as before, and at his death apprehensions were expressed that 'the Royal Society may stagger by the losse of such a proppe and incourager'.[15] Another influential figure who held office only in the Society's earliest years, when he served as President under the system of monthly rotation that preceded the Society's incorporation in 1662, was the mathematician and courtier, Sir Robert Moray, whose death in 1673 so soon after Wilkins's was also seen as 'a great blow'.[16]

[13] Royal Society Miscellaneous Manuscripts 4.72. The 'forthcoming article' has now appeared as 'Towards Solomon's House', above, ch. 6.
[14] Edward Chamberlayne, *Angliae Notitia*, Part I (7th edn, London, 1673), p. 343.
[15] Hunter, 'Social Basis' (n. 3), p. 27. On his role see B. J. Shapiro, *John Wilkins* (Berkeley and Los Angeles, 1969), ch. 7; P. B. Wood, 'Methodology and Apologetics: Thomas Sprat's *History of the Royal Society*', *British Journal for the History of Science*, 13 (1980), 2 – 5; Michael Hunter, 'Early Problems in Professionalizing Scientific Research: Nehemiah Grew and the Royal Society', above, ch. 8.
[16] Hunter, 'Social Basis'(n. 3), p. 27. See Alexander Robertson, *The Life of Sir Robert Moray* (London, 1922), p. 153 note 2, and ch. 8 passim.

Hence it would be wrong to presume that only the President and Secretary of the day mattered. The decisive body in directing the Society's affairs in its early years was its Council, which A.B. likened to the Society's parliament, its twenty-one members elected annually by the Fellows and 'entrusted with the generall care of all our business'. This influential body contained many activists who undoubtedly made a major contribution to the Society's policy. Moreover it is symptomatic that many of its members were continuously involved with the Society but only haphazard holders of office. These included Daniel Colwall, Thomas Henshaw, Abraham Hill, Sir John Hoskins, Sir Robert Southwell and Sir Christopher Wren. All these men were frequently on the Council (and often in attendance at meetings) but — apart from Colwall and Hill, who were successively Treasurers for long periods — they held office only for a few years at a time. And in addition to holding formal office, such men will also be found playing a decisive role at meetings in the capacity of Vice-President, deputizing for the President, especially in the 1690s.

Who Nominated Fellows?

In the absence of any contemporary evidence in favour of the theory, it therefore seems hard to justify an exclusive role for one of the secretaries in the direction of the Society in its early years. There is still less evidence to support the hypothesis that the secretaries were responsible for candidates who presented themselves for election. No contemporary testimony is adduced to support this thesis.

The only evidence as to the nomination of candidates is provided by the names of their 'proposers', which are often (though not always) given in the Society's minutes. Moreover, although candidates were sometimes brought forward by officers, proposers were seen to have a genuine role in securing candidates' election and in being responsible for them once they had joined. Edward Chamberlayne advised his readers that 'the Candidate is proposed by some that know him well', while it is revealing that when the Society expressed concern about arrears of subscriptions in 1666, it was those who had proposed the chief culprits for membership who were given the responsibility for dunning them.[17]

Since such proposals provide concrete evidence either of acquaintance patterns or of organizational roles within the Royal Society, I have included the names of all known proposers in the new version of my paper on the Royal Society's early membership. I have also analysed the evidence they provide, with some interesting results. Oldenburg proposed fifteen Fellows, and of these all but three were correspondents of his, eleven of them living abroad. This bears out the view of Oldenburg as a correspondent and

[17] Chamberlayne, *Angliae Notitia* (n. 14), pp. 343 – 44; Birch, *History*, ii, 118, and cf. D. M. 5.2

journal-editor which has already been indicated (and which has been illustrated by the work of the Halls in publishing his correspondence), but does not suggest the dominant role in the direction of the Society or in recruitment as a whole postulated by the Mulligans.

During the supposed 'Oldenburg' period, on the other hand, two Fellows proposed more candidates than he did: Wilkins, who proposed no fewer than forty-one Fellows up to his death in 1672 (all but eight falling within the period beginning in 1663 postulated by the Mulligans), and Seth Ward, Bishop of Salisbury, who proposed sixteen between 1666 and 1675; in addition, Moray proposed nineteen, all but one of them in the period 1661 – 66. It thus seems as if Wilkins, Moray and Ward were either actively proselytizing on the Society's behalf or acting as a sort of official vetting agency; whichever way, they must have had a decisive influence on the composition of the membership which Oldenburg lacked. Brouncker, it may be noted, proposed fifteen Fellows between 1661 and 1675, while a total of forty-six other Fellows between them proposed the remainder of the candidates for whom a proposer's name is known, none of them accounting for more than eight (Robert Boyle), and most for only one or two. In these cases, it seems as if straightforward acquaintance explains sponsorship: no hypothesis of a recruitment drive by a single officer is needed to explain why these men joined. Least of all are 193 Fellows elected between 1663 and 1677 to be labelled 'Oldenburg's Recruits'.[18]

What happens when we look at the supposed 'Hooke' period from the point of view of proposers? Perhaps the most striking single point arises in the year 1681 and concerns John Houghton, the apothecary and tea-dealer who at this time was just starting to publish his celebrated *Collection of Letters for the Improvement of Husbandry and Trade*, and who had been elected to the Society in 1680. In 1681, Houghton proposed seven Fellows, more than were otherwise proposed by one Fellow in any single year during the whole of the seventeenth century. Indeed, since these recruits included three merchants, among others, Houghton personally made a quite disproportionate contribution to the broadening of the Society's social base.

Not only did Houghton propose these seven elected candidates; he also proposed three who were not elected, including two of fairly low status. This untraditional intensity of recruitment by a new Fellow, including an untraditionally high percentage of men of low status, must help to explain why, early in 1682, the Society's admission policy was changed (a point noted in my paper but ignored by the Mulligans).[19] For on 25 January 1682 reservations were expressed about the ease with which elections were made, and later that year a statute was passed ordering that all candidates should be considered by the Council before they were allowed to stand. This apparently not only explains why, from 1682 onwards, a higher proportion of those elected showed at least some interest in the Society's activities, but may also partly account for the reversal of the trend towards a broader

[18] Mulligan, 'Reconstructing' (n. 1), p. 345.

[19] Hunter, 'Social Basis' (n. 3), pp. 31, 70n., 147 (Russell, Hewer and Joseph Martin were proposed by Houghton); Birch, *History*, iv, 121, 158, 159.

social basis that the Society experienced from about this time.[20] Houghton was never an officer of the Society, though he took an active part in it, promoting technological topics almost single-handedly at meetings in the 1690s:[21] but he clearly played a major role in determining the pattern of recruitment in the early 1680s.

What about Hooke? In note 111 of their article the Mulligans refer to the proposers of a certain number of candidates in the so-called 'Hooke' period, including Hooke and various of 'his friends' (though it is odd that six of Hooke's nominations are ignored, and that of the Fellows said there to have been nominated by his friends, two were in fact proposed by Hooke).[22] In all, Hooke himself proposed ten (or 16 per cent) of the sixty-one candidates elected during his secretaryship for whom a proposer is named (14 per cent of all those elected). Though this is comparable to the percentage proposed by Moray between 1661 and 1666 (17 per cent) or by Ward between 1666 and 1675 (16 per cent), it is significantly less than the percentage proposed by Wilkins during the longer period 1661 – 72 (24 per cent). Moreover it is perhaps revealing that a further candidate proposed by Hooke was never elected.[23] This would suggest that Hooke was responsible for the pattern of recruitment during his secretaryship only to a small extent, and though the Mulligans would argue that the task of proposing was shared around among Hooke's 'friends', it seems to be that this represents an unnecessary multiplication of hypotheses. In fact — in these years, as in the 1660s and earlier 1670s — a large number of Fellows were responsible for bringing forward candidates on an individual basis.[24]

The Role of Hooke

The Mulligans claim to find evidence of Hooke's influence on recruitment in the fact that thirty of the seventy-five Fellows elected between November 1677 and 1682 were 'friends or acquaintances mentioned in his diary' and fifteen more were 'foreigners recruited, at least in part, by him'; elsewhere evidence is said to be offered of a common educational background linking 'Hooke's recruits'.[25] These claims will bear careful scrutiny, since they reveal clearly the weakness of the Mulligans' case.

20 Hunter, 'Social Basis', (n. 3), p. 31, and see above, p. 340.

21 Hunter, *Science and Society*, p. 98.

22 Tyson, Joseph Lane (since this was the only man named Lane elected at this time, I am mystified by the second 'Lane' listed as nominated by Hooke). Clench, Firmin, Heusch, Houghton, Mayow and Moore junior were also proposed by Hooke. There is no evidence that Hooke proposed Slare, as claimed by the Mulligans, though he did 'introduce' him when he was admitted Fellow (Birch, *History*, iv, 61). On the problems of identifying Hooke's 'friends' see below, pp. 346 – 7.

23 Birch, *History*, iv, 104 (joint proposal with Thomas Gale).

24 For the names of all known proposers see the revised Catalogue of Fellows in the 1982 edition of my 1976 paper, which also includes an index of proposers.

25 Mulligan, 'Reconstructing' (n. 1) pp. 332, 346 – 48.

First, let us take the evidence for common education. The reader should note that the 'many' near-contemporaries' at Trinity College, Cambridge, referred to by the Mulligans on page 347 amount to only four out of seventy-five, while it transpires on page 350 that these men were up at Trinity 'at different times'. Though it is added that 'most' of Hooke's recruits were educated at St Paul's or at Westminster School, the number involved is nowhere specified. Nor is the schooling of recruits of other periods analysed, so that it is difficult to know what significance to attach to the statement. Similarly, though the authors mention that sixteen new Fellows of this period attended the Inns of Court (whether in addition to or instead of going to university), comparable figures for other periods are not given, so it is difficult to know whether or not this was atypical, as is here implied.[26] None of this seems to me to prove very much. As for the association with foreigners, I know of no evidence to link Hooke with all but four of the fifteen foreign Fellows whom he is said to have recruited 'at least in part' (one of these four was, it should be noted, not actually foreign).[27]

Then we have the evidence of Hooke's diary, which seems to me to have been employed in a highly questionable way. Hooke's diary is an unusual source which needs to be handled with care, particularly when arguing for Hooke's close contacts with people when no comparable documents exist to assess the 'acquaintance networks' of others at the time. This collection of jottings on Hooke's day-to-day activities and contacts is so voluminous that it can only be said to prove Hooke's close acquaintance with those who appear extremely frequently in its pages, and not with those who appear only a handful of times. London in the 1670s and 1680s was a small place by modern standards, and Hooke could hardly avoid bumping into all sorts of people when walking about it, including even King Charles II.[28] Though Hooke religiously recorded many such occasional encounters, it can hardly be claimed that they account for the presence of the individuals concerned in the Royal Society.

A consideration of those listed by the Mulligans in their note 51 will illustrate the flimsiness of this argument in many cases. The index to the published edition of the diary (which, I grant, is not quite complete) shows that of the thirty people listed as having significant contacts with Hooke one does not appear at all (Wheler), three only once (Chardin, Moore junior, Waller), two twice (Bridgeman, Chetwynd), one three times (Dickenson) and six more less than ten times (Clüver, Deane, Gale, Oliver Hill, Lane and

[26] Ibid., pp. 347 – 48. As n. 82 explains, the percentages of those attending the Inns given in Table 5 excludes those who had already been to university.

[27] Ibid., p. 332 and n. 51. Heusch was proposed by Hooke and Sarotti and Spanhiem each appear twice in the diary: see Hooke, *Diary*, pp. 373, 403, 434, 456. John Bemde, or Van der Bemde — who appears once in the *Diary*, p. 387 — was always designated an English, as against a foreign, Fellow in the Society's annual printed membership lists. For further information on Bemde, see Hunter, *Royal Society*, p. 71.

[28] Hooke, *Diary*, p. 161.

Perry).[29] Since literally scores of other people are mentioned in the diary a handful of times, none of whom was ever suggested for election to the Royal Society, it seems to me that we are hardly closer to 'explaining' why these men joined. In the case of Hill, the Mulligans appear to have confused two different people: Oliver Hill was elected to the Royal Society in 1678 and was briefly active at meetings, Hooke's few comments about him in the diary (like the attitudes of other Fellows revealed in the minutes) showing almost total hostility; Abraham Hill, on the other hand, appears frequently in the diary as a regular coffee-house associate of Hooke's.[30]

This leaves us with a substantially smaller group, and while some of these were certainly friends of Hooke who may well have joined the Society due to his encouragement — such as George Ent junior, whom he proposed — in other cases it is unlikely that Hooke's influence is a necessary hypothesis in explaining why the men in question became Fellows. The Society was thus a natural focus for Edmond Halley, who was proposed by Sir Jonas Moore, and for Robert Plot, whose proposer is not recorded, and, though it is possible that there is a connection between the overthrow of the Brouncker-Oldenburg regime and the disproportionate number of activists recruited in the late 1670s, this requires an explanation transcending the influence of Hooke.

This is not to deny that Hooke played a part in recruitment, along with Houghton and others. In attributing so much to Hooke's personal influence, however, the Mulligans are trying to explain too much too simply. The routes by which candidates came forward were in fact more varied, reflecting the diffuseness of control in the Society's affairs. It is thus notable that, though the increased recruitment of professional scholars is a significant feature of this period, none of the nine elected during Hooke's secretaryship whose proposers are known was nominated by Hooke.[31]

If it is unlikely that Hooke is to be held responsible for more than a minority of those who joined the Society while he was Secretary, it is equally unlikely that he discouraged others from joining, as implied by the Mulligans (although they offer no evidence in support of the suggestion).[32] It is true that figures prominent in politics or at court were unusually infrequent among recruits at this time, and that no member of the

[29] On Waller's later friendship with Hooke see Margaret 'Espinasse, *Robert Hooke* (London, 1956), pp. 138 – 39; but this friendship was probably a result, rather than a cause, of his membership of the Society. Hooke presumably proposed Moore junior (see above, n. 22) because of his friendship with his father, who *is* frequently mentioned in the diary. Lane, whom he also proposed, proved useful to the Society in its legal business (see Birch, *History*, iii, 410; iv, 75, 79, 82, 86, 97, 109) and was perhaps recommended to Hooke for this purpose.

[30] Mulligan, 'Reconstructing' (n. 1), p. 350. The only remotely favourable reference to Oliver Hill is on p. 376 of the *Diary*; but the 'Mr Hill' with whom Hooke records 'much good discourse' on that page was (as usual) Abraham. See also Hunter, *Science and Society*, pp. 19 – 20 and n. 33.

[31] The proposers were Edward Browne, Colwall, Haak, Holder, Hoskins, Mapletoft, Moore (senior) and Wren.

[32] Mulligan, 'Reconstructing' (n. 1), pp. 346, 352.

aristocracy became a Fellow between the Marquis of Halifax in 1675 and the Second Earl of Clarendon in 1684. But there is no reason to connect this alteration with Hooke, who was Secretary only from 1677 to 1682. What we evidently see is a lacuna between the first decade and a half of the Society's history, when it had close links with courtly and aristocratic circles through figures like Brouncker and Moray; and a period, especially in the 1690s, when contacts with the government and with high-ranking circles were re-established through men like Sir Robert Southwell. The temporary lull in recruitment of men of high rank in the interim is apparently to be attributed to the lack of such contacts — and perhaps also to the Society's changing reputation[33] — but certainly not to unproven secretarial influence.

A 'Hooke Period'?

All in all, most of the evidence for attributing the positive and negative features of recruitment in this period to Hooke seems doubtful. This raises the broader question of whether there was a distinctive 'Hooke period' at all. No one would deny Hooke's hostility to the Brouncker-Oldenburg regime in the 1670s, as illustrated by his diary. Moreover Hooke certainly attained greater prominence in the Society in the reshuffle that followed the ousting of Brouncker from the presidency in 1677, when he gained a place on the Council for the first time. But was the Society's character reshaped under Hooke's influence during his secretaryship?

Part of the Mulligans' argument to this effect depends on a contrast between Oldenburg's and Hooke's styles of science which they have exaggerated. While Oldenburg is almost caricatured, Hooke is presented as more single-minded than he actually was.[34] The evidence cited for this evaluation is superficial, and the only detailed instance given has been misinterpreted. In their note 45, the Mulligans tell us that 'Hooke rejected further correspondence with a Mr Young about the grisly details in the life of an ancient woman — such 'trifles' did not interest him'. Yet if the letter cited (dated 9 January 1685) is examined, it transpires that Hooke was re-establishing contact with this correspondent. He refers to a former letter, which had gone astray, in which he had requested further information on this topic, which had proved of great interest to him and other members of the Royal Society. In any case, the man in question was not 'a Mr Young'

33 Hunter, Social Basis' (n. 3), p. 27.
34 Mulligan, 'Reconstructing' (n. 1), pp. 330 – 32, 352. On Oldenburg see Paul B. Wood, 'Francis Bacon and the "Experimental Philosophy": a Study in Seventeenth-century Methodology' (London M.Phil. thesis, 1978), ch. 3. On Hooke see 'Espinasse, *Robert Hooke* (n. 29), pp. 108 – 10 and passim.

but James Yonge, the Plymough surgeon, a 'good friend' of Hooke (whom he had visited in 1678 and 1681); by this time Hooke had published more than one of Yonge's writings.[35]

Furthermore, no radical change under Hooke is suggested by the extant minutes of meetings (which the Mulligans appear not to have consulted for information on this topic). Hooke had, after all, always dominated the Society's scientific business as Curator of Experiments: as Wren put it in a letter to him of 1665, 'I know you are full of employment for the Society which you all-most wholy preserve together by your own constant paines'.[36] This is borne out by Frank's analysis of the minutes, which showed Hooke carrying out a quite disproportionate number of experiments throughout the period he studied, in 1680 as earlier.[37] Frank's analysis, it may be noted, provides no evidence from the minutes for changes like the shift towards applied science postulated by the Mulligans as presided over by Hooke; instead, he finds a familiar range of miscellaneous business.[38] Such continuity is perhaps in some way ironic, since Hooke *was* interested in a radical reform of the Royal Society.[39] Yet none of this came to fruition during Hooke's supposed 'regime'.

Both in the running of the Society, and in activity at meetings, there is plenty of evidence for the influence of others apart from Hooke. Throughout these years — as throughout the Society's early history — important decisions were taken by the Council, and active stalwarts who attended frequently included Colwall, Henshaw, Hill, Hoskins and Wren, who became President in 1680. The Mulligans are thus quite wrong to attribute the large-scale expulsions which took place for the first time in the Society's history in 1682 to the influence of Hooke: Hooke was not even present at the meeting at which the decision to make the expulsions was reached, though he was there when it was subsequently confirmed. The expulsions were due, not to his decisiveness, but to new circumstances brought about by the Society's substantial endowment for the first time by the King in 1682.[40]

35 E.L. H.3.71; F. N. L. Poynter, ed., *The Journal of James Yonge (1647 – 1721) Plymouth Surgeon* (London, 1963), pp. 23, 155, 168; R. Hooke, *Lectures and Collections* (London, 1678), pp. 105 – 12; Hooke, *Lectures de Potentia Restitutiva . . . to which are added some Collections* (London, 1678), pp. 28 – 35.

36 Wren-Hooke, 20 April 1665, E.L. W.3.6.

37 Frank, 'Institutional Structure' (n. 2), p. 86.

38 Ibid., pp. 91 – 94. Frank there discusses the minor shifts in interest that occurred. Compare Birch, *History*, iii, 358 – 520, iv, 1 – 168, and Mulligan 'Reconstructing' (n. 1), pp. 348, 352 – 53.

39 The hitherto unpublished evidence for this, very briefly adumbrated in Hunter, *Science and Society*, p. 57, is expounded more fully in 'Towards Solomon's House', above, ch. 6.

40 Mulligan, 'Reconstructing' (n. 1), p. 332. See Birch, *History*, iv, 159 – 60, 161; Hunter, *Science and Society*, p. 41.

Indeed, so far from being the fruitful period of scientific achievement depicted by the Mulligans, there is in fact evidence for severe problems during the years when Hooke was Secretary. Attempts to improve the quality of the Society's proceedings recurred not only soon after the reshuffle of 1677 but at intervals thereafter (in December 1679, January 1680 and January 1681), while in June 1680 an emergency Council meeting was called to which John Evelyn summoned Samuel Pepys with the assurance that it involved 'the most material concerne of a Society, which ought not to be dissolved for want of a redresse, which is yet certainely in its power'.[41]

Recruitment, it is true, was fairly high in these years, averaging fourteen elections a year in 1677 – 81, by contrast to six a year in the earlier 1670s.[42] But this could as easily be seen as a desperate expedient in hard times, as a symptom of the good health of the Society. Indeed the 1682 decision on membership suggests a reaction against indiscriminate elections. And, though the Mulligans have devoted their attention exclusively to recruitment, other indices show the 'Hooke period' in a less favourable light than the years after 1682. Both the number of meetings held per year and the attendance at the annual anniversary meetings was larger afterwards, while the size of the Society's active nucleus also increased. In addition, the bulk of expulsions at this time occurred in 1685 rather than 1682, while certain novel developments of the 1680s are discussed in the new version of my 1976 paper.[43]

The greatest objection to associating the period too optimistically with Hooke, however, stems from the fact that this seems to have been a problematic time for Hooke himself. In 1678 and especially in 1679 he attempted to undertake the correspondence work that Oldenburg had long carried out for the Society, in addition to displaying experiments at meetings; while from 1679 to 1682 he published his *Philosophical Collections* as a replacement for Oldenburg's *Philosophical Transactions.* As if this double burden was not enough, Hooke also acquired an extra responsibility outside the Society to add to his continuing architectural practice. From 1678 to 1680 he collaborated with the publisher, Moses Pitt, in producing a lavish atlas, and Hooke's part in the project was (in the words of its historian) 'no sinecure'.[44]

[41] Birch, *History*, iii, 365, 513 – 14, 516; iv, 6 – 8, 20, 65; Evelyn, *Diary*, iv, 125 – 26, 205; Evelyn-Pepys, 25 June 1680, *Historical Manuscripts Commission: MSS of J. Eliot Hodgkin* (London, 1897), p. 177.

[42] See Hunter, 'Social Basis' (n. 2), p. 26 (Table 1) (ibid., p. 31, perhaps rather underrates the change). I am unable to explain how the Mulligans have arrived at an average figure of seventeen (Mulligan, 'Reconstructing' (n. 1), p. 332).

[43] Hunter, 'Social Basis' (n. 3), pp. 26 (Tables 3 and 4), 40 (Table 7), 55 – 56. No evidence is cited to support the statement to the contrary concerning attendance in Mulligan, 'Reconstructing' (n. 1), p. 335. On the changes of the 1680s, see Hunter, *Royal Society*, pp. 41 – 2.

[44] E. G. R. Taylor, ' "The English Atlas" of Moses Pitt, 1680 – 83', *Geographical Journal*, 95 (1940), 298. On Hooke's architectural practice see 'Espinasse, *Robert Hooke* (n. 29), ch. 5.

Then, in 1680, there was a crisis in the dispute between Hooke and Sir John Cutler, who was ever reluctant to pay Hooke the emoluments of the lectureship Cutler had established for him in 1664. In the early 1680s there are hints both in Hooke's diary and in his letters of troublesome legal business, 'which took up most of my Leasure time'.[45]

As might be expected, the Royal Society suffered from this overcommitment. In December 1679, the duty of looking after the Society's correspondence was taken away from Hooke — to his chagrin, as recorded in the diary[46] — and given to Thomas Gale. The Mulligans' delicate suggestion that 'perhaps the Society did not endorse Hooke's attitude to his correspondence work'[47] disguises the fact that there had been complaints from foreign correspondents because the task had been neglected. Then, in 1682, there were evidently further complaints, this time about Hooke's performance of his more general secretarial duties, and on 30 November of that year Hooke was more or less given the sack; he was replaced as Secretary by Robert Plot and was not even re-elected to the Council for 1682–83. (It should be noted that the Mulligans have included as 'Hooke recruits' people elected at this very meeting.[48])

Great Men and Institutional History

This does not suggest a triumphant reformist episode presided over by Hooke, contrasting with less successful periods before and after. On the contrary, it suggests that during Hooke's secretaryship the Society experienced the same mixture of normal activity and periodic crisis which is to be found throughout its early history. And it is in general easy to exaggerate the extent to which individual office-holders were the focus of good and bad periods in the Royal Society's early history. Though the significance of the palace revolution of 1677 is undoubted, it is revealing that the two most serious crises of the seventeenth century — in 1673–75 and in 1698–99, when major programmes of reform and reinvigoration were projected — came in the middle, rather than at the beginning, of secretarial eras. They represented a response to declining activity among the membership at large, while the initiative in them was often taken by individuals who were not even officers of the Society, such as Sir William

[45] Hooke-Yonge, 9 January 1685, E.L. H.3.71. See also Hooke-Sir John Hoskins, Last of March 1683, ibid., H. 3. 70, and Hooke, *Diary*, pp. 449–50 and passim thereafter. There are also a few relevant hints in the sections of the diary covering 1681–83 which were omitted by the editors of the printed edition: see City of London Guildhall Library MS 1758. For the background see 'Espinasse, *Robert Hooke* (n.29), p. 83.

[46] Hooke, *Diary*, p. 433. See Birch, *History*, iii, 518, and cf. ibid., iii, 515.

[47] Mulligan, 'Reconstructing' (n. 1), p. 332.

[48] Birch, *History*, iv, 168. See also ibid., iv, 174, 196. Faria and Chardin were elected at this meeting.

Petty and Jonathan Goddard in the early 1670s and William Bridgeman in 1698.[49]

Indeed, a more serious issue is involved in the Mulligans' paper than simply a dubious hypothesis in support of which no satisfactory evidence has been adduced. For behind it lies a misleading view of institutional history, which they share with commentators on the greatest man associated with the Society at this time, Sir Isaac Newton. Recent biographers of Newton have presumed that he played a greater personal role in reshaping the Society during his presidency from 1703 to 1727 than was really the case. In fact, Newton's novelty has been overrated and his influence on the Society's structure was less than has sometimes been thought,[50] and the same is true — for all his scientific brilliance — of Hooke.

The point is that the Royal Society cannot be properly understood by the 'Great Man' theory of institutional history to which the Mulligans evidently subscribe. From the first, the Society developed a complexity which meant that all sorts of interconnecting initiatives went into its day-to-day running, its recruitment, its activities and its reforms. Moreover the Society early gained a sort of momentum of its own, which made it impregnable to reform efforts but at the same time more resilient to crisis than many contemporary commentators realized; its continuity as a symbol of scientific enterprise was thereby ensured.[51] It should also be made quite clear that the Mulligans' misleading stress on secretaries has not — as they claim — countered the case for general trends which no individual could reverse. One such notable trend was the decline in public experimentation at meetings which Frank and others have noted, and which proved inexorable despite repeated efforts at reform, not least during Hooke's secretaryship.[52]

We have come some way from the question of recruitment, but this discussion illustrates the misapprehensions likely to stem from an over-schematic approach to the history of a major institution. I do not believe that there was a 'Hooke period', and almost none of the other periods postulated by the Mulligans seems to me to have the significance they believe their statistics prove. There are significant variations in recruitment to the Royal Society in its first forty years, which the Mulligans and I agree in discerning: but these are only slightly connected with the influence of the Society's Secretaries. Moreover the Mulligans' broader claims have even less to substantiate them. Recruitment is only one of many aspects of the

[49] On the former, see Hunter, Social Basis' (n. 3), pp. 27 – 29, and Lindsay Sharp, 'Sir William Petty and Some Aspects of Seventeenth-century Natural Philosophy' (Oxford D. Phil. thesis, 1977), pp. 255 – 73. The latter is discussed in Hunter, *Royal Society*, pp. 45 – 6.

[50] See F. E. Manuel, *A Portrait of Isaac Newton* (Cambridge, Mass., 1968), ch. 13; R. S. Westfall, *Never at Rest* (n. 12), ch. 13; Hunter, *Royal Society*, 'Conclusion'.

[51] Hunter, *Royal Society*, 'Conclusion'.

[52] Mulligan 'Reconstructing' (n. 1), p. 353; Frank, 'Institutional Structure' (n. 2), pp. 94 – 98; see also R. K. Bluhm, 'Remarks on the Royal Society's Finances 1660 – 1768', *Notes and Records* 13 (1958), 91, and R. P. Stearns, *Science in the British Colonies of America* (Urbana, Ill., 1970), pp. 96 – 98. On Hooke's secretaryship, see the references in n. 41.

Society's development which need to be scrutinized before legitimate conclusions can be drawn about the Society's activities and significance, and — as I have indicated in the Appendix — undue emphasis on recruitment can actually impede a true understanding of the Society's corporate character.

If this Response seems negative, this is not because I am hostile to the enterprise in which the Mulligans are engaged. Quite the reverse is true, for institutional history is a crucial topic in the history of seventeenth-century science. The Royal Society has already suffered greatly from overschematized views of its origins and nature.[53] It is essential that in studying institutions we do justice to their full historical complexity.

NOTES

ON THE MULLIGANS' STATISTICS

This Appendix is intended partly to help to explain discrepancies between statistics on topics which both the Mulligans and I have studied, but also to explain my reservations about their statistics on matters which I have not pursued. It concludes with some remarks on the limitations of exclusive attention to recruitment in reconstructing the Society's character.

First, the Mulligans have included foreign members in their analysis of Fellows' occupation, education, and so on, whereas I kept English and foreign members separate and did not go further than itemizing all the latter as 'foreign'. This has provided some information which readers will value, but it also accounts for some divergences between our statistics. By the late 1680s and 1690s, foreign members accounted for some 24 per cent of all recruits: the considerable number of foreign correspondents with the degree of MD who were invited to become nominal Fellows in that period apparently accounts for the more marked and consistent increase in the recruitment of medical men at that time in the Mulligans' figures than is in evidence in the tables in the extended version of my 1976 paper.[54] This conflation must affect all the Mulligans' statistics to some extent — they even cite 'foreign' illustrations of social trends in the text[55] — but since foreign members constitute an even more random sample of European scientific enthusiasts than do English ones, who by definition could do even

53 Perhaps particularly in Margery Purver's *The Royal Society: Concept and Creation* (London, 1967).

54 Hunter, *Royal Society*, appendix 10 and passim; Mulligan, 'Reconstructing' (n. 1), Table 2, pp. 333, 351, and n. 56.

55 For example, Mulligan, 'Reconstructing' (n. 1), p. 346 (Hannisius).

less for the Society once elected, I am inclined to feel that they are best analysed separately.

Secondly, there is the question of accuracy, though this can only be tested on a random basis since most of the Mulligans' statistics are inscrutably anonymous with (generally) a respectable proportion of unknown cases. This also concerns topics which they cover and I did not, such as education. The problems involved are individually small, but cumulatively important. Thus whereas I calculate that 551 Fellows were elected between 1660 and 1700, the Mulligans have only 542, and I simply do not believe that my 1976 Catalogue of Fellows and its new extension to 1700 contains nine ghosts. Again, Sloane's secretaryship is deemed by the Mulligans to have begun in 1692, whereas in fact he was not elected to the office until 30 November 1693:[56] this must raise doubt as to whether the eight Fellows elected in 1693 should be included in the supposed 'Sloane' period. In some cases where it has been possible to check figures included in the tables, the information given seems incorrect. For example, far more Fellows in the 'post-Hooke' period were alumni of the Inns of Court (but not of any university) than appear in Table 5: according to my investigations, some eleven Fellows elected after 1682 were educated at one or other of the Inns, a number quite at odds with the figures given there, which suggest that only one or two Fellows attended.[57] In addition, the fact that biographical misidentifications appear even in the text does not inspire confidence in those which have been silently subsumed into the statistics: there are also various inaccuracies of nomenclature and citation.[58]

Elsewhere I am dubious of the precision which the tables claim. Table 1, on social status, includes no unknown cases, yet I know from my own experience of the impossibility of identifying or classifying certain Fellows that some of these categorizations must be based on guesswork of which the reader is not warned. Table 6 similarly has no unknown cases and therefore presumes a completeness of information which seems extremely dubious in the case of lesser 'contributors', since interests and activities are frequently evidenced only by haphazard hints: for instance, I have recently come across a letter suggesting that William Hammond, a Fellow on whom information is scarce, was 'a great lover of Mathematics',[59] and such cases could cumulatively quite alter the statistics given. Moreover the sheer breadth of

[56] Ibid., p. 334.

[57] The men involved were: Byrd, Chadwick, Copley, Day, Foley, Hammond, Isted, Meres, Pitfield, Earl of Radnor, Sydenham. Some doubt remains, however, about the identification of Chadwick and Day: see Hunter, *Royal Society*, pp. 72, 73 – 4.

[58] See, for instance, a comment in Mulligan, 'Reconstructing' (n. 1), p. 350, showing that Edward Stillingfleet, FRS, has been misidentified with E.S. (1635 – 99), Bishop of Worcester, whereas in fact the Fellow was the bishop's son, Gresham Professor of Physic (1660? – 1708). Nomenclature: for example, Thomas Gale is referred to variously as 'Robert Gale' (n. 94), as 'Edward Gale' (p. 350) and as 'Cole' (n. 49). Citation: for instance, MS volume numbers are omitted in nn. 58, 61, while the quotation on p. 335 has been garbled (cf. Hunter, 'Social Basis' (n. 3), p. 45).

[59] John Collins-John Wallis, 14 February 1677, Royal Society Miscellaneous Manuscripts 6.86. On Hammond see Hunter, 'Social Basis' (n. 3), p. 43.

interest of many virtuosi of the time suggests that some of this table's categories are rather artificial.[60]

It should also be pointed out, for the benefit of the unwary reader, that Table 6 conflates Fellows' activities inside the Royal Society with interests evidenced outside it. Even men who were totally inactive in the Society are included in the figures (such as, presumably, John Mayow, a 'systematic scholar' elected in 1678, who had nothing to do with the Society at any stage). The value of this statistical exercise is not obvious, since it seems to show very little of significance about the composition of the Society or the way in which its activities changed over the period.

Table 8 is based on the analysis of Fellows' activity on my 1976 paper, and it illustrates quantitatively the influx of new activists that the Society experienced in the late 1670s and early 1680s, virtually for the first time since the 1660s. But whereas in my Catalogue of Fellows I made clear how long individuals supported the Society, the Mulligans have conflated together all those I described as 'active', whether their activity lasted for a few months or for forty years. Yet this clearly makes a great difference to the contribution of the Fellow involved to the Society's vitality.

Indeed, a major limitation of the Mulligans' paper is its restriction to recruitment. The Society was at any one time only partly composed of new recruits: in 1700, for instance, 27 per cent of those who played an active role in the Society had been elected in the past five years, 15 per cent in the previous seven, while the rest were of longer standing.[61] Exclusive attention to the newest members is therefore likely to give a misleading view of the state of the Society at any time: that is why both Frank and I analysed the whole membership, or the whole activity, of the Society in sample years, while I tried to accompany this by a more narrative view of how the movement of Fellows both into and out of the Society made up the texture of its institutional life. This depended on an examination of the Society's minutes (and its accounts) on a consecutive basis, as would any attempt to assess the Society's vitality. Yet the Mulligans have apparently used the minutes only to assess the interests of each recruit rather than to gauge the Society's character at successive stages. They have thus not actually established whether, or how far, the concerns of new recruits had any impact on the Society's areas of interest: hence their claim to have analysed the 'activities' of the Society as well as 'changes in the nature of the membership' is largely unfounded.[62]

60 See, for instance, Michael Hunter, *John Aubrey and the Realm of Learning* (London, 1975). Readers will also note that, though this table includes interests in the humanities, it entirely ignores Fellows' occult concerns.

61 Hunter, *Royal Society*, Table 9 (pp. 120 – 1).

62 Mulligan, 'Reconstructing' (n. 1), pp. 329, 352 – 53.

BIBLIOGRAPHICAL ESSAY

Recent Studies of the Early Royal Society and its Milieu

One feature of my book *Science and Society in Restoration England* (Cambridge, 1981) which was universally well-received was the Bibliographical Essay which concluded it (pp. 198 to 219). Since that was compiled in 1979 – 80, there has been much activity in this field, and the object of this essay is to indicate what has been written on related topics since. The focus will be on the Royal Society itself, as the central theme of this book, recent writings on which will be described as completely as possible. But I will also draw attention to what seem to me the more important trends in recent work on the Society's context in its early years, though space precludes quite as comprehensive an account of such material.[1]

Starting, then, with the Royal Society, we may first note a review essay on the Society's origins and early history which also includes some intriguing reflections on its Harringtonian legacy, Anna M. Strumia, 'Vita Instituzionale della Royal Society Seicentesca in Alcuni Studi Recenti', *Rivista Storica Italiana*, 98 (1986), 500 – 23. On the institutional side of things, I should explain that the article which I have included here as an appendix is a response to a lengthy analysis by Lotte and Glenn Mulligan: 'Reconstructing Restoration Science: Styles of Leadership and Social Composition of the Early Royal Society', *Social Studies of Science*, 11 (1981), 327 – 64. On one aspect of the Society's organization a paper has appeared which glimpses at this period and places it in the context of later developments, Maurice Crosland, 'Explicit Qualifications as a Criterion for Membership of the Royal Society: a Historical Review', *Notes and Records of the Royal Society*, 37 (1983), 167 – 87, while on a range of themes my monograph, *The Royal Society and its Fellows, 1660 – 1700: the Morphology of an Early Scientific Institution* (Chalfont St Giles: British Society for the History of Science, 1982) wholly supercedes the earlier recension of the same study, published in the Society's *Notes and Records* in 1976. Not only does it extend the coverage from 1685 to 1700; it also incorporates information on the earlier period not to be found in the former version, dealing at length with topics like Fellows' proposers and the Society's foreign membership. I have to point out that the 1985 reprint of this work is marginally preferable to the original edition of 1982, since I had the opportunity to correct a few

[1] For a briefer and more general survey, see also my 'Reading History: Science in Seventeenth-century England', *History Today*, February 1984, 39 – 41.

misprints and to add a 'Note' on relevant material among the Petty Manuscripts at Bowood House.

The Society's foreign relations have been the subject of various studies. On its links with Italy we have had both Marta Cavazza, 'Bologna and the Royal Society in the Seventeenth century', *Notes and Records of the Royal Society*, 35 (1980), 105 – 23, and M. B. Hall, 'The Royal Society and Italy, 1667 – 1795', ibid., 37 (1982), 63 – 81, while much light is shed on the Society's continental contacts by the twelfth and thirteenth — and final — volumes of *The Correspondence of Henry Oldenburg*, edited by A. R. and M. B. Hall, which have now at last appeared (London: Taylor and Francis, 1986). On the Society's overseas links as a whole, a lengthy and elaborate account appears in: S. Gomes, A. Milian, C. Moreno and M. J. Pascual, 'Las Relaciones Internacionales de la Royal Society of London (1660 – 1700)', *Sylva Clius: Revista de Historia de la Ciencia*, 1 (1987), 21 – 107, the product of a lively centre for studies of English science in this period in Madrid. A traveller's account which provides some intriguing hints about the Society in its early years is: W. E. K. Middleton, ed., *Lorenzo Magalotti at the Court of Charles II: his 'Relazione d'Inghilterra' of 1668* (Waterloo: Wilfrid Laurier University Press, 1980).

On the Society's institutional background, a few items may be noted. The London College of Physicians, with which the Society enjoyed a somewhat stressful relationship in its early years, has been the subject of a valuable assessment by H. J. Cook in *The Decline of the Old Medical Regime in Stuart London* (Ithaca: Cornell University Press, 1986). Also relevant is Cook's 'The Society of Chemical Physicians, the New Philosophy and the Restoration Court', *Bulletin of the History of Medicine*, 61 (1987), 61 – 77, while important background material is provided by R. G. Frank jr.'s study of the flourishing of extra-curricular activities among members of the College of Physicians, 'The Physician as Virtuoso in Seventeenth-century England', in Barbara Shapiro and R. G. Frank, jr., *English Scientific Virtuosi in the Sixteenth and Seventeenth Centuries* (Los Angeles: William Andrews Clark Memorial Library, 1979), pp. 57 – 114. An interesting account of an abortive attempt to emulate the Society north of the border is given in Roger Emerson, 'Sir Robert Sibbald, Kt., the Royal Society of Scotland and the Origins of the Scottish Enlightenment', *Annals of Science*, 45 (1988), 41 – 72, while a broader view of the Royal Society, and particularly of its history and influence in the eighteenth century, is offered by J. E. McClellan in his *Science Reorganized: Scientific Societies in the Eighteenth Century* (New York: Columbia U.P., 1985); the research for this book is, however, somewhat less recent than its date of publication would imply. McClellan has also contributed a prosopographical study of 'The Académie Royale des Sciences, 1699 – 1793: A Statistical Portrait', *Isis*, 72 (1981), 541 – 67.

On more miscellaneous aspects of the Society's early institutional life, C. A. Rivington's 'Early Printers to the Royal Society, 1663 – 1708', *Notes and Records of the Royal Society*, 39 (1984), 1 – 27, is useful, while A. D. C. Simpson's 'Newton's Telescope and the Cataloguing of the Royal Society's Repository', ibid., 38 (1984), 187 – 214, is broader in scope than its title implies. By way of background to the repository, it is worth noting recent work on the early years of the other principal 'public' collection of rarities in England at this time, the Ashmolean Museum at Oxford: of this there has been a plethora, inspired largely by that institution's tercentenary. Most substantial is Arthur MacGregor, ed., *Tradescant's Rarities: Essays on the Foundation of the Ashmolean Museum, 1683, with a Catalogue of the Surviving Early Collections* (Oxford: Clarendon Press, 1983); in addition, there are the relevant sections of A. V. Simcock's excellent *The Ashmolean Museum and Oxford Science, 1683 – 1983* (Oxford: Museum of the History of Science, 1983); Michael Hunter, *Elias Ashmole, 1617 – 1692: the Founder of the Ashmolean Museum and his World* (Oxford: Ashmolean Museum, 1983); A. G. MacGregor and A. J. Turner, 'The Ashmolean Museum', in L. S. Sutherland and L. G. Mitchell, eds., *The History of the University of Oxford, vol. 5: The Eighteenth Century* (Oxford: Clarendon Press, 1986), pp. 639 – 58; and the early chapters of R. F. Ovenell, *The Ashmolean Museum, 1683 – 1894* (Oxford: Clarendon Press, 1986).

As for scientific activities carried out under the Royal Society's direct aegis, a valuable essay on the period of Newton has been published by J. L. Heilbron, *Physics at the Royal Society during Newton's Presidency* (Los Angeles: William Andrews Clark Memorial Library, 1983); this also has some material on the earlier period. A thematic article dealing with the pre-Newtonian years is Penelope M. Gouk's 'Acoustics in the early Royal Society, 1660 – 80', *Notes and Records of the Royal Society*, 36 (1982), 155 – 75, while an unusual study which focusses on the exchanges that occurred at the Society's meetings is Rhoda Rappaport, 'Hooke on Earthquakes: Lectures, Strategy and Audience', *British Journal for the History of Science*, 19 (1986), 129 – 46. In this connection one might also note an account of a topic that aroused a good deal of curiosity among Fellows in the Society's earliest years, Laurel Brodsley, Sir Charles Frank and J. W. Steeds, 'Prince Rupert's Drops', *Notes and Records of the Royal Society*, 41 (1986), 1 – 26. An overview is provided by Marie Boas Hall's 'Salomon's House Emergent: the Early Royal Society and Co-operative Research' in Harry Woolf, ed., *The Analytic Spirit: Essays in the History of Science in Honor of Henry Guerlac* (Ithaca: Cornell University Press, 1981), pp. 177 – 94, while on one dispute arising from work carried out in close association with the Society, see A. D. Farr, 'The First Human Blood Transfusion', and A. R. and M. B. Hall, 'The First Human Blood Transfusion: Priority Disputes', *Medical History*, 24 (1980), 143 – 62, 461 – 5.

The work of individual scientists will be considered below, but it may be appropriate to include here recent publications on a topic central to the broad aspirations for the Society's repository surveyed elsewhere in this book, language-planning. A number of studies have appeared which have made more explicit than hitherto the broad epistemological context of such schemes, namely: M. M. Slaughter, *Universal Languages and Scientific Taxonomy in the Seventeenth Century* (Cambridge University Press, 1982); Sidonie Clauss, 'John Wilkins's Essay towards a Real Character: its Place in the Seventeenth-century Episteme', *Journal of the History of Ideas*, 43 (1982), 531 – 53; and Paolo Rossi, 'Universal Languages, Classifications, and Nomenclatures in the Seventeenth century', *History and Philosophy of the Life*

Sciences, 6 (1984), 119 – 31. It is perhaps also worth noting that Vivian Salmon's essay on Wilkins' critics and continuators is now available in her *The Study of Language in Seventeenth-century England* (Amsterdam: Benjamins, 1979), pp. 191 – 206, while Hans Aarsleff's excellent *Dictionary of Scientific Biography* account of Wilkins has been reprinted in his *From Locke to Saussure: Essays on the Study of Language and Intellectual History* (Mineapolis: University of Minnesota Press, 1982), pp. 239 – 77 (it is preceded by a reprint of his account of Sprat, pp. 225 – 38).

The Royal Society's interest in the 'History of Trades' is dealt with by K. H. Ochs in 'The Royal Society of London's History of Trades Programme: an Early Episode in Applied Science', *Notes and Records of the Royal Society*, 39 (1985), 129 – 58, though readers should be warned that some of this author's statements and assessments are questionable. Technological aspects of the new science are also the subject of a whole volume of Clark Lectures edited by J. G. Burke, *The Uses of Science in the Age of Newton* (Berkeley and Los Angeles: University of California Press, 1983), which includes valuable essays by A. Rupert Hall, Albert van Helden and others, though it is rather paradoxical that much of a book with this title should be devoted to denying close links between science and the world of practice. A critical note concerning the approach adopted by some of these studies, which one hopes will be developed more fully in due course, is to be found in J. A. Bennett's 'The Mechanics' Philosophy and the Mechanical Philosophy', *History of Science*, 24 (1986), 1 – 28.

On the more general issue of innovation in the late seventeenth-century economy, various recommendations can be made. Agricultural change and its stimuli are now covered by Joan Thirsk's chapters on 'Agricultural Policy: Public Debate and Legislation' and 'Agricultural Innovations and their Diffusion' in Joan Thirsk, ed., *The Agrarian History of England and Wales: vol. 5, 1640 – 1750: part 2: Agrarian Change* (Cambridge University Press, 1985), pp. 298 – 388, 533 – 89; some of this material will also be found in her 'Plough and Pen: Agricultural Writers in the Seventeenth Century', in T. H. Aston, P. R. Coss, C. Dyer and J. Thirsk, eds., *Social Relations and Ideas: Essays in Honour of R. H. Hilton* (Cambridge University Press, 1983), pp. 295 – 318. A crucial text on one aspect of contemporary technology to which scientists devoted attention has been published in Brian Lavery, ed., *Deane's Doctrine of Naval Architecture, 1670* (London: Conway Maritime Press, 1981), while for a definitive account of the patent system at the time, see Christine MacLeod, 'The 1690s Patent Boom: Invention or Stock-Jobbing?', *Economic History Review*, 2nd series 39 (1986), 549 – 71.

Turning to the broader social and political affiliations of the new science in contemporary society, this is a topic that has attracted much attention. On the place of science in the universities and in English society more generally in the late sixteenth and earlier seventeenth centuries, see Mordechai Feingold, *The Mathematicians' Apprenticeship: Science, Universities and Society in England, 1560 – 1640* (Cambridge University Press, 1984), together with J. A. Bennett's review essay, 'The Mathematicians' Apprenticeship', *British Journal for the History of Science*, 18 (1985), 212 – 8. An interesting attempt to rehabilitate the Merton thesis by arguing that most of those who have considered it have interpreted its thrust too narrowly is made by G. A. Abraham, 'Misunderstanding the Merton Thesis: A Boundary Dispute between History and Sociology', *Isis*, 74 (1983), 368 – 87. Abraham may be right, though in my view the argument for the role of 'cultural values' in encouraging science would be better applied to the eighteenth century than the seventeenth, as I have already pointed out in connection with Joseph Ben-David's *The Scientist's Role In Society*, which, incidentally, was reissued by the University of Chicago Press in 1984 with a new introduction.[2] I might also note here an item drawn to my attention since I compiled my earlier bibliographical essay: L. S. Feuer, 'Science and the Ethic of Protestant Asceticism: A Reply to Professor Robert K. Merton', *Research in Sociology of Knowledge, Sciences and Art*, 2 (1979), 1 – 23.

On the affiliations of science in the Interregnum, see Lotte Mulligan, 'Puritans and English Science: A Critique of Webster', *Isis*, 71 (1980), 456 – 69, which, though a little negative, rightly points out the ambiguities of Charles Webster's use of the term 'Puritan'. A further position-statement by Webster is to be found in his 'Puritanism, Separatism and Science', in D. C. Lindberg and R. L. Numbers, eds., *God and Nature: Historical Essays on the Encounter between Christianity and Science* (Berkeley and Los Angeles: University of California Press, 1986), pp. 192 – 217. Also interesting is P. W. Elmer, 'Medicine, Science, and the Quakers: the "Puritanism-Science Debate" Reconsidered', *Journal of the Friends' Historical Society*, 54 (1981), 265 – 86, while a refreshingly different view of the uses to which scientific ideas might be put is offered by John Henry in 'Atomism and Eschatology: Catholicism and Natural Philosophy in the Interregnum', *British Journal for the History of Science*, 15 (1982), 211 – 39.

If the Interregnum has continued to receive attention, it could be argued that the liveliest field of study over the past few years has concerned the broad ideological import of the new science in the period after 1660 and the controversies in which the Royal Society's members (and, often, its reputation) were involved. By far the most interesting such work is Steven Shapin's and Simon Schaffer's study of the controversy between Boyle and Hobbes in *Leviathan and the Air-pump: Hobbes, Boyle, and the Experimental Life* (Princeton University Press, 1985). This throws new light on experimental practice and its context in the period, even if it is sometimes in danger of attaching undue portentousness to rather obvious points, and of overarguing its case concerning the ideological dimension of science and the role of Hobbes. A valuable pendant to it is provided by Steven Shapin's 'Pump and Circumstance: Robert Boyle's Literary Technology', *Social Studies of Science*, 14 (1984), 481 – 520; see also Simon Schaffer, 'Wallifaction: Thomas Hobbes on School Divinity and Experimental Pneumatics', *Studies in the History and Philosophy of Science*, 19 (1988), 275 – 98. Comparable preoccupations are pursued with worthwhile results in J. V. Golinski's 'Robert Boyle: Scepticism and Authority in Seventeenth-century Chemical Discourse', in Andrew Benjamin, Geoffrey Cantor and John Christie, eds., *The Figural and the Literal: Problems of Language in the History of Science and*

[2] See Hunter, *Science and Society*, p. 193.

Philosophy, 1630–1800 (Manchester University Press, 1987), pp. 58–82, and in Peter Dear's '*Totius in Verba*: Rhetoric and Authority in the Early Royal Society', *Isis*, 76 (1985), 145–61. The latter interestingly illustrates the new techniques of argument that the Society's apologists advocated, though Dear's account is marred by an inexplicable ignorance of Paul Wood's study of Sprat, with which his essay partially overlaps.[3]

A further recent book-length study is J. R. Jacob's *Henry Stubbe, Radical Protestantism and the Early Enlightenment* (Cambridge University Press, 1983); a related article by the same author appears as 'Aristotle and the New Philosophy: Stubbe versus the Royal Society', in M. P. Hanen, M. J. Osler and R. G. Weyant, eds., *Science, Pseudo-Science and Society* (Waterloo: Wilfrid Laurier University Press, 1980), pp. 217–36. Jacob certainly makes Stubbe a much more interesting figure than he has often been thought, though whether his contemporary position has been altogether accurately captured is more doubtful. It seems likely that both this and Jacob's earlier study of Robert Boyle will continue to be controversial, as is illustrated by the debate over the case of Valentine Greatrakes, on which Jacob's two books should be read in conjunction with Eamon Duffy, 'Valentine Greatrakes, the Irish Stroker: Miracle, Science, and Orthodoxy in Restoration England', *Studies in Church History*, 17 (1981), 251–73, Nicholas Steneck, 'Greatrakes the Stroker: the Interpretations of Historians', and Barbara Kaplan, 'Greatrakes the Stroker: the Interpretations of his Contemporaries', *Isis*, 73 (1982), 161–77 and 178–85. An interesting broader context to this episode is provided by the early chapters of R. M. Burns, *The Great Debate on Miracles: From Joseph Glanvill to David Hume* (Lewisburg: Bucknell University Press, 1981), while an important local dimension to the debate between Stubbe and Glanvill is opened up in N. H. Steneck, '"The Ballad of Robert Crosse and Joseph Glanvill" and the Background to *Plus Ultra*', *British Journal for the History of Science*, 14 (1981), 59–74.

The amount of attention lavished on Stubbe has perhaps detracted from the other principal critic of the new science, Meric Casaubon. However, M. R. G. Spiller's edition of his hitherto unpublished tract 'Of Learning' has now appeared, together with a reprint of Casaubon's *Letter. . . to Peter du Moulin* (1669) and introductory material, in *'Concerning Natural Experimental Philosophie': Meric Casaubon and the Royal Society* (The Hague: Nijhoff, 1980); this should be read in conjunction with my essay review, 'Ancients, Moderns, Philologists, and Scientists', *Annals of Science*, 39 (1982), 187–92. On the long-standing 'querelle' to which my title referred and with which, in my view, the controversy over the claims of the new science only partially overlaps, there is a recent contribution by J. M. Levine in 'Ancients and Moderns Reconsidered', *Eighteenth-century Studies*, 15 (1981), 72–89. Further related material appears in the same author's essay-collection,

[3] P. B. Wood, 'Methodology and Apologetics: Thomas Sprat's *History of the Royal Society*', *British Journal for the History of Science*, 13 (1980), 1–26.

Humanism and History: Origins of Modern English Historiography (Ithaca: Cornell University Press, 1987).

Arguably the most important facet of anxiety about the new science was its perceived religious implication. I have written a paper on this topic with the title 'Science and Heterodoxy: An Early Modern Problem Reconsidered', which, though completed in 1983, still remains unpublished: I understand, however, that it will be appearing imminently in R. S. Westman and D. C. Lindberg, eds., *Reappraisals of the Scientific Revolution* (Cambridge University Press, 1989 or 1990). I have made some related points in 'The Problem of "Atheism" in Early Modern England', *Transactions of the Royal Historical Society*, 5th series 35 (1985), 135 – 57, while anxiety about 'atheism' is also discussed along with other themes in N. C. Gillespie, 'Natural History, Natural Theology, and Social Order: John Ray and the "Newtonian Ideology" ', *Journal of the History of Biology*, 20 (1987), 1 – 49. Other relevant recent publications include Keith Hutchison's 'Supernaturalism and the Mechanical Philosophy', *History of Science*, 21 (1983), 297 – 333, and John Henry's account of Glisson, who may well turn out to be the true target of Ralph Cudworth's attack on 'hylozoic' atheism in his *True Intellectual System*: 'Medicine and Pneumatology: Henry More, Richard Baxter and Francis Glisson's *Treatise on the Energetic Nature of Substance*', *Medical History*, 31 (1987), 15 – 40. Another figure who is crucial to debates about heterodoxy in the period and the role of natural philosophy in relation to this is John Toland, on whom two books have been published: S. H. Daniel, *John Toland: His Methods, Manners, and Mind* (Montreal and Kingston: McGill-Queen's University Press, 1984) and R. E. Sullivan, *John Toland and the Deist Controversy: A Study in Adaptations* (Harvard University Press, 1982), which, as its title promises, also gives a valuable overview of the problem of Deism.

A relevant earlier study omitted from my former bibliographical essay is E. M. Klaaren, *Religious Origins of Modern Science: Belief in Creation in Seventeenth-century Thought* (Grand Rapids: Eerdmans, 1977), which is mainly focussed on Boyle. For essays on related topics, see D. C. Lindberg and R. L. Numbers, eds., *God and Nature: Historical Essays on the Encounter between Christianity and Science* (Berkeley and Los Angeles: University of California Press, 1986), while a further recent contribution is Amos Funkenstein, *Theology and the Scientific Imagination from the Middle Ages to the Seventeenth Century* (Princeton University Press, 1986).

The question of the new philosophy's religious implications is inextricably bound up with the relationship between science and magic, as is illustrated by various studies of the reaction against 'enthusiasm' in the period. A useful review of the latter theme is provided by Michael Heyd in 'The Reaction to Enthusiasm in the Seventeenth Century: Towards an Integrative Approach', *Journal of Modern History*, 53 (1981), 258 – 80; see also his 'The New Experimental Philosophy: A Manifestation of "Enthusiasm" or an Antidote to it?', *Minerva*, 25 (1987), 423 – 40. In an essay on Sprat's *History*, Brian Vickers has stressed how Sprat's views on prose style make sense as part of a polemical — and somewhat disingenuous — assault on the 'enthusiasm' of the Interregnum: 'The Royal Society and English Prose

Style: A Reassessment', in Brian Vickers and Nancy S. Struever, *Rhetoric and the Pursuit of Truth* (Los Angeles: William Andrews Clark Memorial Library, 1985), pp. 1 – 76.

The denominational affiliations of the rejection and retention of magic are explored by Michael MacDonald in 'Religion, Social Change and Psychological Healing in England, 1600 – 1800', *Studies in Church History*, 19 (1982), 101 – 25; this develops comparable ideas to be found in the concluding pages of his *Mystical Bedlam: Madness, Anxiety and Healing in Seventeenth-century England* (Cambridge University Press, 1981), while a not dissimilar approach to views on witchcraft will be found in T. H. Jobe, 'The Devil in Restoration Science: the Glanvill-Webster Witchcraft Debate', *Isis*, 72 (1981), 343 – 56.

As far as astrology is concerned, the political overtones of contemporary attitudes towards the art are stressed by Patrick Curry in 'Saving Astrology in Restoration England: "Whig" and "Tory" Reforms', in Patrick Curry, ed., *Astrology, Science and Society: Historical Essays* (Woodbridge: the Boydell Press, 1987), pp. 245 – 59, though readers may wish to consider the broader range of factors invoked in my own essay in the same volume, 'Science and Astrology in Seventeenth-century England: an Unpublished Polemic by John Flamsteed', pp. 260 – 300. A further relevant essay in *Astrology, Science and Society* is Simon Schaffer's 'Newton's Comets and the Transformation of Astrology', pp. 219 – 43, while the same author relates developments in natural philosophy more directly to the controversy over atheism and enthusiasm in his 'Godly Men and Mechanical Philosophers: Souls and Spirits in Restoration Natural Philosophy': this appears in the first issue of a new journal, *Science in Context*, 1 (1987), 55 – 86, which also includes the first English translation of writings from the 1930s by Franz Borkenau and Henryk Grossmann on 'The Emergence of Modern Science out of the Production Process'.

Echoes of such ideas are to be found in a series of recent ambitious attempts to place the rejection of magic in the context of a broad view of political change and expropriation: Brian Easlea, *Witch-hunting, Magic and the New Philosophy: An Introduction to Debates of the Scientific Revolution, 1450 – 1750* (Brighton: Harvester Press, 1980); Caroline Merchant, *The Death of Nature: Women, Ecology and the Scientific Revolution* (San Francisco: Harper and Row, 1980); Morris Berman, *The Reenchantment of the World* (Ithaca: Cornell University Press, 1981); and David Kubrin, 'Newton's Inside Out: Magic, Class Struggle, and the Rise of Mechanism in the West', in Harry Woolf, ed., *The Analytic Spirit: Essays in the History of Science in Honor of Henry Guerlac* (Ithaca: Cornell University Press, 1981), pp. 96 – 121. Quite apart from their proneness to rather simplistic correlations, however, such accounts suffer from taking for granted a clear distinction between magical and mechanical ideas which is now seeming increasingly untenable. From this point of view, a brief compendium of what are now virtually orthodox views is Charles Webster's *From Paracelsus to Newton: Magic and the Making of Modern Science* (Cambridge University Press, 1982), which indicates the vitality which magic retained into the late seventeenth century. This may be supplemented by the latest more detailed studies, including various essays in the useful collection edited by Brian Vickers, *Occult and Scientific Mentalities in the Renaissance* (Cambridge University Press, 1984), though there is a slight disharmony between the thrust of most of the contributions to this volume and the views of its editor, as is pointed out in a review essay by Patrick Curry, 'Revisions of Science and Magic', *History of Science*, 23 (1985), 299 – 325. One thing that is clear concerning the rejection of 'occult' ideas and the onset of mechanization is the need to avoid oversimple categorization, as is well-illustrated by two important studies: John Henry, 'Occult Qualities and the Experimental Philosophy: Active Principles in Pre-Newtonian Matter Theory', *History of Science*, 24 (1986), 335 – 81, and Keith Hutchison, 'What Happened to Occult Qualities in the Scientific Revolution?', *Isis*, 73 (1982), 233 – 53.

The same message comes across from recent studies of the work of individual scientists. The Newton industry continues apace, its most recent products including the first volume of Newton's *Optical Papers*, edited by A. E. Shapiro (Cambridge University Press, 1984), and J. E. McGuire and Martin Tamny's edition of *Certain Philosophical Questions: Newton's Trinity Notebook* (Cambridge University Press, 1983); see also their article, 'Newton's Astronomical Apprenticeship: Notes of 1664/5', *Isis*, 76 (1985), 349 – 65. A number of studies of aspects of Newton's ideas and related topics appear in the festschrift for Richard S. Westfall, M. J. Osler and P. L. Farber, eds., *Religion, Science and Worldview* (Cambridge University Press, 1985), while the proceedings of a discussion meeting on Newton held at the Royal Society in June 1987 have been published both as part 1 of volume 42 of *Notes and Records of the Royal Society* and separately as D. G. King-Hele and A. R. Hall, eds., *Newton's 'Principia' and its Legacy* (London: the Royal Society, 1988). Other recent studies include B. J. T. Dobbs, 'Newton's Alchemy and his Theory of Matter', *Isis*, 73 (1982), 511 – 28; Henry Guerlac, 'Theological Voluntarism and Biological Analogies in Newton's Physical Thought', *Journal of the History of Ideas*, 44 (1983), 219 – 29; R. S. Westfall, 'Newton's Marvelous Years of Discovery and their Aftermath: Myth versus Manuscript', *Isis*, 71 (1980), 109 – 21, and 'Alchemy in Newton's Library', *Ambix*, 31 (1984), 97 – 101; and Paolo Casini, 'Newton: the Classical Scholia', *History of Science*, 22 (1984), 1 – 58. On Newton's debate with Leibniz, readers may sample both A. Rupert Hall, *Philosophers at War: The Quarrel between Newton and Leibniz* (Cambridge University Press, 1980) and Steven Shapin, 'Of Gods and Kings: Natural Philosophy and Politics in the Leibniz-Clarke Disputes', *Isis*, 72 (1981), 187 – 215, and 'Licking Leibniz', *History of Science*, 19 (1981), 293 – 305, while an attempt at an overview of Newton's impact is provided by I. B. Cohen, *The Newtonian Revolution* (Cambridge University Press, 1980), generalized still further in his *Revolution in Science* (Harvard University Press, 1985).

Newton has also been the subject of a major biography, R. S. Westfall, *Never at Rest: A Biography of Isaac Newton* (Cambridge University Press, 1980). At least one aspect of this, however — separately divulged in R. S. Westfall, 'Isaac Newton in Cambridge: The Restoration University and Scientific Creativity', in Perez Zagorin, ed., *Culture and Politics From Puritanism to the Enlightenment* (Berkeley and Los Angeles: University of

California Press, 1980), pp. 135 – 64 — has already to be fundamentally revised: see John Gascoigne, 'The Universities and the Scientific Revolution: the Case of Newton and Restoration Cambridge', *History of Science*, 23 (1985), 391 – 434. A further full-length biography has appeared since Westfall's, Gale E. Christianson, *In the Presence of the Creator: Isaac Newton and his Times* (New York: Free Press, 1984), while one might also draw attention here to A. R. Hall's 'Further Newton Correspondence', *Notes and Records of the Royal Society*, 37 (1982), 7-35.

As for Boyle, a number of studies have already been referred to above in connection with their stress on the ideological implications of his ideas, but one might also note here two books omitted from my 1981 listing — Angelo Campodonico, *Filosophia dell'Esperienza ed Epistemologia della Fede in Robert Boyle* (Florence: Felice le Monnier, 1978) and M. A. Stewart, ed., *Selected Philosophical Papers of Robert Boyle* (Manchester University Press, 1979) — and two new studies: J. G. Lennox, 'Robert Boyle's Defense of Teleological Inference in Experimental Science', *Isis*, 74 (1983), 38 – 52, and M. B. Hall, 'Boyle's Method of Work: Promoting His Corpuscular Philosophy', *Notes and Records of the Royal Society*, 41 (1987), 111 – 43.

Hooke has received the attention of J. A. Bennett in 'Robert Hooke as Mechanic and Natural Philosopher', ibid., 35 (1980 – 1), 33 – 48; of P. M. Gouk in 'The Role of Acoustics and Music Theory in the Scientific Work of Robert Hooke', *Annals of Science*, 37 (1980), 573 – 605; and of David Oldroyd in 'Some "Philosophicall Scribbles" attributed to Robert Hooke', *Notes and Records of the Royal Society*, 35 (1980), 17 – 32, 'Some Writings of Robert Hooke on Procedures for the Prosecution of Scientific Inquiry, including his "Lectures of Things Requisite to a Natural History" ', ibid., 41 (1987), 145 – 67, and (with J. C. Kassler) 'Robert Hooke's Trinity College "Musick Scripts", his Music Theory and the Role of Music in his Cosmology', *Annals of Science*, 40 (1983), 559-95. Hooke also appears along with other thinkers in Steven Pumfrey, 'Mechanizing Magnetism in Restoration England: the Decline of Magnetical Philosophy', *Annals of Science*, 44 (1987), 1 – 21, and in Paolo Rossi, *The Dark Abyss of Time: the History of the Earth and the History of Nations from Hooke to Vico* (English translation, University of Chicago Press, 1984). Relevant to the latter is a further study of geological ideas at the time, Frances Willmoth's 'John Flamsteed's Letter concerning the Natural Causes of Earthquakes', *Annals of Science*, 44 (1987), 23 – 70.

Studies of other influential thinkers include J. A. Bennett, *The Mathematical Science of Christopher Wren* (Cambridge University Press, 1982), the published version of his 1974 thesis, and Sabina Fleitmann, *Walter Charleton (1620 – 1707), 'Virtuoso': Leben und Werk* (Frankfurt am Main: Peter Lang, 1986). C. E. Raven's *John Ray, Naturalist* (second edition, Cambridge University Press, 1950) has been reissued in paperback with a new introduction. In addition, one might note here Walter Pagel's crucial study of the influential *Joan Baptista Van Helmont: Reformer of Science and Medicine* (Cambridge University Press, 1982), and the English translation of Alessandro Roncaglia's *Petty: the Origins of Political Economy* (University College Cardiff Press, 1985). An interesting source is made accessible in

E. H. Cohen and J. S. Ross, 'The Commonplace Book of Edmond Halley', *Notes and Records of the Royal Society*, 40 (1985), 1 – 41, while a useful research tool is provided by Wilbur Applebaum in 'A Descriptive Catalogue of the Manuscripts of Nicolaus Mercator, F.R.S. (1620 – 1687) in Sheffield University Library', ibid, 41 (1986), 27 – 37.

On one leading figure of the Royal Society in its earliest years, Sir Robert Moray, one can recommend David Stevenson's 'Masonry, Symbolism and Ethics in the Life of Sir Robert Moray, F.R.S.', *Proceedings of the Society of Antiquaries of Scotland*, 114 (1984), 405 – 31, while an equally important figure of the later years of the century, Sir Hans Sloane, is dealt with by Maarten Ultee in 'Sir Hans Sloane, Scientist', *British Library Journal*, 14 (1988), 1 – 20. On Denis Papin, who was employed to execute experiments for the Society in the 1680s, see Charles-Armand Klein, *Denis Papin: Illustre Savant Blaisois* (Chambray-les-Tours: Editions C.L.D., 1987), while Richard Waller, Sloane's co-secretary of the Royal Society in the 1690s and subsequently the editor of Hooke's *Posthumous Works*, has received definitive treatment in M. J. E. Ezell, 'Richard Waller, S.R.S.: "In the Pursuit of Nature" ', *Notes and Records of the Royal Society*, 38 (1984), 215 – 33. Equally definitive coverage is being given to the crucial John Beale, the subject of a full account by Mayling Stubbs which currently stops just short of his association with the Society, but which is due to be extended to take in his career during his association with it: 'John Beale, Philosophical Gardener of Herefordshire. Part I. Prelude to the Royal Society (1608 – 1663)', *Annals of Science*, 39 (1982), 463 – 89.

Rather slighter studies have appeared of two other men who both deserve more adequate accounts than they have yet had: Stephen Passmore, 'Thomas Henshaw, F.R.S. (1618 – 1700)', *Notes and Records of the Royal Society*, 36 (1982), 177 – 88, and S.Mendyck, 'Robert Plot: Britain's "Genial Father of County Natural Histories" ', ibid., 39 (1985), 159 – 77. The Bristol customs officer and virtuoso, William Cole, is dealt with by Anthony Turner in 'A Forgotten Naturalist of the Seventeenth Century: William Cole of Bristol and his Collections', *Archives of Natural History*, 11 (1982), 27-41, while a somewhat similar figure who came very briefly into contact with the Royal Society in the 1690s is the subject of Michael Hunter and Annabel Gregory, *An Astrological Diary of the Seventeenth Century: Samuel Jeake of Rye, 1652 – 1699* (Oxford: Clarendon Press, 1988). On John Webster, the former critic of the universities whose *Displaying of Supposed Witchcraft* was published with the Royal Society's imprimatur in 1677, valuable data is to be found in Peter Elmer's *The Library of Dr John Webster: the Making of a Seventeenth-century Radical* (*Medical History, Supplement No. 6*, London: Wellcome Institute, 1986).

A handful of more thematic studies that have not already been referred to may be noted here. Albert van Helden's important study, *Measuring the Universe: Cosmic Dimensions from Aristarchus to Halley* (University of Chicago Press, 1985), includes a good deal of material on English scientists of this period. On the other hand, it has to be recorded that Lesley Murdin's *Under Newton's Shadow: Astronomical Practices in the Seventeenth century* (Bristol: Adam Hilger, 1985) is a considerable disappointment: though informative on a number of minor scientific enthusiasts, it is often inaccurate and gives a rather confused picture of scientific circles at the time. It is perhaps also worth pointing out that the introductory chapters of J. L. Heilbron's *Electricity in the Seventeenth and Eighteenth Centuries* (1979) have been separately reprinted as *Elements of Early Modern Physics* (Berkeley and Los Angeles: University of California Press, 1982); that A. R. Hall has updated his older textbook on the Scientific Revolution under the title, *The Revolution in Science, 1500 – 1750* (London: Longman, 1983); and that Brian Vickers has produced an introductory anthology of texts illustrative of *English Science, Bacon to Newton* (Cambridge University Press, 1987). Barbara Shapiro has brought out *Probability and Certainty in Seventeenth-century England: A Study of the Relationships between Natural Science, Religion, History, Law and Literature* (Princeton University Press, 1983); a preliminary version of part of this portentous and slightly disappointing book had appeared as 'History and Natural History in Sixteenth- and Seventeenth-century England: An Essay on the Relationship between Humanism and Science', in Barbara Shapiro and R. G. Frank, jr., *English Scientific Virtuosi in the Sixteenth and Seventeenth Centuries* (Los Angeles: William Andrews Clark Memorial Library, 1979), pp. 1 – 55. Useful bibliographical essays which include this period will be found in Pietro Corsi and Paul Weindling, eds., *Information Sources in the History of Science and Medicine* (London: Butterworth, 1983), while another such work referred to as forthcoming in my earlier essay has now been published, the Royal Commission on Historical Manuscripts' guide to *The Manuscript Papers of British Scientists, 1600 – 1940* (London: Her Majesty's Stationery Office, 1982).

Lastly, what has been happening in the eighteenth century? Here we have had an ambitious attempt by Margaret Jacob to extend the ideas that she had adumbrated earlier about English Newtonianism in *The Radical Enlightenment: Pantheists, Freemasons and Republicans* (London: Allen and Unwin, 1981). The research of others, however, has tended not to second her view either of this or the earlier period. For a critique of her more recent book, see G. C. Gibbs, 'The Radical Enlightenment', *British Journal for the History of Science*, 17 (1984), 67 – 81, while the ideas expressed in her *The Newtonians and the English Revolution* (1976) need at least to be modified, as is suggested by such studies as Anita Guerrini, 'The Tory Newtonians: Gregory, Pitcairne and their Circle', *Journal of British Studies*, 25 (1986), 288 – 311, J. E. Force, *William Whiston, Honest Newtonian* (Cambridge University Press, 1985) and Colin Russell, *Science and Social Change, 1700 – 1900* (London: Macmillan, 1983), ch. 4. A more subtle and accurate approach to the way in which the new science became increasingly culturally dominant will be found in John Gascoigne, 'Politics, Patronage and Newtonianism: the Cambridge Example', *Historical Journal*, 27 (1984), 1 – 24, and 'Mathematics and Meritocracy: the Emergence of the Cambridge Mathematical Tripos', *Social Studies of Science*, 14 (1984), 547 – 84. While these studies deal with the universities, the place of science in the metropolis has been illuminated by Larry Stewart in 'The Selling of Newton: Science and Technology in Early Eighteenth-century England', *Journal of British Studies*, 25 (1986), 178 – 92, and 'Public Lectures and Private Patronage in Newtonian England', *Isis*, 77

(1986), 47 – 58; a related study is Simon Schaffer, 'Natural Philosophy and Public Spectacle in the Eighteenth Century', *History of Science*, 21 (1983), 1 – 43.

An excellent introductory text on natural philosophy in this later period is provided by T. L. Hankins in *Science and the Enlightenment* (Cambridge University Press, 1985), while a valuable overview is given by the essays in G. S. Rousseau and Roy Porter, eds., *The Ferment of Knowledge: Studies in the Historiography of Eighteenth-century Science* (Cambridge University Press, 1980). The latter, however, is already being overtaken by more recent detailed work, both on the content of scientific ideas — as, for instance, in Geoffrey Cantor, *Optics after Newton: Theories of Light in Britain and Ireland, 1704 – 1840* (Manchester University Press, 1983) — and, as already noted, on their context. I have already referred to McClellan's *Science Reorganized*, which gives a worthwhile view of the broader scientific community and the Royal Society's place in it in the eighteenth century. It may be appropriate to end with an intriguing study which, although brief, illustrates clearly the way in which science came to permeate cultural values in the eighteenth century, C. B.Paul, *Science and Immortality: the Eloges of the Paris Academy of Sciences (1699 – 1791)* (Berkeley and Los Angeles: University of California Press, 1980). Studies like the present one are significant in illuminating the formative — and sometimes rather fumbling — steps towards the state of affairs which books like Paul's and McClellan's document.

INDEX

www.ingramcontent.com/pod-product-compliance
Ingram Content Group UK Ltd.
Pitfield, Milton Keynes, MK11 3LW, UK
UKHW020231290926
15470UKWH00002B/483

9 780851 155067